Radiowave Propagation and Antennas for Personal Communications

Third Edition

DISCLAIMER OF WARRANTY

The technical descriptions, procedures, and computer programs in this book have been developed with the greatest of care and they have been useful to the author in a broad range of applications; however, they are provided as is, without warranty of any kind. Artech House, Inc. and the author and editors of the book titled *Radiowave Propagation and Antennas for Personal Communications, Third Edition* make no warranties, expressed or implied, that the equations, programs, and procedures in this book or its associated software are free of error, or are consistent with any particular standard of merchantability, or will meet your requirements for any particular application. They should not be relied upon for solving a problem whose incorrect solution could result in injury to a person or loss of property. Any use of the programs or procedures in such a manner is at the user's own risk. The editors, author, and publisher disclaim all liability for direct, incidental, or consequent damages resulting from use of the programs or procedures in this book or the associated software.

For a complete listing of the *Artech House Mobile Communications Library*, turn to the back of this book.

Radiowave Propagation and Antennas for Personal Communications

Third Edition

Kazimierz Siwiak
Yasaman Bahreini

ARTECH HOUSE
BOSTON | LONDON
artechhouse.com

Library of Congress Cataloging-in-Publication Data
A catalog record of this book is available from the Library of Congress.

British Library Cataloguing in Publication Data
A catalogue record of this book is available at the British Library.

ISBN 13: 978-1-59693-073-5

Cover design by Igor Valdman

685 Canton Street
Norwood, MA 02062

10 9 8 7 6 5 4 3 2 1

To my family for their support and guidance,
my mom for teaching me patience,
and my dad for teaching me perseverance.
—Yassi

Moim Rodzicom Janowi i Bronisławie
– Kai

Contents

Preface to the First Edition

In the late spring of 1990, the director of training at the Paging Products Group of Motorola asked me to "think about what would make a good graduate course in antennas—something practical." That request drew a paper napkin response listing a telecommunications problem that explores the radio links between fixed-site antennas, includes propagation into and within urban and suburban environments, and considers the problem of small antennas next to, or in the hands of, people. The paper napkin outline evolved into a well-attended 1990 summer session graduate elective entitled "Antennas and Propagation" at Florida Atlantic University. The objective of that course was to expose graduate practicing radio engineers to antennas and electromagnetics problems that are appropriate to the personal telecommunications industry. These radio engineers are not necessarily experts in electromagnetic theory or in antennas but can often be, for example, experts in traffic and queuing theory tasked with designing complex personal radio communications systems. The course became a three day intensive workshop on "Radiowave Propagation and Antennas for Personal Communications" in late 1993, and this book is the outgrowth of the course and the workshop notes.

The book differs from other available texts on antennas and propagation in that there is an emphasis on three distinct communications problems: fixed-site antennas, radiowave propagation, and small antennas proximate to the body. These three problems form the basis for the investigation of the personal communication path link. The emphasis and the style of approaching these problems have an eye toward the radio engineer tasked with radio system analysis and design. There is a marked tendency to present material in a manner that is suitable for practical implementation.

The objective of this book is to provide an introduction to the antenna and propagation problems for radio communications engineers and to supplement the many excellent available theoretical texts on these subjects. Another objective is to reach mechanical and reliability engineers tasked with designing small antennas for personal portable products who wish to comprehend more fully the trade-offs between physical implementations of antennas, the fundamental limitations in small antennas, and system issues relative to radiowave propagation. The text attempts to bridge the gap between the purely theoretical and general treatments of electromagnetics and the purely practical empirical approaches associated with radio system design. Material presented here, for example, includes a treatment of the receiver-sensitivity measurement problem using simulated human body devices, as well as a detailed analysis of receiver-field-strength sensitivity test sites.

The text is sprinkled with illustrative examples worked out using Mathcad (Mathcad is a trademark of MathSoft, Inc.) templates. The Mathcad examples are marked in the text with a diskette symbol (🖫), which indicates that the example is available as a template on a companion floppy diskette. Readers of the textbook, especially students, educators, and engineers, are encouraged to generate additional templates dedicated to the public and to submit them to the author care of the publisher or directly by e-mail at k.siwiak@ieee.org. Submitted templates that are relevant to the text will be considered for inclusion, with appropriate credits, in any subsequent editions of the diskette. The formulas and equations in this book have been formulated with numerical evaluation in mind. All of the presented examples can be worked out using Mathcad, as implied by the Mathcad templates.

The contents of this book fall into three main parts, which follow the personal communications path link. Chapters 1 and 2 present fundamental concepts, define the personal communications link, and explore the fixed-site antennas of that link. Chapters 3 to 7 concentrate on radiowave propagation problems. Chapters 8 to 11 explore the nature of waves in the vicinity of personal communications devices and explore small antennas suitable for small telecommunications devices. The book can also be categorized by personal communications system design considerations. Chapter 2 deals with the performance of fixed-site antennas, while Chapters 5, 7, and 8 present the radiowave propagation problem. Chapters 9 to 11 consider the field-strength performance and measurement of body-proximate personal telecommunications devices, and Chapter 12 presents the complete personal communications system as an application of the communications path link. This book is written to be applicable to a special-topics course at the graduate level targeted at radio and communications engineers. It is also a suitable reference text for a professional short course or workshop on antennas and propagation.

The contents of the book are subdivided as follows. Chapter 1 defines the personal telecommunications problem, gives a historical perspective, and introduces the fundamentals of radiation and antennas. Definitions of terms commonly used in antennas and radiowave propagation are provided. Chapter 2 introduces the antennas that are typically employed at fixed sites in a telecommunications infrastructure. The material concentrates on the radiation characteristics of fixed-site antennas that directly affect radio-link performance, including beam shaping and "smart antenna" technology. The effects of distortions of fixed-site antenna patterns on the path link encountered when these antennas are placed in the vicinity of radio towers and other antennas are also included.

Chapter 3 introduces the radio communication channel in terms of guided and radiated waves. The performance and characteristics of transmission lines and of the basic free space propagation law are presented in the context of their impact on personal communications system designs. Basic radiowave propagation behavior is presented as the Friis transmission formula. Finally, the polarization characteristics of radiated waves and of antennas, along with polarization mismatch losses, are explored.

Chapter 4 presents an overview of the radio frequency spectrum and briefly surveys some of the behavior of electromagnetic waves in the various radio frequency bands in view of their potential application to personal communications. It is shown that the primary range of interest in personal communications lies in the frequency spectrum between 30 MHz and 3 GHz.

Chapter 5 investigates some of the special problems associated with personal communications using earth-orbiting satellites in contrast with the more familiar fixed-geometry link of the geostationary satellites. Chapter 6 introduces the two-ray model for radiowave propagation and develops a foundation for the analysis of open-field antenna test sites that are suitable for measuring the field-strength characteristics of personal communications receivers. The radiowave propagation models developed in this chapter are applicable to the open-field antenna testing ranges that are so critical in personal communications device development. The test range analysis explores the effects of range geometry and fixed antenna arrangements with the aim of analyzing measurement errors and test range capabilities.

Chapter 7 studies the urban and suburban radiowave propagation paths in the radio frequency spectrum (30 MHz to 3 GHz) that is most important to personal communications services. Theoretical urban propagation models based on an assumed regularity in the urban environment are presented, along with empirical models that rely on measured parameters describing the urban and suburban environment. Propagation within, into, and in the vicinity of buildings is studied. In Chapter 8, the wave behavior in the large-scale area and in the local vicinity of the personal communications device is modeled statistically, and a system design strategy is developed.

Chapters 9 and 10 present the basics of receiver-sensitivity measurements using statistically based measuring methods. Open-field antenna test ranges, especially their accuracy as measurement devices, are explored. The simulated body devices used in receiver-sensitivity testing are presented here and analyzed in detail. Chapter 11 investigates the fundamental properties, limitations, and performance of small antennas that are used with personal communications devices. Finally, Chapter 12 describes a complete radio communication system in terms of the personal communication path link problem.

Acknowledgments are extended to Robert J. Schwendeman, Dr. Tadeusz M. Babij, Phil Macnak, Dr. Lorenzo Ponce de Leon, Dr. Huey-Ru Chuang, and the Artech reviewers for their constructive criticism of this book.

Preface to the Second Edition

The second edition is enlarged, and chapters are enhanced to provide more detail about some topics and to cover new material relevant to the personal communications device (PCD) in a personal communications system (PCS). The growth in PCS has been spectacular over the last two years. Paging has seen the introduction of the high-speed FLEX signaling protocol family and in just two years placed 25 million subscribers on the air. In the same time frame, two-way paging using the ReFLEX protocol has emerged as the most popular two-way messaging system, surpassing all other two-way radio messaging systems combined. The ERMES VHF high-speed paging signaling protocol was introduced and enjoys over a million subscribers across Europe and the Middle East. GSM has emerged as the world-dominant cellular telephony standard command half the world's subscribers, while IS-95 CDMA vies for the digital cellular business in the United States, stimulated by newly available PCS spectrum. Personal communications have staked a presence in space with the launching on May 5, 1997, of the first 5 of 66 Iridium satellites. Test broadcasts of the EUREKA-147 system have made digital voice broadcasting a reality in Europe.

Meanwhile, this book has found its way into university classrooms, initially as a reference and then as a course textbook. Problems sets are now included at the end of each chapter, many with answers or solutions. Additional Mathcad templates are added onto the diskette that accompanies the book. The Mathcad template examples in the text are marked with a 🖫 symbol and now include the template file name, such as "[1-3a.mcd]" to help readers find the relevant template. The diskette is enhanced: the original version 2.5 Mathcad files are retained, and version 6 renditions of those templates are added; all new templates are presented in version 6 of Mathcad. Some chapter problem solutions are also included as templates. The FORTRAN listings of the dipole analysis in Appendix A and the loop analysis in Appendix B are moved to the diskette in source code that is easily complied, and compiled executable code (PC-DOS) for each of the FORTRAN programs is included. Appendixes A and B now contain annotated listings of dialogs with the FORTRAN programs to illustrate their usage.

New material has been added that includes multiple access techniques in view of their radiowave propagation channel characteristics. Specifically, the FLEX protocols, ERMES system, and POCSAG signaling code are used to illustrate paging and notification services, and the FLEX protocol for two-way paging (ReFLEX) and FLEX protocol for voice paging (InFLEXion) illustrate two-way paging and messaging characteristics, as well as complex QAM modulation. A new Appendix C illustrates several digital code renditions, including those that appear in paging. Real-time voice system characteristics are illustrated by AMPS (analog), GSM (time and frequency multiplexing), and IS-95 (code division multiple access) systems. The EUREKA EU-147 protocol illustrates a modern approach to digital voice broadcasting, and finally packet radio access techniques are studied in view of their use in PCD channel access. System examples of one-way messaging are enlarged and include two-way systems.

The problems in radiowave propagation and antennas are not limited to just simple link margin studies. Increasingly, total system behavior dominates the picture. System subscriber capacity, coverage, and system economics are tightly intertwined, and the cost of implementing complex solutions is rapidly decreasing. Radiowave propagation and antennas is evolving into radiowave propagation channel engineering. New material on diversity reception and transmission techniques is added to study ways of improving radio channel performance. The sections on receiver-sensitivity measurements have been enlarged to include transmitting device characterization. In August 1996 the U.S. Federal Communications Commission issued regulations limiting exposure of people to radio frequency electromagnetic fields; hence, new material has been added about compliance with exposure guidelines and regulations.

Based on my experience teaching this material at Johns Hopkins University, Whiting School of Engineering, Organizational Effectiveness Institute, the textbook material can be presented in the classroom starting with a general picture of the propagation channel engineering problem and leading to a detailed explanation of the channel components. Starting with radio system design, Chapter 1 introduces historical perspectives, a background on personal communications, and electromagnetics fundamentals. Chapter 4 then introduces the radio frequency spectrum and includes propagation channel engineering characteristics of several multiple access signaling protocols. Chapter 12 shows radio systems with an emphasis on one- and two-way messaging. Chapter 8 provides a level of detail into wave behavior in the multipath environment and introduces diversity reception techniques. The next level of detail, antennas in personal communications systems, begins with fixed-site antennas described in Chapter 2. The chapter has been enlarged to better cover mutual coupling and to treat Yagi array antennas. Chapter 11 then introduces the personal communications device antennas, and Chapter 9 extends this to consider body proximity effects, which leads to an examination of compliance with radio frequency exposure guidelines and regulations. Chapter 10 introduces simulated body devices used in PCD testing. The last detailed topic, radiowave propagation in personal communications, starts in Chapter 3 with transmission line and spherical radiation. The detail is extended to the communications satellite propagation problem in Chapter 5 and to two-ray propagation models of open-area test sites in Chapter 6. Finally, the urban and suburban propagation paths are studied in Chapter 7.

The objective of the second edition, like the first, remains to provide an introduction to the antenna and propagation problems associated with personal communications for radio communications engineers and to supplement the many excellent available theoretical texts on these subjects.

Acknowledgments are extended to Robert J. Schwendeman, Dr. Tadeusz M. Babij, Phil Macnak, Dr. Lorenzo Ponce de Leon, Dr. Huey-Ru Chuang, and the Artech reviewers for their constructive criticism of this edition of the book.

Preface to the Third Edition

This third edition benefits from a coauthor who provides new materials on ultrawideband (UWB) technology and a focus on mobile/cellular in the sections on personal communications systems (PCSs), as well as an introduction to wireless local area networks (W-LANs) and personal area networks (W-PANs). We are in an age where within a few years, one-third of the occupants of the planet will have accessed a network wirelessly. Paging, on the other hand, is demised as relatively unimportant; thus, the sections on ERMES VHF high-speed paging, which is seeing the deallocation of its frequency bands, have been eliminated.

A special emphasis is placed on UWB technology throughout this edition. The unique properties of UWB pulse radiation, pulse propagation, and reception are presented in comparison with conventional radio systems. Elementary loop and dipole radiation are now additionally discussed for UWB pulses, and circular polarization for UWB pulses is explained in Chapter 1. Chapter 2 now includes W-LANs and W-PANs under the IEEE 802.11 and 802.15 families of standards, as well as indoor UWB technology. Multiple Input Multiple Output (MIMO) techniques are also introduced. Treatment of the Friis transmission formula has been moved into Chapter 3. New high frequency (HF) band propagation software has been included on the CD-ROM to support an enhanced look at HF propagation in Chapter 4. The chapter has also been expanded to reflect the importance of mobile/cellular telephony and wireless local and personal area networks, including UWB technology. Chapter 5 has been updated to reflect the economic failure of low-earth-orbit voice communications systems and shows that data and media transmissions from satellites are gaining economic footing. Chapter 6 has been expanded to cover the unique propagation properties of UWB pulses in free space and over a plane earth. In Chapter 7, the

Siwiak-Bertoni-Yano (SBY) propagation model relates multipath and propagation power law, and an in-room deterministic ray-tracing model is added in support of local and personal area networks. Chapter 8 includes new material on UWB pulses in multipath as well as a background on MIMO technology. Chapter 9 now includes the methodologies for measuring mobile phone receiver sensitivities and transmitter radiation. Chapter 10 describes an anthropomorphic simulated head used with mobile phone measurements. Chapter 11 explains the radiation properties (as distinct from the propagation properties) of short UWB pulses. Chapter 12 has an improved treatment of radio frequency noise and includes new system design examples for a local area wireless network and for a very-high-speed UWB personal area system.

Problems sets have been enhanced, especially in mobile and cellular systems and in UWB technology. Mathcad template examples in the text are marked with a "🖫" symbol, as are several software examples of HF propagation, which are illustrated using NTIA HF propagation software, now included in the CD-ROM and explained in Appendix D. All of the Mathcad templates are now additionally rendered in PDF so the Mathcad program is not needed, except for interactive use. An in-room ray-tracing propagation model is added as a Mathcad template and is also rendered in PDF. The included CD-ROM is organized as follows:

In root directory:

\DISCLAIM.TXT

\HOW-THIS-DISK-IS-ORGANIZED.TXT

\FORTRAN

\DIPOLE

\LOOP

\HF-prop

\In-building-prop

\Mcad-problems

\Mcad-problems-PDF

\Mcad60

\Mcad60-PDF

- FORTRAN\DIPOLE contains the FORTRAN source code for dipole analysis and DOS executables for that FORTRAN code explained in Appendix A.

- FORTRAN\LOOP contains the FORTRAN source code for loop analysis and DOS executables for that FORTRAN code explained in Appendix B.
- HF-prop contains HF propagation software explained in Appendix D.
- In-building-prop contains the Mathcad 6.0 template IN-ROOM.mcd for the ray-tracing in-building propagation model, as well as PDF-rendition-of-IN-ROOM.pdf.
- Mcad60 has all of the third edition templates, which have been tested in Mathcad 6.0.
- Mcad60-PDF has a PDF rendition of all the Mathcad files in Mcad60.
- Mcad-problems contains Mathcad template solutions for some chapter problems.
- Mcad-problems-PDF has PDF renditions of the Mathcad files in Mcad-problems.

The objective of the third edition, like the first and second, remains to provide an introduction to the antenna and propagation problems associated with personal communications for radio communications engineers and to supplement the many excellent available theoretical texts on these subjects.

The authors wish to acknowldege Gregory Hand for supplying the HF propagation software itshfbc_060418.exe, Alex Shovkoplyas for supplying the HamCAP GUI interface included on the disk with this edition. We also acknowledge the kind support of our families, who persevered through the development of this edition of the book.

Kazimierz Siwiak
Yasaman Bahreini

1

Introduction

1.1 Introduction and Historical Perspective

In 1864, James Clerk Maxwell, in his chair at the University of Edinburgh, placed the concept of electricity and magnetism into the language of mathematics in the form of his equations of electromagnetism. His theory stated that energy can be transported through materials and through space at a finite velocity by the action of electric and magnetic waves, each transferring energy, one to the other, in an endless rhythmic progression through time and space. Maxwell never validated his theory by experiment, and his results were opposed at the time. It was left to Heinrich Rudolf Hertz, starting 22 years later, to put into practice what Maxwell had spelled out in mathematics in a remarkable set of historical experiments [1] spanning the years 1886 to 1891. Hertz demonstrated communications over several-meter distances experimentally with his spark gap apparatus. The era of "wireless" had begun. We will review Maxwell's equations and their boundary conditions in this chapter.

It was left to the pioneer inventors like Nikola Tesla and Alexander Popov to develop the radio arts to the point of practical usability at the turn of the century with their grasp of resonant transmitter and receiver circuits. In 1901 Guglielmo Marconi put to use the innovations of his predecessors and bridged the 3,000 km distance [2] between St. John's Newfoundland and Cornwall, on the southwest tip of England, using Morse transmission of the letter "S." Morse code (see Appendix C) is a form of digital communications using on-off keying with characteristics that can be likened to a Huffman code. In the United States, the Radio Act of July 23, 1912, attempted unsuccessfully to mitigate the mutual interference caused by wideband spark emissions. Finally the Radio Act of 1927 established the Federal Radio Commission (FRC), which evolved into the

Federal Communications Commission (FCC) with the Communications Act of 1934. Transmitting stations were to be licensed, and unfiltered spark "damped wave emissions" were prohibited, services were separated by wavelength, and wireless marched toward narrow bandwidths. Wireless communications were becoming a practical reality, and the radio frequencies of interest to personal communications were steadily evolving into voice communications using analog modulations. By the mid-1930s, the era of two-way radio communications in the low very-high-frequency (VHF) range (30 to 40 MHz) had become a reality. Edwin Armstrong discovered the advantages of wideband frequency modulation (FM) in 1933, eventually ushering in noise-free FM broadcasting. Meanwhile, Dan Nobel championed the use of narrowband FM for land-mobile communications, effectively opening up the VHF bands for economical communications systems. Wireless had become radio. By the mid-1940s, radio frequencies for land-mobile communications were allocated in the 150-MHz range, followed by frequencies in the 450-MHz range during the 1960s. Today, the most significant growth in personal communications is taking place at frequencies up to and beyond 3 GHz. Voice and data are coded digitally; the new era of "digital wireless" has begun. We will introduce several examples of modern digital communications in view of their characteristics in propagation channels.

Thought the 1960s, 1970s, and 1980s, experimentation in wideband pulse transmissions for radar and communications demonstrated the practicality of modern low-power-impulse radio. On April 14, 2002, the FCC published a report and order permitting unlicensed ultrawideband (UWB) operation under Part 15 of the rules, further noting that "the prohibition ... against Class B damped wave emissions does not apply to UWB devices" [3]. UWB transmissions are permitted at levels below the unintentional emissions of electronic devices and other radio transmissions; thus, they can coexist with other users of the same spectrum. The unique properties of UWB pulse radiation, propagation, and reception will be presented in comparison with conventional radio systems.

From the modest beginnings of land-mobile communications to today's explosive growth, personal communications spans a remarkably short historical time of just a few decades. Figure 1.1 shows a global estimate of personal communications users in paging, mobile telephone, and wireless local-area and personal area networks (WLANs and WPANs). The number of paging subscribers had been growing an order of magnitude per decade. Mobile telephone and Personal Communications Systems (PCS) users have, in recent years, been growing two orders of magnitude per year. WLAN users have increased dramatically faster than pager or mobile phone users. Each new personal communications evolution grows faster and climbs higher than its predecessor, and all are headed to a convergence of capabilities.

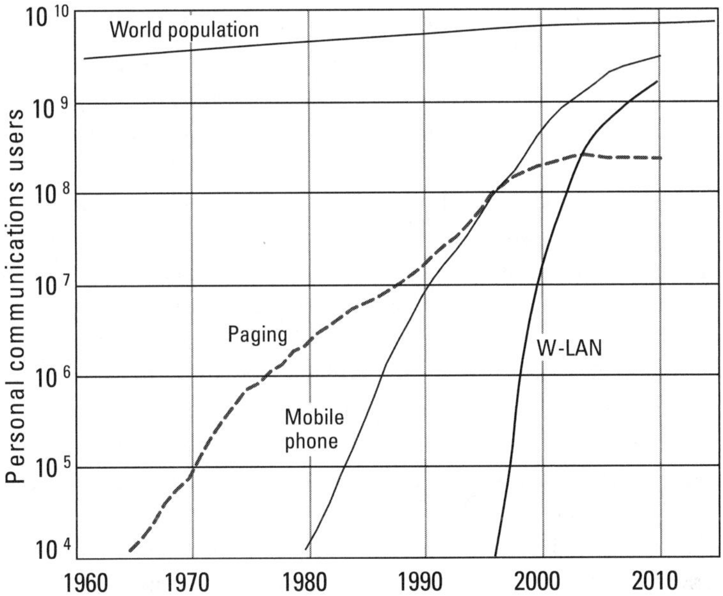

Figure 1.1 Worldwide personal communications users, current and projected. (*Sources:* U.S. Bureau of the Census; MTA-EMCI, Inc.; Computer Industry Almanac).

What the figure does not show is that the message size that a personal communications user expects is also growing rapidly. A paging message of the 1960s was a simple alert, equivalent to a few thousand bits per user each month. Today, in addition to the greatly increased numbers of users, there is explosive growth in the volume of wireless data being transferred by each user. Future personal communications, for which additional frequency spectrum in the 1–6-GHz range is just now being utilized, will provide additional, and no doubt dramatic, growth. UWB in the 3.1–10.6-GHz band in the United States, and typically from around 7 to 10 GHz in Europe and Asia, will usher in a spectacular growth in very high-rate digital communications on the personal-area scale. The future personal communications user base will not only be greatly larger but will routinely expect billions of bits per month in services that include voice, notification services, and multimedia transmissions.

1.2 Personal Communications

The radio path link for personal communications will be covered by studying three separate significant electromagnetics problems that are encountered in radio engineering. The first involves fixed sites where the antennas interact with the immediate environment. The second involves the propagation channel.

Finally, the third involves small antennas, including interactions with the human body and with the immediate environment. In the case of UWB pulse systems, we additionally subdivide the "radiation" and the "propagation" processes to study some of the unique properties of UWB pulses. Since we are dealing with personal communications devices (PCDs) that may be receivers, additional topics include measurement techniques for body-mounted receiver devices and the analysis and practice of using simulated human body devices in radio testing. The radio communications path link, as seen in Figure 1.2, can be identified for both the "downlink" path, the path from a transmitting fixed site to a receiving portable device, and for the "uplink" path from the portable device to the receiver at a fixed site. Multiple antennas may be employed at either end of the link to exploit characteristics of channel impairments to improve the link. Such exploitation will be studied as diversity techniques.

Fixed-site antennas are generally large antennas located at prime sites. These are not always clear sites, and there is often coupling between proximate fixed-site antennas, which results in distortions to the radiation patterns. The kinds of antennas used at fixed sites will be studied in Chapter 2. The distortions in antenna patterns caused by scattering from nearby tower structures or other antennas collocated at the site will also be analyzed. The antenna pattern distortion can result in significant distortions of the coverage area compared to the usual assumption of omnidirectional antenna coverage.

Radiowave propagation in free space is a relatively simple phenomenon. For the most common emitters, the waves are spherical, and power density remains constant for a given solid angle on a sphere expanding about the origin of the wave. In the presence of the Earth, however, the problem rapidly becomes complex. Even with the simplifying assumption of a planar Earth, we additionally use plane wave approximations to model waves reflected from the Earth. In the case of obstacles such as buildings, hills, and mountains, we are forced to use additional approximations because the geometric description of the physical problem can quickly outpace our computational capabilities. We will deal with the suburban propagation problem by making simplifying assumptions based on a perceived regularity in the city environs. The multipath scattering problem will be handled using a statistical description of the waves appropriate to radio system designs. Narrowband systems are most conveniently analyzed employing harmonic wave solutions (steady-state sine waves) to Maxwell's equations. UWB

Figure 1.2 The communications link.

transmissions, on the other hand, are more transient and can best be viewed in the time rather than the frequency domain.

The antennas usually associated with personal communications devices are small, both electrically and physically, and are very close to people and to host devices. In fact, personal communications devices are generally worn on the body or held in the hand. The electromagnetics problem of antenna coupling with the body of interest to the communications link is the external radiating problem, as compared with the bioeffects problem of energy deposition into body tissue. We will concentrate on the external problem. Since receiver performance of personal communications devices is often of interest, we will explore the techniques for measuring receiver performance of body-proximate devices.

1.3 Electromagnetics Fundamentals

This section introduces Maxwell's laws, along with the boundary conditions that allow solutions to those equations for cases of interest to radio communication problems. It is usual in the communications industry, and in this textbook, to represent the electric E and magnetic H fields by their root mean square (RMS) fields values, not their peak values as is common in many other electromagnetics texts. This matters only when the fields quantities are related to power or power density.

Maxwell's equations form the basis for the solution of all problems involving the motion of charges, or currents, on conductors, which gives rise to traveling electromagnetic (EM) waves. Figure 1.3 illustrates in summary form a range of concerns in the radiowave propagation and antenna path link problem of this text. In Figure 1.3(a), our concern is with currents on structures at fixed sites that generate those waves and focus them around a radio system coverage area. We are further concerned with scattering and the resulting distortions to the "focusing patterns" that are, in fact, antenna patterns, as represented in Figure 1.3 by the building-mounted antennas and their distorted radiation patterns. Subsequently, the waves propagate near, and are reflected and diffracted by, the ground, hills, and buildings and other structures, as represented by the several propagation rays. We will thus study the problem of propagation in the suburban and urban area using approximate methods, as the boundaries are too cumbersome and the geography too extensive to solve exactly. The waves in the vicinity of a personal communications device on the body or in the hands of a person will be studied. The waves in the local vicinity of the person are represented in Figure 1.3 by the interference pattern resulting from vector addition of waves from multiple-ray paths. Because the precise geometry of the local vicinity is rarely specified accurately enough (the exception is at open-field test sites) to calculate the exact wave interference behavior, the problem is solved in terms of

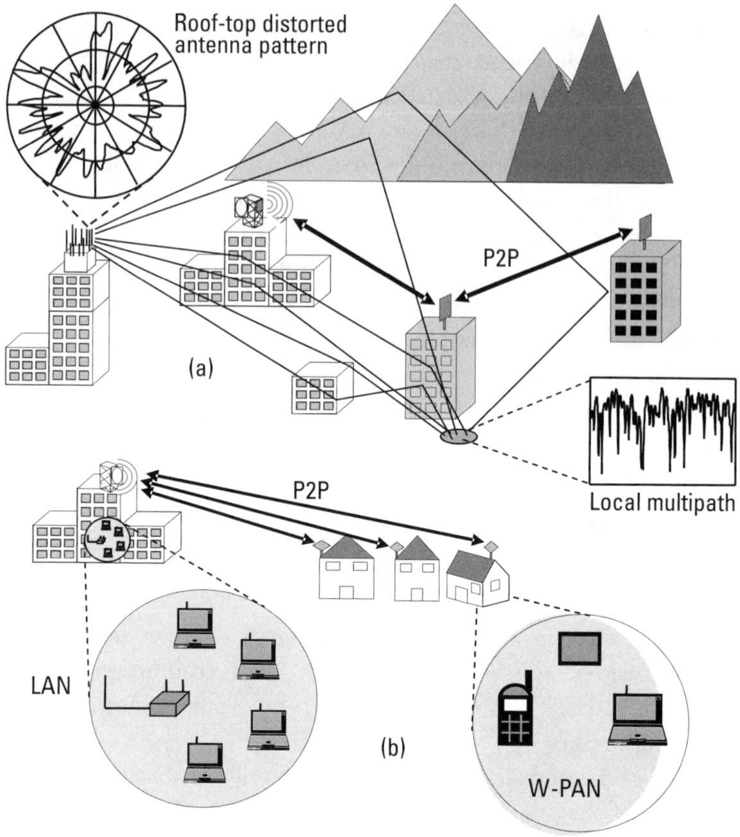

Figure 1.3 Radiowave propagation and antennas in (a) wide, and (b) local/personal areas.

the *statistics* of the wave amplitudes in a local region. The reverse link problem is similar. A transmitter located among a scattering environment of buildings and other objects proximate to the transmitter appears as a diffuse source to the tower or building-mounted antennas even on some point-to-point (P2P) links. We will see later in Chapter 8 how the characteristics of the signal radiated by this diffuse source can be exploited to improve reception using two receiving antennas spaced slightly in the horizontal plane. Finally—and this is not specifically identified in Figure 1.3—small antennas interacting with the human body will be studied. In Figure 1.3(b) we see the local- and personal-area network scale where primarily handheld devices interact with other handheld and stationary devices. This link is typified by very high data rate and can involve UWB transmissions.

The form of Maxwell's equations given here is valid over the range of velocities and the range of temperatures normally encountered by people. With relativistic corrections, the range of validity can be extended for velocities approaching that of light. Although Maxwell presented his equations describing the behavior of electromagnetic waves relatively recently in history, they have been suitable for describing most electromagnetic phenomena [4, 5] for a considerably longer time, as Table 1.1 depicts. Table 1.1 is a thumbnail sketch of the evolution of the cosmos. It serves to illustrate that Maxwell's equations describe macroscopic phenomena of physics under certain constraints of physical temperature. The equations are an approximation, which is valid in a physical system where the four fundamental forces of nature (the weak and strong nuclear forces, gravity, and electromagnetic forces) are decoupled. *Approximation* is the operative word, and this text presents many additional approximations that enable solutions to the complex geometries of the personal communications path link.

There are relatively few electromagnetics problems for which closed-form solutions can be found. This is because the geometries of most problems of interest cannot be described in terms of a geometric coordinate system for which we know a set of orthogonal basis functions. Problems in rectangular

Table 1.1
Timeline of Electromagnetics Phenomena

Time (sec)	Event	Effect
0	"Big Bang" takes place.	Four fundamental forces are coupled.
10^{-43}	Gravity is frozen out.	Weak and strong nuclear and electromagnetic forces are still coupled.
10^{-35}	Strong nuclear forces are frozen out.	Weak nuclear and electromagnetic forces are still coupled.
10^{-6}	Protons are able to form	The universe is cooling.
1	Weak nuclear and electromagnetic forces dissociate.	Maxwell's equations are adequate to describe macroscopic field behavior.
4.31×10^{17}	Dinosaurs vanish from Earth.	Earth becomes ready for man.
4.32×10^{17}	Maxwell's equations are written.	"Wireless" is discovered, beginning the era of digital (Morse code) communications.
$4.3200000010 \times 10^{17}$	30 years pass since the era of Maxwell.	The era of "radio," analog voice, and invention in the radio arts begins.
$4.3200000031 \times 10^{17}$	100 years pass since the era of Maxwell.	PCS and a new era of "digital wireless" begin.

geometries, like rectangular waveguides and rectangular cavities, have solutions in terms of the rectangular trigonometric functions, the harmonic sines and cosines. Problems in cylindrical coordinates have solutions in terms of Bessel and Hankel function harmonics. The cylindrical problems are two dimensional, and approximations are made when problems involve truncated cylinders, such as dipoles of finite thickness and cylindrical scatters. Problems expressed in spherical coordinates have solution that are spherical Bessel harmonics and elliptical functions that are solutions in elliptical coordinates. Often, the boundaries are just too numerous and perhaps even transitory, so we employ statistical methods to arrive at engineering solutions.

This section, in effect, summarizes the radio wave propagation and antenna problem at the lowest level of abstraction of the behavior of electromagnetic waves. We are introduced to fundamental laws of the theory governing the interactions between electric and magnetic fields, as well as the sources that give rise to those fields. Later chapters develop higher-level solutions to problems more specific to the personal communications path link.

1.3.1 Maxwell's Equations

Maxwell's equations provide us with the relationship between electric and magnetic fields, as well as sources that give rise to the fields, and allow us to determine the way in which these fields (radio waves) interact with the environment. This section lists the laws of Michael Faraday (1791–1867), André-Marie Ampere (1775–1836), and Carl Friedrich Gauss (1777–1855), along with the continuity relationships.

The four vector field quantities with which this book deals are the electric field $\mathbf{E}$ (V/m), the electric displacement field $\mathbf{D}$ (C/m^2), the magnetic field intensity $\mathbf{H}$ (A/m), and the magnetic flux $\mathbf{B}$ (tesla). The equations are initially written in their time-varying form and are valid without bandwidth restrictions. Symbols in bold are vector quantities. The sources are current density $\mathbf{J}$ (A/m^2) and charge density ρ (C/m^2), which obey Maxwell's equations in the following manner. The law of induction, discovered experimentally by Faraday, states that the time rate of change of magnetic flux is given by the curl of the electric field

$$\nabla \times \mathbf{E} = -\frac{\partial \mathbf{B}}{\partial t} \tag{1.1}$$

The law of Ampere, generalized by Maxwell to include the conduction or convection current $\mathbf{J}$, states that the time rate of change of the electric displacement field is equal to the curl of the magnetic intensity

$$\nabla \times \mathbf{H} = \frac{\partial \mathbf{D}}{\partial t} + \mathbf{J} \tag{1.2}$$

Equations (1.1) and (1.2) are mathematical duals of each other; their mathematical forms are identical when a magnetic current –**M**, in volts per square meter, is added to (1.1), as will be shown later. Gauss's electric law relates the divergence of the electric displacement field to the charge density:

$$\nabla \cdot \mathbf{D} = \rho \tag{1.3}$$

Gauss's magnetic law expresses the continuity of magnetic flux:

$$\nabla \cdot \mathbf{B} = 0 \tag{1.4}$$

These equations are not independent; (1.1) and (1.4) are coupled, as are (1.2) and (1.3). It can be shown that only six independent scalar equations can be written from Maxwell's four relationships. We assume in solving electromagnetic field problems that the current and charge sources are given and are related by the continuity equation:

$$\nabla \cdot \mathbf{J} = -\frac{\partial \rho}{\partial t} \tag{1.5}$$

The continuity equation states that the flow of current out of a differential volume equals the rate of decrease of electrical charge in the volume; that is, electrical charge is conserved.

There are 12 scalar field variables in Maxwell's equations (1.1) to (1.4), one for each of the three components of **E**, **H**, **D**, and **B**, but there are only 6 independent scalar equations. Maxwell's equation alone are therefore insufficient to solve for the 12 unknowns. Six additional scalar equations, known as the *constitutive relations*, are needed. In a vacuum, or free space, these relate the three components of the electric displacement field,

$$\mathbf{D} = \varepsilon_0 \mathbf{E} \tag{1.6}$$

and the three components of magnetic flux density with the magnetic field,

$$\mathbf{B} = \mu_0 \mathbf{H} \tag{1.7}$$

where ε_0 (F/m) is the permittivity, and μ_0 (H/m) is the permeability of free space.

In materials that are homogeneous, the relations (1.6) and (1.7) are $\mathbf{D} = \varepsilon\mathbf{E}$, where $\varepsilon = \varepsilon_0\varepsilon_d$ is the complex permittivity, and $\mathbf{B} = \mu\mathbf{H}$, where $\mu = \mu_0\mu_d$ is the complex permeability. Often, the lossy component of the relative permittivity due to the finite conductivity σ of the lossy dielectric material is written $\varepsilon_i = \sigma/j\omega$ for harmonic wave solutions (sine waves) to Maxwell's equations, and a conduction current $\mathbf{J}_c = \sigma\mathbf{E}$ can be included along with $\mathbf{J}$ in (1.2).

The analysis of electromagnetics and radiation problems deals often with sinusoidal time-varying fields usually specified as phasors with the explicit time-dependence $e^{j\omega t}$ suppressed. This means that the time derivative operator $\partial/\partial t$ can be replaced by the operator $j\omega$. The consequence of this assignment is that the subsequent solutions are valid only for narrowband systems in the sense that the solution is for a particular frequency ω. It follows that the phase progresses along a positive axis with an increasing negative magnitude. *The explicit time-dependence suppression is justified only for those problems for which the fields and waves are sinusoidal or nearly sinusoidal.* We will see later, however, that some problems of interest to us deal with modulation periods commensurate with time differences of arrival between multiply reflected copies of the waves, and others deal with short-pulse UWB signals. Thus, care must be taken to account properly for the time dependence.

1.3.2 Boundary Conditions

The environment within which radio waves are generated and propagate includes a complicated geometry of scatterers and reflectors, which result in multiple instances of boundaries between materials having different electromagnetic properties. Maxwell equations (1.1) to (1.7) must be solved in these regions of inhomogeneity. The equations are, hence, complemented by a set of boundary conditions. Such problems will be considered when dealing with scattering from objects near antennas, such as other antennas, radio towers, and the human body. At the surface of a perfect electric conductor, the tangential component of the electric field vanishes; hence,

$$\mathbf{n} \times \mathbf{E} = 0 \tag{1.8}$$

where $\mathbf{n}$ is the unit vector normal to the surface. Since magnetic flux cannot penetrate a perfect electric conductor, the normal component of the magnetic field must vanish at the surface. So,

$$\mathbf{n} \cdot \mathbf{H} = 0 \tag{1.9}$$

The value of the magnetic field tangential to the surface is equal to the surface current $\mathbf{J}_s$ whose flow of charges is perpendicular to the tangential magnetic field vector. So,

$$\mathbf{n} \times \mathbf{H} = \mathbf{J}_s \tag{1.10}$$

Since there is no electric field within a perfect conductor, electric flux lines terminate on surface charges ρ_s; hence,

$$\mathbf{n} \cdot \mathbf{D} = \rho_s \tag{1.11}$$

At the boundary of two dielectrics with permittivity ε_1 and ε_2, the tangential components of the field components are equal on both sides. So,

$$\mathbf{n} \times \mathbf{E}_1 = \mathbf{n} \times \mathbf{E}_2 \tag{1.12}$$

$$\mathbf{n} \times \mathbf{H}_1 = \mathbf{n} \times \mathbf{H}_2 \tag{1.13}$$

Finally, the electric flux is continuous across the boundary. So,

$$\mathbf{n} \cdot \mathbf{D}_1 = \mathbf{n} \cdot \mathbf{D}_2 \tag{1.14}$$

1.3.3 Vector and Scalar Potentials

The wave equation, in the form of the inhomogeneous Helmholtz equation, is presented here with most of the underlying vector arithmetic omitted (see [6–8] for the details). Since the magnetic flux **B** is always solenoidal (the field lines do not originate or terminate on sources)—that is, in (1.4) the divergence is zero—it can be represented by the curl of an arbitrary vector **A**. So,

$$\mathbf{B} = \mu_0 \mathbf{H} = \nabla \times \mathbf{A} \tag{1.15}$$

where **A** is called the vector potential and obeys the vector identity $\nabla \cdot \nabla \times \mathbf{A} = 0$. Substituting (1.15) into (1.1) with (1.6) and with the vector identity $\nabla \times (-\nabla \Phi) = 0$, where Φ represents an arbitrary scalar function of position, it follows that for the harmonic wave solution (sine waves),

$$\mathbf{E} = -\nabla \Phi - j\omega \mathbf{A} \tag{1.16}$$

Taking the curl of both sides of (1.8) and for a homogeneous medium, along with (1.2) and (1.6), after some manipulation and with (1.16), we get

$$\nabla^2 \mathbf{A} + k^2 \mathbf{A} = -\mu \mathbf{J} + \nabla(\nabla \cdot \mathbf{A} + j\omega\mu\varepsilon\Phi) \tag{1.17}$$

where k is the wave number, and $k^2 = \omega^2\mu\varepsilon$. Although (1.15) defines the curl of **A**, the divergence of **A** can be independently defined. ∇^2 is the Laplacian operator given by

$$\nabla^2 = \frac{\partial^2}{\partial x^2} + \frac{\partial^2}{\partial y^2} + \frac{\partial^2}{\partial z^2}$$

The *Lorentz condition* is chosen:

$$j\omega\mu\varepsilon\Phi = -\nabla \cdot \mathbf{A} \tag{1.18}$$

Substituting the simplification of (1.18) into (1.17) leads to the inhomogeneous Helmholtz equation:

$$\nabla^2 \mathbf{A} + k^2 \mathbf{A} = -\mu \mathbf{J} \tag{1.19}$$

Similarly, by using (1.18) and (1.16) in (1.3), we see that

$$\nabla^2 \Phi + k^2 \Phi = -\rho / \varepsilon_0 \tag{1.20}$$

For time-varying fields, the charge ρ is not an independent source term, as seen from (1.5). Consequently, it is not necessary to solve for the scalar potential Φ. Using (1.16) and the Lorentz condition (1.18), we can find the electric field solely in terms of the vector potential **A**. The utility of that definition becomes apparent when we consider the case of current source aligned along a single vector direction, for example, $\mathbf{J} = \mathbf{z}J_z$, for which the vector potential is $\mathbf{A} = \mathbf{z}A_z$, where **z** is the unit vector aligned in the z-axis direction, and (1.19) becomes a scalar equation.

1.3.4 Radiation from a Sinusoidally Excited Current Element

A spherical wave solution to the wave equation (1.19) is presented here, again with the details suppressed. This will be seen in later chapters to be the fundamental form of radio wave propagation in free space. Here, the results are used to derive the radiation properties of the infinitesimal sinusoidal excited current element.

The infinitesimal current element $\mathbf{J} = \mathbf{z}J_z$ located at the origin satisfies a one-dimensional, hence scalar, form of (1.19). At points excluding the origin, where the infinitesimal current element is located, (1.19) is source free and can be written as a function of radial distance r,

$$\nabla^2 A_z(r) + k^2 A_z(r) = \frac{1}{r^2}\frac{\partial}{\partial r}\left[r^2 \frac{\partial A_z(r)}{\partial r}\right] + k^2 A_z(r) = 0 \tag{1.21}$$

and can be reduced to

$$\frac{d^2 A_z(r)}{dr^2} + \frac{2}{r}\frac{dA_z(r)}{dr} + k^2 A_z(r) = 0 \tag{1.22}$$

Since A_z is a function of only the radial coordinate, the partial derivative of (1.21) is replaced with the ordinary derivative. For sinusoidal excitation, (1.22) has a solution

$$A_Z = C_1 \frac{e^{-jkr}}{r} \tag{1.23}$$

There is a second solution where the exponent of the phasor quantity is positive; however, we are interested in outward traveling waves, so we discard that solution. The phasor quantity is unity in the static case. Constant C_1 is related to the strength of the source current and is found by integrating (1.19) over the volume including the source. The infinitesimal current density J_z over the cross-sectional area dS and length dl is equal to a current filament I of length dl, which finally gives

$$C_1 = \frac{\mu_o}{4\pi} J_Z dS\, dl = \frac{\mu_o}{4\pi} I\, dl \tag{1.24}$$

The solution for the vector potential is in the **z** unit vector direction,

$$\mathbf{A} = \frac{\mu_o}{4\pi} I\, dl \frac{e^{-jkr}}{r}\mathbf{z} \tag{1.25}$$

which is an outward propagating spherical wave with increasing phase delay (increasingly negative phase) and with amplitude decreasing as the inverse of distance. We may now solve for the electric fields of an infinitesimal current element by inserting (1.25) into (1.16) with (1.18). The fields, after sufficient manipulation, and for $r >> dl$, are

$$E_r = -j\frac{Idl}{2\pi}e^{-jkr}\eta_o k^2\left[\frac{j}{(kr)^2}+\frac{1}{(kr)^3}\right]\cos(\theta) \quad (1.26)$$

$$E_\theta = -j\frac{Idl}{4\pi}e^{-jkr}\eta_o k^2\left[-\frac{1}{kr}+\frac{j}{(kr)^2}+\frac{1}{(kr)^3}\right]\sin(\theta) \quad (1.27)$$

Similarly, from (1.25) inserted into (1.15), the magnetic field is

$$H_\phi = j\frac{Idl}{4\pi}e^{-jkr}k^2\left[\frac{1}{kr}-\frac{j}{(kr)^2}\right]\sin(\theta) \quad (1.28)$$

Equations (1.26) to (1.28) describe a particularly complex field behavior for what is a very idealized selection of sources: a simple current element I of infinitesimal length dl. This is the case of the ideal (uniform current) infinitesimal dipole. Expressions (1.26) to (1.28) are valid only in the region sufficiently far ($r >> dl$) from the region of the current source I.

The solution to the radiation problem of the half-wave dipole, particularly one of finite wire thickness, is inordinately complex and has been extensively investigated [9–13]; most often, approximate solutions [6–8] are satisfactory. Appendix A presents the FORTRAN code based on [13] for computing the currents on, and the fields close to (within a wire diameter), dipoles and helical dipoles having finite thickness. For most purposes here, especially when the detail of the fields very near the dipole conductor are not of interest, the half-wave dipole may be modeled by a sinusoidal current on a very thin wire, as seen in Chapter 11.

1.3.5 Duality in Maxwell's Equations

Once a solution for an electromagnetics problem is found, a related solution may be found by exploiting the dual nature of Maxwell's equations. For example, (1.1) and (1.2) are mathematical duals of each other, and their variables are dual quantities when a "magnetic current" **M** is subtracted from the right-hand side of (1.1). It must be emphasized that the magnetic current **M** is a mathematical abstraction, which enables us to study Maxwell's equations as dual quantities. No magnetic currents are presently know to exist in nature, but equivalent magnetic currents arise when we apply volume or surface equivalence theorems. Table 1.2 lists the dual equations for electric- and magnetic-current harmonic

Table 1.2
Electric and Magnetic Dual Equations for Harmonic Waves

Electric Sources, J	Magnetic Sources, M
$\nabla \times \mathbf{E} = -j\omega\mu_o \mathbf{H}$	$\nabla \times \mathbf{H} = j\omega\varepsilon_o \mathbf{E}$
$\nabla \times \mathbf{H} = j\omega\varepsilon_o \mathbf{E} + \mathbf{J}$	$-\nabla \times \mathbf{E} = -j\omega\mu_o \mathbf{H} + \mathbf{M}$
$\nabla^2 \mathbf{A} + k^2 \mathbf{A} = -\mu_o \mathbf{J}$	$\nabla^2 \mathbf{F} + k^2 \mathbf{F} = -\varepsilon_o \mathbf{M}$
$\mathbf{A} = \frac{\mu_o}{4\pi} \frac{e^{-jkr}}{r} \mathbf{J}$	$\mathbf{F} = -\frac{\varepsilon_o}{4\pi} \frac{e^{-jkr}}{r} \mathbf{M}$
$\mu_o \mathbf{H} = \nabla \times \mathbf{A}$	$\varepsilon_o \mathbf{E} = -\nabla \times \mathbf{F}$
$\mathbf{E} = -j\omega \mathbf{A} - \frac{1}{\omega\mu_o\varepsilon_o} \nabla(\nabla \cdot \mathbf{A})$	$\mathbf{H} = -j\omega \mathbf{F} - \frac{1}{\omega\mu_o\varepsilon_o} \nabla(\nabla \cdot \mathbf{F})$

wave sources. Care must be exercised when the sources are not sinusoidal, as will be shown later.

The dual quantities are given in Table 1.3. The wave number k remains invariant in the dual equations.

The principal of duality is sometimes applied to problems of radiation from apertures such as open-ended waveguides, radiating slots, and edges of microstrip patch antennas. The surface across which the radiating electric field exists is replaced by a perfect conductor across which a magnetic current **M** exists, oriented in a direction perpendicular to the electric field. This surface current then gives rise to radiated fields, which are found using the dual equations in the same way as would they would be determined from an electric surface current **J**.

Table 1.3
Electric and Magnetic Dual Quantities and Variables

Electric Sources, J	Magnetic Sources, M
E	**H**
H	**–E**
J	**M**
A	**F**
ε_o	μ_o
μ_o	ε_o
η_o	$1/\eta_o$

1.3.6 Current Loop for Sinusoidal Excitation

Applying the principal of duality, the infinitesimal current loop consisting of a circulating current I enclosing a surface area S is solved by analogy to the infinitesimal dipole problem solved earlier. The dipole current Idl is replaced by a magnetic current equal to $M_z\, dS = Idl$, and when the surface area $S = dl/k$, the fields due to the infinitesimal loop are then given by

$$H_r = \frac{kIS}{2\pi} e^{-jkr} k^2 \left[\frac{j}{(kr)^2} + \frac{1}{(kr)^3} \right] \cos(\theta) \tag{1.29}$$

$$H_\theta = \frac{kIS}{4\pi} e^{-jkr} k^2 \left[-\frac{1}{kr} + \frac{j}{(kr)^2} + \frac{1}{(kr)^3} \right] \sin(\theta) \tag{1.30}$$

$$E_\phi = \eta_o \frac{kIS}{4\pi} e^{-jkr} k^2 \left[\frac{1}{kr} - \frac{j}{(kr)^2} \right] \sin(\theta) \tag{1.31}$$

where $\eta_0 = c\mu_0$ is the intrinsic impedance of free space. The electric field equations (1.26) and (1.27) of the infinitesimal dipole have exactly the same form as the magnetic field equations for the infinitesimal loop, while the magnetic field of the dipole (1.28) has exactly the same form as the electric field of the loop—when the excitation is a sinusoidal current. In the case where the loop moment kIS equals the previously presented dipole moment Idl, and the loop and dipole are superimposed, the fields in all space will be circularly polarized.

The field of arbitrary-sized circular loop antennas was investigated as early as 1897 with H. C. Pocklington's [9] study of thin wire loops excited by plane waves. Later, E. Hallén [10] and J. E. Storer [14] considered driven antennas. The close near fields of wire loops are quite complex, especially if the loop wire is fat [15]. Appendix B contains the FORTRAN computer code for a detailed analysis of the current distribution on, and the close near field of (within a fraction of a wire diameter), the fat wire loop antenna for sinusoidal excitation. The computer code, based on [15], provides some detail of the current density around the circumference of the loop wire.

1.3.7 Radiation of a UWB Elementary Dipole and Loop

For UWB signals, the fields surrounding an infinitesimal current element of length Δh can be derived from the Boit-Savart law, which relates a static current distribution to a magnetostatic field, and generalized by Oleg Jefimenko to time-varying solutions. Some details are given in [16], and the magnetic field, here written in terms a current $I(t)$ oriented along the z-axis, is

$$\mathbf{H}_D(r,t) = \frac{\Delta h \sin(\theta)}{4\pi r}\left[\frac{I(\tau)}{r} + \frac{1}{c}\frac{\partial}{\partial t}I(\tau)\right]\hat{\boldsymbol{\varphi}} \tag{1.32}$$

The corresponding electric field is

$$\mathbf{E}_D(r,t) = \frac{\Delta h}{4\pi\varepsilon_0 r}\left[\left(\frac{\int I(\tau)dt}{r^2} + \frac{I(\tau)}{cr}\right)\left(2\cos(\theta)\hat{\mathbf{r}} + \sin(\theta)\hat{\boldsymbol{\theta}}\right) + \frac{\sin(\theta)}{c^2}\frac{\partial}{\partial t}I(\tau)\hat{\boldsymbol{\theta}}\right] \tag{1.33}$$

where $\tau = t - r/c$ is the time-distance retardation term. The $1/r^3$ dependent electrostatic term involves an integration of the current in time and, hence, is a charge accumulation—the expected result for a static electric doublet (capacitor). For the duality to be complete, the magnetic current moment must have the same functional form as the electric dipole moment; thus, we make the substitution

$$\frac{1}{c}\frac{\partial}{\partial t}I(\tau)\Delta S \Rightarrow \Delta h I(\tau) \tag{1.34}$$

where ΔS is the infinitesimal loop area in the *x-y* plane surrounded by the loop current. Thus, the infinitesimal loop fields *for the same source current in the loop as in the dipole* are

$$\mathbf{E}_L(r,t) = \frac{\Delta S \sin(\theta)}{4\pi\varepsilon_0 rc}\left[\frac{\partial}{\partial t}\frac{I(\tau)}{rc} + \frac{\partial^2}{\partial t^2}\frac{I(\tau)}{c^2}\right]\hat{\boldsymbol{\varphi}} \tag{1.35}$$

and the corresponding electric field is

$$\mathbf{H}_L(r,t) = \frac{\Delta S}{4\pi r}\left[\left(\frac{I(\tau)}{r^2} + \frac{\partial}{\partial t}\frac{I(\tau)}{cr}\right)\left(2\cos(\theta)\hat{\mathbf{r}} + \sin(\theta)\hat{\boldsymbol{\theta}}\right) + \sin(\theta)\frac{\partial^2}{\partial t^2}\frac{I(\tau)}{c^2}\hat{\boldsymbol{\theta}}\right] \tag{1.36}$$

Only the $1/r$ terms survive into the far-field radiation region. The induction region fields have $1/r^2$ behavior. The $1/r^3$ term in (1.36) is the magnetostatic term, which would produce a static solenoidal magnetic field if $I(t)$ were a constant direct current. We see that (1.32) to (1.36) are the general case of (1.26) to (1.28) and (1.29) to (1.31).

The dipole and loop fields are not only in *space quadrature*, but their representations in time are also orthogonal since they are related by a time derivative. This has interesting consequences. The superposition of an infinitesimal loop and dipole with the same current excitation produces a rotating electric field vector everywhere in space for the duration of time that the field vector exists;

hence, it defines circular polarization for UWB pulse transmission. Figure 1.4 portrays the electric far field of a collocated infinitesimal loop and dipole pair with $I(t)$ as an odd function of time. The pulse expands in all radial directions, but viewed along the y-axis, the projection of the E-field on the y-z plane is an even function, while its projection on the x-y plane is an odd function, and the polarization vector appears to rotate in space as it passes a fixed observation point on the y-axis.

Figure 1.5 shows the dipole and loop electric field strengths, along with the total amplitude of the circularly polarized signal. Comparing Figures 1.4 and 1.5, we can see that as the signal passes a stationary point in its propagation along the y-axis, it appears to an observer to rotate in a counterclockwise direction. As it does so, the E-field vector magnitude is given by the dotted trace in Figure 1.5.

Figure 1.6 shows the shape of the signal that would be received by a linearly polarized antenna for different rotational orientations. The signal shape changes with the rotation angle. This behavior is not as obvious with a time-harmonic signal. In that case, the *relative phase* of the signal shifts as the receive dipole is rotated, but the signal remains a sinusoid. One of the

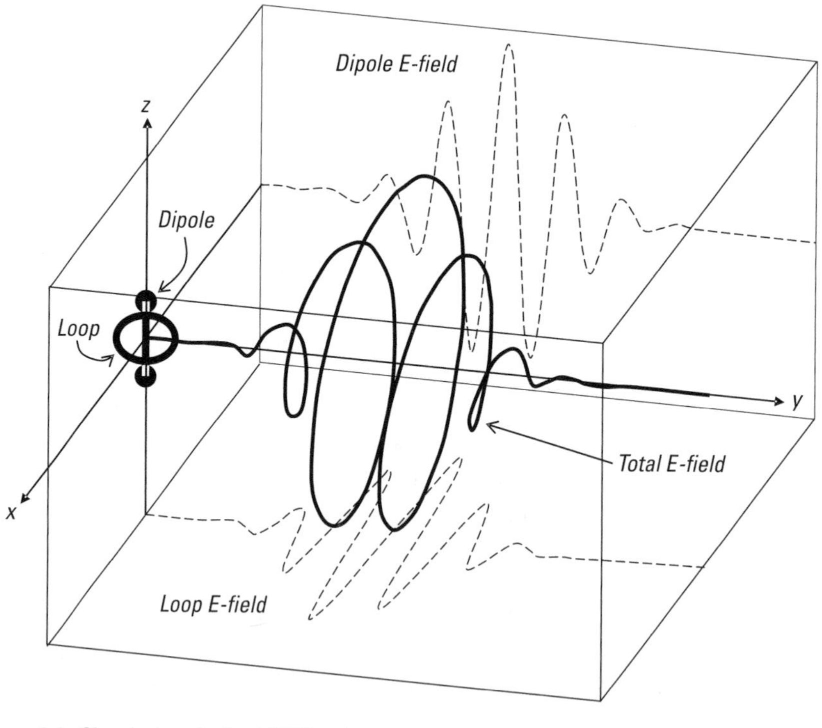

Figure 1.4 Circularly polarized UWB pulse.

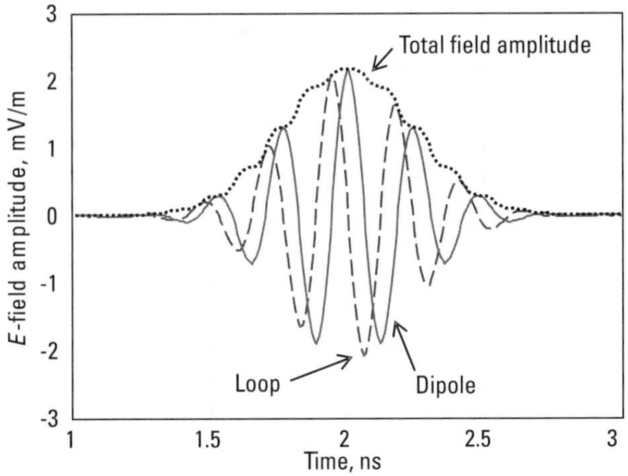

Figure 1.5 UWB circularly polarized field components.

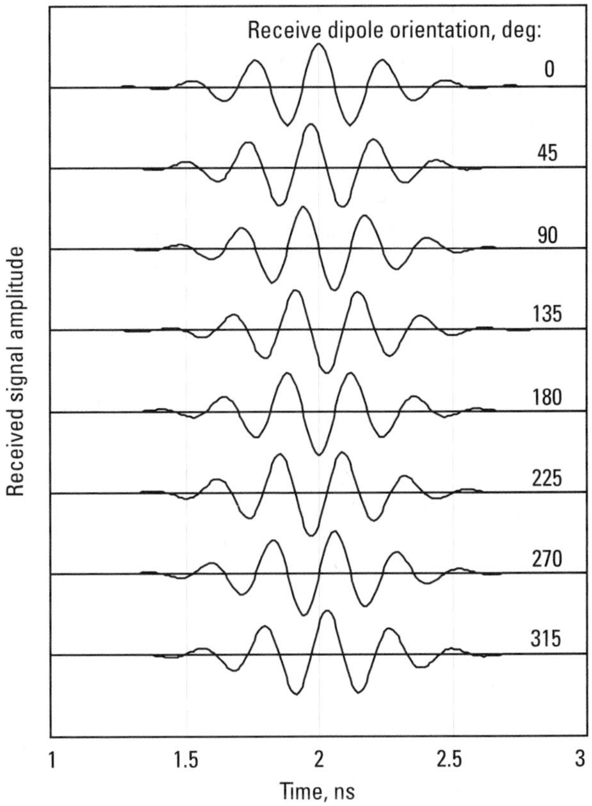

Figure 1.6 Circularly polarized UWB signal received by linearly polarized antenna.

consequences of "suppressing" $e^{j\omega t}$ in the time-harmonic solution is that we tend to suppress our understanding of the physics as well.

Equations (1.32) and (1.33) reduce to (1.26) to (1.28) and (1.38) and (1.36) reduce to (1.29) to (1.31), when a sinusoidal current $Ie^{j\omega t - jkr}$ is used in place of $I(\tau)$ in (1.32) to (1.36). UWB current pulses are shown in Chapter 6 and described in [16, 17].

1.3.8 Radiation Zones

Inspection of (1.32), (1.33), and (1.26) to (1.28) for the dipole and (1.35), (1.36), and (1.29) to (1.31) for the loop reveals a very complex field structure. Components of the fields vary as the inverse third power of distance r, inverse square of r, and the inverse of r. In (1.32), we see that there is a transition point where the energy associated with a pulse $I(t)$ equals the energy associated with r/c times the energy associated with the time derivative of the pulse. We will see in Chapter 6 that the energy associated with the time derivative of a UWB pulse is very nearly equal to ω_c^2 times the energy associated with a UWB pulse, where frequency ω_c is the geometrical center frequency of the UWB pulse energy. In fact, as the pulse becomes narrower and narrower in bandwidth (its time representation approaches a sinewave), the two become equal, which is the harmonic wave solution (see [17]). Our discussion, then, continues for the sine wave solution.

In the near field or induction region of the idealized infinitesimal loop (but for $r >> Sk$ for the loop and $r >> dl$ for the dipole), for kr<<1 the magnetic fields vary as the inverse third power of distance. The region where kr is nearly unity is part of the radiating near field of the Fresnel zone. The inner boundary of that zone is commonly taken to be $r^2 > 0.38 D^3/\lambda$ [8], and the outer boundary is $r < 2D^2/\lambda$, where D is the largest dimension of the antenna. The outer-boundary criterion is based on a maximum phase error of $\pi/8$. There is a significant radial component of the field in the Fresnel zone.

The far-field, or Fraunhofer, zone is region of the field for which the angular radiation pattern is essentially independent of distance. That region is usually defined as extending from $r > 2D^2/\lambda$ to infinity, and the field amplitudes there are essentially proportional to the inverse of distance from the source. The far-zone behavior will later be identified with the basic free space propagation law.

We can study the "induction zone" in comparison to the far field by considering induction zone coupling, which was investigated by L. A. Hazeltine [18] and which applied to low-frequency radio receiver designs of his time. Today, that solution might be applied to the design of a miniature radio module where inductors must be oriented for minimum coupling. The problem is one of finding the geometric orientation for which two loops have minimum coupling in the induction zones of their near fields and that illustrates that "near-field" behavior differs fundamentally and significantly from "far-field" behavior. To solve the problem, we invoke the principle of reciprocity, which states

$$\int_V \left[\mathbf{E}^b \cdot \mathbf{J}^a - \mathbf{H}^b \cdot \mathbf{M}^a\right] dV \equiv \int_V \left[\mathbf{E}^a \cdot \mathbf{J}^b - \mathbf{H}^a \cdot \mathbf{M}^b\right] dV \qquad (1.37)$$

That is, the reaction on antenna (a) of sources (b) equals the reaction on antenna (b) of sources (a). For two loops, we want to find the angle θ for which the coupling between the loops vanishes; that is, both sides of (1.37) are zero. In the case of the loop, there are no electric sources in (1.37), so $\mathbf{J}_a = \mathbf{J}_b = 0$, and $\mathbf{M}_a$ here is aligned with **z**, the unit vector parallel to the *z*-axis. For now, we orient the second loop moment so that $\mathbf{M}_b = \cos(\theta - \theta')\mathbf{r} - \sin(\theta - \theta')\theta$. That is, $\mathbf{M}_b$ is tilted at an angle θ' with respect to the *z*-axis, but both θ and θ' are in the same plane. Retaining only the inductive field components and clearing common constants, (1.29) and (1.30) are placed into (1.37). To minimize the loop coupling, we require that $\mathbf{H}_a - \mathbf{M}_b = 0$, or

$$\left[H_r\mathbf{r} + H_\theta\boldsymbol{\theta}\right] - \left[\cos(\theta - \theta')\mathbf{r} - \sin(\theta - \theta')\boldsymbol{\theta}\right] = 0 \qquad (1.38)$$

It is easy to show that (1.38) reduces to

$$\left(2\cos^2\theta - \sin^2\theta\right)\cos\theta' + 3\cos\theta\sin\theta\sin\theta' = 0 \qquad (1.39)$$

The loops will not couple when the plane of one loop is tangent to the solenoidal magnetic fields of the other loop. Two especially useful cases emerge. One special case was investigated by Hazeltine [18]: $\theta' = 0$, and the loops are in parallel planes, so we are left with $2\cos^2(\theta) - \sin^2(\theta) = 0$, for which $\theta = 54.736°$. As shown in Figure 1.7, two loops parallel to the *x-y* plane, whose centers are displaced by an angle of 54.736° with respect to the *z*-axis, will not couple in their near fields; the plane of loop *b* there is tangent to the magnetic field of loop *a*. In the second interesting case, the loops are mutually orthogonal ($\theta' = 90°$) so that $3\cos\theta\sin\theta = 0$ and, thus, will not couple when $\theta = 0°$ or 90° (shown as *a* to *c* in Figure 1.7).

🖫 [1-3a.mcd] Using (1.37), prove that in the near fields of two loops in parallel planes, the orientation of the loops for which the coupling between the loops vanishes is the one shown in Figure 1.7.

The loop-coupling problem can now be investigated in the far field by applying (1.37) with (1.29) and (1.30), retaining only the $1/r$ components of

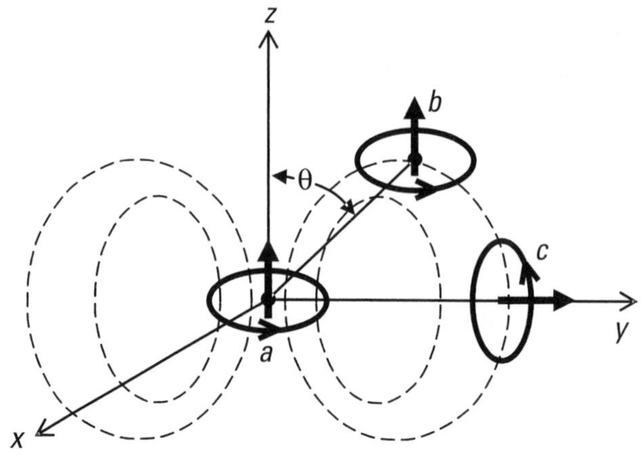

Figure 1.7 One loop in the plane of the other loop magnetic field (θ = 54.736° for loops in parallel planes, 90° for perpendicular planes) will not couple.

the fields. Only the H_θ term of the magnetic field survives into the far region, and by inspection of (1.30), the minimum coupling occurs for θ = 0° or 180°. Figure 1.8 compares the coupling (normalized to their peak values) for loops in parallel planes as a function of angle θ for the induction-zone case (kr = 0.001), an intermediate region (kr = 2), and the far-field case (kr = 1,000). The patterns are fundamentally and significantly different.

[1-3b.mcd] Using (1.37), show that in the far fields of two loops in parallel planes, the orientation of the loops for which the coupling between the loops vanishes is θ = 0° or 180°.

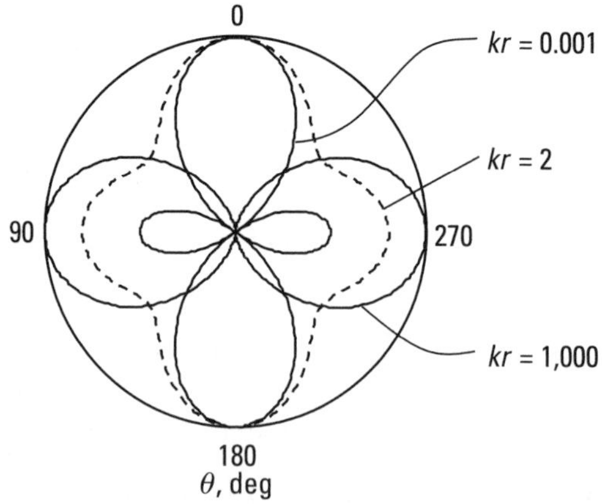

Figure 1.8 Normalized induction-zone, intermediate-zone, and far-zone coupling between loops in parallel planes.

1.4 Basic Radiowave and Antenna Parameters

Terms and definitions commonly encountered in the study of antennas and radio wave communication can be found in Institute of Electrical and Electronics Engineers (IEEE) [19, 20] and European Telecommunications Standards Institute (ETSI) [21] standards, as well as in FCC rules [3] and ITU-R reports and recommendations [22]. Here, we will present some of the parameters and definitions that are especially relevant and specific to the problems in this textbook.

Antenna factor (AF) in a receiving antenna is the ratio of electric field strength to the voltage across the terminating impedance connected to the antenna. By common usage in the telecommunications industry [23], the antenna factor is sometimes stated under conditions where the field incident on the antenna is not uniform along the antenna. Antenna factor is therefore, by common usage, a parameter that is measurement site dependent.

Antenna illumination efficiency, sometimes called the *aperture illumination efficiency*, is the ratio of the directivity D of an antenna to its reference directivity D_{ref}, given by

$$D_{ref} = \frac{4\pi A_{area}}{\lambda^2} \tag{1.40}$$

written in terms of the wavelength λ and aperture area A_{area}.

Antenna pattern lobes are defined as follows:

- *Back lobe* is the radiation lobe whose axis makes an angle of 180° with respect to the axis of the main lobe. The back lobe is not defined for omnidirectional vertical collinear array antennas.
- *Front-to-side-lobe ratio* is the ratio of the maximum directivity of an antenna to the peak radiation in a specified side-lobe direction.
- *Major lobe*, or the *main lobe*, is the antenna pattern radiation lobe containing the direction of maximum radiation.
- *Side lobes* are antenna pattern radiation lobes in any direction other than that of the main or major antenna pattern lobe.

Aperture distribution is the field over the antenna aperture described by amplitude, phase, and polarization distributions. For collinear arrays of dipole antennas, the aperture distribution is given approximately by the dipole excitations.

Beamwidth is the angular width of the main lobe of an antenna far-field radiation pattern as measured between the amplitude points on the main pattern lobe that are 3 dB below the peak of the main lobe. The beamwidth in a plane varies with the inverse of the effective antenna dimension in that same plane. Beamwidth and directivity are also inversely related. See Section 2.5 for further discussion of collinear antenna vertical beamwidth and directivity.

Bandwidth is commonly defined by the frequency band of a rectangular power spectral density (PSD) that integrates to the actual signal power having the same peak PSD.

- *3-dB bandwidth:* For simple filters, the bandwidth equals the frequency separation between the 3-dB down points of the filter response.
- *99% power bandwidth:* This is the bandwidth containing 99% of the total signal power.
- *Fractional bandwidth:* This is the bandwidth divided by the arithmetic average of the upper and lower bandwidth defining points.
- *UWB bandwidth:* See *UWB.*

Collinear array antenna is a linear arrangement of radiating elements that are usually polarized in the direction of the antenna axis and exhibit a nominally omnidirectional pattern in the plane perpendicular to the antenna axis.

Directive gain (see also *gain*) is also a far-field quantity and is defined as the ratio of radiation density in a particular angular direction to the radiation density of the same power P_{rad} radiated isotropically. Directive gain can be found from the far-field radially directed Poynting vector in ratio to the average Poynting vector over the radian sphere.

$$D(\theta,\phi) = \frac{\left|(\mathbf{E}\times\mathbf{H}^*)\cdot\mathbf{r}\right|}{\frac{1}{4\pi}\int_0^{2\pi}\int_0^{\pi}\left|(\mathbf{E}\times\mathbf{H}^*)\cdot\mathbf{r}\right|\sin(\theta)d\theta d\phi} \tag{1.41}$$

It is easy to show that the directive gain of the infinitesimal current element considered earlier can be found using (1.41) to be $D = 1.5\sin^2(\theta)$ by noting that the functional form of the product of E and H is simply $\sin^2(\theta)$ in the far field and by carrying out the simple integration in the denominator of (1.41). From duality, the directive gain of the infinitesimal loop is the same as that of the infinitesimal dipole.

Directivity D is the directive gain in the direction of maximum radiation density and can be expressed as

$$D = \frac{P_d 4\pi r^2}{P_{rad}} \tag{1.42}$$

As defined in (1.42), directivity D is quoted relative to isotropic radiation. By applying (1.41) to the infinitesimal dipole or loop, we find that the directivity is 1.5, which expressed in decibels is 1.76 dBi (decibels relative to an isotropic radiator). The application of (1.41) and (1.42) to a resonant half-wave dipole antenna [7] results in a directivity of 2.15 dBi [see (2.42)].

Effective aperture A_e of an antenna is defined under matched polarization conditions as

$$A_e = \frac{\lambda^2}{4\pi} G \tag{1.43}$$

where λ is the wavelength. It relates the gain G of an antenna to the "power-collecting" aperture area of an antenna. The product of effective aperture A_e and radiation density P_d gives the received power of an antenna. From (1.43) and with $G = 1$, we can see that the effective aperture of the isotropic radiator is $\lambda^2/4\pi$. In the case of aperture-type antennas, like flat plate antennas and parabolic reflector antennas, A_e is directly proportional to the aperture area but is usually smaller because the actual antenna illumination is nonuniform; hence, illumination efficiency less than 100%.

Gain G of an antenna is usually stated as the peak gain; hence, it is the product of directivity D and efficiency η:

$$G = D\eta \tag{1.44}$$

The efficiency does not include losses arising from impedance and polarization mismatches. Gain is stated in decibels relative to an isotropic radiator (dBi) or, more commonly in the telecommunications industry, in dBd, gain relative to a lossless resonant half-wave dipole gain of 2.15 dBi.

Impedance with regard to transmission lines and fields is defined in terms of the following:

- *Characteristic impedance* refers to the ratio of the phasor voltage to phasor current on an infinite two-conductor transmission line. The voltage is the integral of the electric field along a path between the conductors, while the current is equal to the integral of the magnetic field around one of the conductors.
- *Intrinsic impedance* refers to the ratio of the fields E and H for a plane wave in unbounded medium. That ratio is $\eta_0 = c\mu_0\Omega$ in free space and can be defined in terms of the square root of the ratio of permeability to permittivity for any medium.

- *Wave impedance* refers to the ratio of an electric field component to a magnetic field component at the same point of the same wave. For a plane wave in unbounded space, the wave and intrinsic impedances are the same. In the presence of higher-order modes, there can be as many wave impedances as there are combinations of electric and magnetic fields.

The relationships between these impedances can be appreciated in the operation of a transverse electromagnetic (TEM) cell [24], which is a test device developed for producing calibrated EM fields in a shielded environment. The device is essentially a physical expansion of a coaxial transmission line exhibiting a *characteristic impedance* of $R = 50\Omega$, yet it provides a *wave impedance* in its operating range equal to η_0, the free space *intrinsic impedance.* In a TEM cell, the outer conductor is expanded to a rectangular shape of height $2h$, and the center conductor is flattened and centered between the top and bottom walls so that in a match-terminated TEM cell operating within its frequency range, the relationship between the supplied power P and field strength E within the test volume is

$$P = (Eh)^2 \;/\; R \tag{1.45}$$

and between the transmission line voltage V and field strength, it is

$$E = V \;/\; h \tag{1.46}$$

At frequencies above the operating ranges, higher-order transmission-line modes are present, and other *wave impedances* are present.

Omnidirectional antenna is an antenna having an essentially constant directivity in a given plane of the antenna and a directional pattern in any orthogonal plane. A vertical dipole or vertical collinear array of dipoles has essentially an omnidirectional pattern on the horizon but a vertical or elevation pattern with varying directivity.

Polarization of a radiated wave is the far-field orientation of the electric field vector as it travels along the direction of propagation. In general, the far-zone electric field will contain E_θ and E_φ components, which are possibly related by a time derivative (or phase difference for time-harmonic waves).

For time-harmonic waves, a complex vector $\mathbf{h}_a$ is used to denote the amplitude and relative phase of the orthogonal polarization components in terms of the components h_θ in the $\boldsymbol{\theta}$ unit vector orientation and h_φ in the $\boldsymbol{\phi}$ unit vector orientation. The general expression for the polarization vector is

$$\mathbf{h}_{\mathbf{a}} = h_\theta \boldsymbol{\theta} + h_\phi \boldsymbol{\phi} \tag{1.47}$$

When h_θ and h_ϕ are in phase, the polarization is linear; when there is a phase difference between h_θ and h_ϕ, the polarization is elliptical. Two special cases of elliptical polarization are the two senses of circular polarization. Right-hand circular polarization (RHCP) is defined as

$$\mathbf{h}_{\mathbf{RHC}} = \frac{\boldsymbol{\theta} - j\boldsymbol{\phi}}{\sqrt{2}} \tag{1.48}$$

and left-hand circular polarization (LHCP) is defined as

$$\mathbf{h}_{\mathbf{LHC}} = \frac{\boldsymbol{\theta} + j\boldsymbol{\phi}}{\sqrt{2}} \tag{1.49}$$

The expressions (1.48) and (1.49) are normalized to unity, but that is neither general nor necessary. It is evident that LHCP and RHCP are orthogonal because the dot product of $\mathbf{h}_{\mathrm{RCH}}$ and the complex conjugate of $\mathbf{h}_{\mathrm{LCH}}$ is zero. See Section 3.4 for further discussions of polarization; a more complete treatment is available in [25].

Polarization axial ratio (AR) is defined as the ratio of the major axis to the minor axis of the polarization ellipse. In terms of the polarization vector quantities,

$$AR = \left[\frac{1 + \left| \dfrac{h_\phi}{h_\theta} \cos\left[\arg(h_\phi) - \arg(h_\theta)\right] \right|^2}{\left| \dfrac{h_\phi}{h_\theta} \sin\left[\arg(h_\phi) - \arg(h_\theta)\right] \right|} \right]^{\pm 1} \tag{1.50}$$

where arg(...) is the phase angle of the complex quantity. The sign of the exponent is chosen so that AR is between 0 and 1.

Polarization tilt angle τ_{pol} is defined here as the angle in the plane perpendicular to the propagation direction between a reference vector and the major axis of the polarization ellipse.

UWB elliptical polarization is defined by the super position of the elementary dipole and an elementary loop fed with the same UWB current pulse $I(t)$, so the far-field energies are proportional to

$$W_\theta = \left(\frac{\eta_0 \sin(\theta)}{4\pi r}\right)^2 \int\left(\frac{\Delta h}{c}\frac{\partial}{\partial t}I(t)\right)^2 dt \tag{1.51a}$$

and

$$W_\phi = \left(\frac{\eta_0 \sin(\theta)}{4\pi r}\right)^2 \int\left(\frac{\Delta S}{c^2}\frac{\partial^2}{\partial t^2}I(t)\right)^2 dt \tag{1.51b}$$

Hence,

$$\frac{E_D(r,t)^2}{W_\theta} + \frac{E_L(r,t)^2}{W_\varphi} = 1 \tag{1.51c}$$

Referring to (1.33) and (1.35), $E_D(r,t)\boldsymbol{\theta} + E_L(r,t)\boldsymbol{\phi}$ traces out an ellipse in space-time. The polarization is circular when $W_\theta = W_\phi$.

Poynting vector **S** is the complex flux density obtained by the vector cross-product of the electric and magnetic field vector quantities

$$\mathbf{S} = \mathbf{E} \times \mathbf{H}^* \tag{1.52}$$

The Poynting vector is defined over all regions of space. In the close near field, **S** is largely reactive, indicating that the near-field energy is largely stored and exchanged between electric and magnetic fields. In the far zone, the ratio of the electric to the magnetic field becomes η_0, the intrinsic impedance of the medium, and the Poynting vector represents a real power-density flow in the radial direction **r**.

Radiation efficiency is the ratio of total power radiated by the antenna to the net power accepted by the antenna from the connected feed line.

Radiation power density P_d is defined as power radiated per square meter. It is a far-field quantity and can be expressed as the dot product of the Poynting vector with the unit radial vector $\mathbf{S} \cdot \mathbf{r}$ or in terms of the RMS electric field strength E or the RMS magnetic field strength H.

$$P_d = \frac{E^2}{\eta_o} = H^2\eta_o \tag{1.53}$$

Receiver-sensitivity test site is an open-field antenna test site, generally complemented with a simulated human body test device (see SALTY and SALTY-LITE) and with transmitting height and measurement distance arranged so that the simulated test device is illuminated by no more than a single lobe formed by the direct and ground-reflected waves originating from the transmitting antenna. The simulated human body test device rests on Earth ground and supports a personal communications device at a nominal height of 1m above ground.

SALTY is a simulated human body test device consisting of a 1.7m-tall dielectric-walled cylinder filled with saline water with sodium chloride added in a concentration of 1.5g per liter of water. The standard SALTY cylinder is 0.305m in diameter. See Chapter 10 for more detail.

SALTY-LITE is a simulated human body test device consisting of concentric dielectric-walled cylinders with a 1.32m-tall column of saline water having sodium chloride added in a concentration of 4.0g per liter of water, filling a 4-cm radially thick space between the concentric cylinders. The standard SALTY-LITE cylinder is 0.305m in outer diameter. See Chapter 10 for more detail.

UWB, or *ultrawideband* [3, 22], is a technology for short-range radio communications, involving the intentional generation and transmission of radio frequency (RF) energy that spreads out over a very large frequency range and may overlap several frequency bands allocated to radio communication services. Devices using UWB technology typically have intentional radiation from the antenna that at any point has a fractional bandwidth equal to or greater than 0.20 or has a UWB bandwidth equal to or greater than 500 MHz, regardless of the fractional bandwidth.

- *UWB bandwidth* is frequency band bounded by the points that are 10 dB below the highest radiated emission, as based on the complete transmission system, including the antenna. The upper boundary is designated f_H and the lower boundary is designated f_L. The frequency at which the highest radiated emission occurs is designated f_M. The center frequency f_C equals $(f_H + f_L)/2$.
- UWB fractional bandwidth equals $2(f_H - f_L)/(f_H + f_L)$.
- *UWB impulse* is a surge of unidirectional polarity that is often used to excite a UWB band-limiting filter whose output, when radiated, is a UWB pulse.
- *UWB pulse* is a radiated, short, transient UWB signal whose duration is nominally the reciprocal of its bandwidth.
- *UWB transmitter* [3] is an intentional radiator that at any point has a fractional bandwidth equal to or greater than 0.20 or has a UWB bandwidth equal to or greater than 500 MHz, regardless of the fractional bandwidth.

1.5 Summary

We started with a short history of communications and an overview of Maxwell's equations. We introduced the personal communications problem in terms of the radio link, including a fixed-site antenna, propagation to and within an urban or suburban environment, and a small antenna that might be close to a human body. We also viewed the personal communications problem in the personal-area space. The fundamentals of electromagnetics were presented, and we noted that most electromagnetics problems of interest to the communications path link can be solved only in approximation. Both the narrowband time-harmonic and UWB pulse solutions to communications problems were introduced. The concepts of radiation and propagation were introduced, particularly in view of the telecommunications problems that will be considered in detail in subsequent chapters. We defined some important antenna and propagation parameters.

Problems

1.1 In Cartesian coordinates, the vector *curl operation* on a vector **A** is

$$\nabla \times \mathbf{A} = \mathbf{x}\left[\frac{\partial}{\partial y}A_z - \frac{\partial}{\partial z}A_y\right] + \mathbf{y}\left[\frac{\partial}{\partial z}A_x - \frac{\partial}{\partial x}A_z\right] + \mathbf{z}\left[\frac{\partial}{\partial x}A_y - \frac{\partial}{\partial y}A_x\right]$$

The *divergence operation* on a vector **A** is

$$\nabla \cdot \mathbf{A} = \frac{\partial}{\partial x}A_x + \frac{\partial}{\partial y}A_y + \frac{\partial}{\partial z}A_z$$

The *gradient operation* on a scalar F is

$$\nabla F = \mathbf{x}\frac{\partial}{\partial x}F + \mathbf{y}\frac{\partial}{\partial y}F + \mathbf{z}\frac{\partial}{\partial z}F$$

Find $\nabla \times \mathbf{A}$ and $\nabla \cdot \mathbf{A}$ and ∇F

when $\mathbf{A} = \mathbf{y}\cos(x)$ and $F = \cos(x)\sin(y)$

and when $\mathbf{A} = 3\mathbf{x} + 4\mathbf{y} + 5\mathbf{z}$ and $F = xyz$

1.2 Can the magnetic flux density $\mathbf{B} = \mathbf{x}\sin(x) + \mathbf{y}\sin(y)$ exist in the region defined by $-1 \le x \le 1$ and $-1 \le y \le 1$?

Ans:

B violates (1.4) because it is not identically zero in the specified region; hence, it cannot exist.

1.3 In a source-free region, **E** = sin(z) **z**. Is **B** time varying?

Ans:

No. Apply (1.1) and the cross-product is zero. This is not a time-harmonic solution, so use the time derivative of **B**. Hence, $-\partial \mathbf{B}/\partial t = 0$.

1.4 Find the charge density in a region where **D** = 5 y**y**.

Ans:

Charge density $\rho = \nabla \cdot \mathbf{D}$, so $\rho = 5\ \mathrm{C/m^2}$.

1.5 A charge qC exists on a small spherical surface of radius $R << 1$m with charge density ρ C/m^2. Sketch the electric field and give its magnitude. What is the potential at a distance of 1m?

1.6 A point charge of $+q$C is surrounded by a perfect electric conducting (PEC) sphere at a radius of 1m from the charge. (a) Sketch the electric fields within the sphere and outside the sphere. (b) The charge is now displaced a small amount ($+\delta << 1$m) along the z-axis, indicate the changes to the electric field outside the PEC sphere. (c) A second charge, $-q$C, is introduced within the PEC sphere at $z = -\delta$. Sketch the fields inside and outside the PEC sphere.

1.7 Consider the system in Figure 1.P1 of two capacitors, $C_1 = 100\ \mu$f and $C_2 = 300\ \mu$f, connected in a series arrangement with a voltage source of $V_1 = 100$V as in (a). After the system has reached steady state, the circuit is reassembled with the voltage source absent and with the capacitors

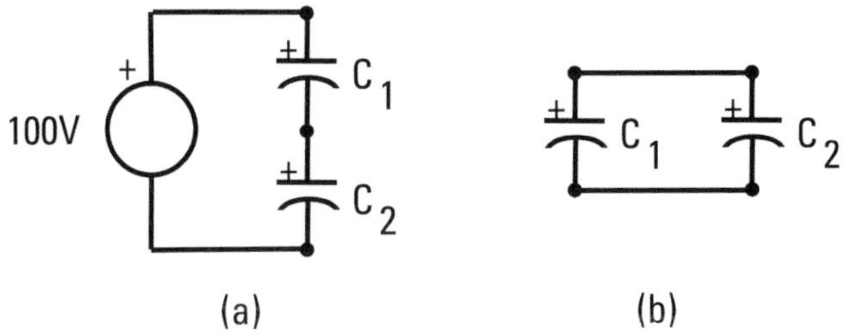

Figure 1.P1 C_1 and C_2 are charged as in (a) and reconnected as in (b).

arranged in parallel as in (b). The stored energy in a capacitor is $W = 1/2CV^2$, and the stored charge is $q = CV$.

Find the stored energy and stored charge in each capacitor in the first arrangement (a) and in the reassembled arrangement (b). Has energy been conserved? Has charge been conserved? Explain.

1.8 Given electric field $\mathbf{E} = \mathbf{x}\ 0.5\ e^{-jkz}$, where k is a constant, find the associated magnetic field **B**.

Ans:

Using (1.1), $\mathbf{B} = \mathbf{y}\ (0.5k/\omega)\ e^{-jkz}$. Note that **B** satisfies (1.4).

1.9 Take Maxwell's equations in a source-free region (where $\mathbf{J} = \rho = 0$), and using the free space constitutive relations, show that taking the curl of (1.1) and substituting (1.2), then applying the vector identity

$$\nabla \times (\nabla \times \mathrm{E}) = \nabla (\nabla \cdot \mathrm{E}) - \nabla^2 \mathrm{E}$$

results in a wave equation for **E**.

1.10 Show that the wave equation $\nabla^2\mathbf{E} + k^2\mathbf{E} = 0$ has the solution $\mathbf{E} = \mathbf{x}\ E_0\ e^{-jkz}$ for the simple case where **E** is parallel to the x-axis and is a function of only the z-coordinate. Find the corresponding magnetic field **H**.

1.11 Starting with the magnetic field strength of a current element (1.28), use the radiated power density (1.42) and directive gain (1.41) to show that the radiation resistance of a small current element of length h is

$$R_{rad} = (kh)^2\eta_0/6\pi$$

1.12 Show that the maximum directivity of an infinitesimal current element whose fields are described by (1.26) to (1.28) is 1.5, or 1.76 dBi.

1.13 The antenna factor (AF) is the ratio of electric field strength to voltage developed across the terminating load for the antenna. Assuming a lossless system, calculate the AF for a system of a small current element of length h ($h << \lambda$) and a lossless matching network terminating in 50Ω match.

Ans:

We note that A_e is given by (1.43) with $G = 1.5$. Then,

(Power density × aperture area) $= E^2A_e/\eta_0 = V^2/50$

so $E/V = 7.94/\lambda$.

1.14 Use (1.42) to calculate the total power radiated by the sun. Assume the sun radiates energy isotropically and that the power density is 1.4 kW/m^2. What is the total power incident on Earth. The Earth's radius is 6,368 km, and its distance from the sun is 1.5×10^8 km.

Ans:

$P_{total} = (1.4\ \text{kW/m}^2)(4\pi)(1.5 \times 10^8\ \text{km} \times 1{,}000\text{m/km})^2 = 4 \times 10^{23}\ \text{kW}.$

$P_{Earth} = (1.4\ \text{kW/m}^2)(\pi)(6{,}368\ \text{km} \times 1{,}000\text{m/km})^2 = 1.8 \times 10^{14}\ \text{kW}.$

1.15 Show that right-hand circular (RHC) polarized waves and left-hand circular (LHC) polarized waves are orthogonal.

Ans:

$\mathbf{h}_{RHC} \cdot \mathbf{h}_{LHC}{}^{*} = 0.$

1.16 A right-hand circular polarized wave reflects from a perfectly conducting plane. What is the polarization after reflection?

Ans:

The boundary condition for an $\mathbf{h}_{RHC}$ incident on a perfect conductor states that both orthogonal components must be zero at the interface, so the reflected incident electric field vectors point in the same direction, but their directions of travel are opposite. Hence, $\mathbf{h}_{RHC}$ becomes $\mathbf{h}_{LHC}$ traveling in the opposite direction. This has implications for satellite-to-mobile communications.

1.17 What is the axial polarization ratio of the time-harmonic wave $\mathbf{E} = 2\mathbf{x} + 3\mathbf{y}$?

Ans:

The x and y components are in phase, the polarization is linear, and 1/AR = 0.

1.18 What is the axial polarization ratio AR of the time-harmonic wave $\mathbf{E} = 5\mathbf{x} - j5\mathbf{y}$?

Ans:

The x and y components are in quadrature and of equal amplitude, the polarization is right-hand circular, and from (1.50), AR = 1.

1.19 A resonant half-wave dipole is 0.322m long and 1 mm thick. What is its effective aperture at 931 MHz?

Ans:

Use (1.43) with $G = 1.64$ (2.15 dBi), $A_e = 0.00825\ G\,\text{m}^2 = 0.0135\ \text{m}^2$.

1.20 A short dipole of total length 0.322m is used as an electric field probe at 33 MHz. What is its effective aperture and directivity at 33 MHz and at 10 MHz?

Ans:

Use (1.43), and assume that the antenna is decoupled from any attached feed line. The dipole is small compared to a wavelength, so directivity is $G = 1.5$, and $A_e = 6.57\ G\,\text{m}^2 = 9.85\ \text{m}^2$ at 33 MHz, and $A_e = 71.52\ G\,\text{m}^2 = 107.3\ \text{m}^2$ at 10 MHz.

1.21 A TEM cell has *characteristic impedance* of 50Ω and separation between the flat horizontal center conductor and the top wall of 0.4m. If the TEM cell is match-terminated at one end and fed with a matched generator at the other end, show that the relationship between (a) supplied power and field strength within the test volume is $P = E^2/312.5$, and (b) the transmission line voltage and field strength is $E = 2.5\,V$.

1.22 An engineer requires a field strength of 100 V/m at 150 MHz to perform radiation immunity testing on a small (5-cm maximum dimension) radio device. Design a procedure using (a) a test range with a dipole antenna, and (b) a TEM cell with center-conductor-to-top-wall separation of h = 0.12m to accomplish this test. Comment on the two results.

Ans:

(a) From (1.42) and (1.53) $E^2 = 30\ PG/r^2$ and $G = 1.641$. Using at least one-wavelength separation between the dipole and the test device, the transmitter power needed is $P = (100 \times 2)^2/(30 \times 1.641) = 812.5\text{W}$.

(b) The TEM cell field is $E = V/h$, and the supplied power is $P = V^2/50$, so the required power is $P = (Eh)^2/50 = 2.88\text{W}$.

The TEM cell has advantages in shielding and in power needed for the test.

1.23 Show that the near-field coupling is minimized between two loops displaced and angle θ from each other, and whose loop planes are tilted with respect to each other by an angle θ', when the following relationship derived from $\mathbf{H}_a \cdot \mathbf{M}_b = 0$ is satisfied:

$$(2\cos^2\theta - \sin^2\theta)\cos\theta' + 3\cos\theta\,\sin\theta\,\sin\theta' = 0$$

1.24 The magnetic field in (1.32) reduces to the harmonic wave solution when the current is chosen to be $I(t,r) = \exp(j\omega t - r/c)$. Show that (1.32) reduces to (1.28).

1.25 Show that the far-field wave impedance for a UWB pulse emitted by an infinitesimal dipole or an infinitesimal loop equals the intrinsic free space impedance.

1.26 A uniform current pulse $I(t)$ of 1 ns duration is applied to an infinitesimal dipole.

(a) Show that **E** eventually becomes the classical static solution for an electric doublet:

$$E = \frac{q\Delta h}{4\pi\varepsilon_0 r^3}\left[2\cos(\theta)\hat{\mathbf{r}} + \sin(\theta)\hat{\theta}\right]$$

(b) For the same current, what is the magnetic field at 1m from the doublet after 5 ns?

References

[1] Bryant, J. H., *Heinrich Hertz—The Beginning of Microwaves,* New York: IEEE Press, 1988.

[2] DeSoto, C. B., *Two Hundred Meters and Down,* Newington, CT: The American Radio Relay League, 1936.

[3] U.S. 47 C.F.R. Part15, Ultra-Wideband Operations FCC Report and Order, April 22, 2002.

[4] Burns, J. O., "Very Large Structures in the Universe," *Scientific American,* July 1986, pp. 38–47.

[5] Guth, A. H., and P. J. Steinhardt, "The Inflationary Universe," *Scientific American,* May 1984, pp. 116–128.

[6] Balanis, C. A., *Advanced Engineering Electromagnetics,* New York: John Wiley and Sons, 1989.

[7] Collin, R. E., *Antennas and Radiowave Propagation,* New York: McGraw-Hill, 1985.

[8] Jordan, E. C., and K. G. Balmain, *Electromagnetic Waves and Radiating Systems,* 2nd ed., Englewood Cliffs, NJ: Prentice Hall, 1968.

[9] Pocklington, H. C., "Electrical Oscillations in Wires," *Proc. of the Cambridge Phys. Soc.,* Vol. 9, 1897, pp. 324–333.

[10] Hallén, E., "Theoretical Investigation into Transmitting and Receiving Qualities of Antannae," *Nova Acta Regiae Soc. Ser. Upps.,* Vol. 2, November 4, 1938, pp. 1–44.

[11] Balzano, Q., O. Garay, and K. Siwiak, "The Near Field of Dipole Antennas, Part I: Theory," *IEEE Transactions on Vehicular Technology*, Vol. VT-30, No. 4, November 1981, pp. 161–174.

[12] Balzano, Q., O. Garay, and K. Siwiak, "The Near Field of Dipole Antennas, Part II: Experimental Results," *IEEE Transactions on Vehicular Technology*, Vol. VT-30, No. 4, November 1981, pp. 175–181.

[13] Balzano, Q., O. Garay, and K. Siwiak, "The Near Field of Omnidirectional Helices," *IEEE Transactions on Vehicular Technology*, Vol. VT-31, No. 4, November 1982, pp. 173–185.

[14] Storer, J. E., "Impedance of Thin-Wire Loop Antennas," *Transactions of AIEE*, Vol. 75, November 1956, pp. 606–619.

[15] Balzano, Q., and K. Siwiak, "The Near Field of Annular Antennas," *IEEE Transactions on Vehicular Technology*, Vol. VT-36, No. 4, November 1987, pp. 173–183.

[16] Schantz, H., *The Art and Science of Ultrawideband Antennas*, Norwood, MA: Artech House, 2005.

[17] Siwiak, K., and D. McKeown, *Ultra-wideband Radio Technology*, London: John Wiley and Sons, 2004.

[18] Hazeltine, L. A., "Means for Eliminating Magnetic Coupling between Coils," U.S. Patent 1,577,421, March 16, 1926.

[19] *IEEE Standard Definitions of Terms for Antennas*, IEEE Std 145-1993, SH16279, March 18, 1993.

[20] *IEEE Standard Definitions of Terms for Radio Wave Propagation*, IEEE Std 211-1990, SH13904, 1990.

[21] *Radio Equipment and Systems (RES); Improvement of Radiated Methods of Measurement (Using Test sites) and Evaluation of the Corresponding Measurement Uncertainties* (Draft), ETR Version 0.1.0, European Telecommunications Standards Institute (ETSI), Valbonne, France, October 1994.

[22] *Report of the Sixth Meeting of ITU-R Task Group 1/8, Geneva, meeting of October 12–20, 2005*, Document 1-8/443-E, January 17, 2006 Geneva, Switzerland.

[23] Smith, A. A., Jr., R. F. German, and J. B. Pate, "Calculation of Site Attenuation from Antenna Factors," *IEEE Transactions on Electromagnetic Compatibility*, Vol. EMC-24, No. 3, August 1982, pp. 301–312.

[24] Crawford, M. L., "Generation of Standard EM Fields Using TEM Transmission Cell," *IEEE Transactions on Electromagnetic Compatibility*, Vol. EMC-16, No. 4, November 1974, pp. 189–195.

[25] Stutzman, W. L., *Polarization in Electromagnetic Systems*, Norwood, MA: Artech House, 1993.

2

Fixed-Site Antennas

2.1 Introduction

We are exploring radio links that involve fixed-site antennas, propagation into and within urban and suburban environments, and small antennas next to, or in the hands of, people. In this chapter, we investigate the kinds of antennas and the kinds of localized impairments that occur at the fixed sites. Today, fixed sites include access points for wireless local-area networks (WLANs), and these further include ultrawideband (UWB) antennas, so we'll consider those scenarios also. For the roof- or tower-mounted case, we'll consider some of the radiation characteristics especially for vertically polarized telecommunications antennas. Beam shaping and "smart antenna" technology will be introduced as a design tool for emerging PCS designs. We will explore the interactions between fixed-site antennas and their local environments, and we will consider the distortions fixed-site antenna patterns encounter when these antennas are placed in the vicinity of radio towers and other antennas. The effect on range of antenna pattern distortions will be calculated. We will see that in some cases the interactions of driven antennas with scatterers, specifically Yagi-Uda antennas and corner reflectors, can be controlled to optimize pattern shape.

Antennas, unlike active circuits, do not produce gain. They are passive devices that can couple mutually and exhibit directivity and dissipative losses. In the telecommunications industry, vertically polarized systems are encountered almost exclusively, largely because of the standardization to simple, vertical, quarter-wave-long "whip" antennas in land-mobile applications. As a result, vertically polarized collinear arrays of radiating elements form the basic fixed-site antenna of the telecommunications industry. Also, vertically polarized collinear antennas do not require excessive rooftop "real estate" for mounting and exhibit modest wind-loading surface area compared with panel array antennas. They are the mainstay of the telecommunications industry because omnidirectional directivity (at the expense of elevation plane pattern compression) can be achieved in a relatively narrow profile structure in any frequency band of interest to personal communications. Sectored coverage antennas, on the other hand, achieve azimuth directionality by virtue of their horizontal physical dimension, or width. Sufficient width to achieve narrowed azimuth coverage is practical only at the higher frequencies (typically above 800 MHz). Site managers at many fixed-site locations are reluctant to lease space for excessively large antennas because their wind loading may be problematic and because they tend to shadow other proximately located antennas.

For body-mounted and handheld radio applications, the choice of vertical polarizations is particularly advantageous for operation at VHF. The body exhibits a whole-body resonance to vertical polarization, which, as will be shown in Chapter 10, can contribute significantly to VHF radio system link margin. In Chapter 5, we will see that in orbiting satellite applications, circular polarization is a better choice because of the higher frequencies planned (900 to 3,000 MHz) and because of geometrical considerations.

We will explore combining basic dipole radiating elements in collinear vertical omnidirectional antennas to shape the vertical, or elevation plane, radiation pattern to increase signal strength near the ground, where the other (user) antennas are located. The relationship to Fourier transforms will be shown, as will the relation to beamwidth. The consequences of directive patterns on field strengths near the ground will be shown. We will show how shaping the radiation patterns improves radio system coverage.

Recently, panel array antennas have emerged as components in communications system designs. The simplest panels are used to provide sectored azimuth coverage in cellular systems. Azimuth sectoring provides a method dividing a coverage area into zones or cells from the edge or a corner of a cluster of cells. Omnidirectional antennas, in contrast, are located at the centers of cells. Consequently, when the system designer has such a choice, sectored versus omnidirectional antennas allow system and economic trade-offs in coverage as a function of cell size, system capacity, and frequency-reuse possibilities. More complex array designs combine the multibeam technologies developed and refined by the defense industries of the early 1960s. When used with modern digital signal processing capabilities, they enable new system design strategies in emerging personal communications technologies to exploit spatial diversity, improved multipath fading performance, and increased system capacities.

2.2 Antennas as Arrays of Current Sources

The basic radiating element in large antennas is the half-wave dipole. Dipoles have been analyzed in excruciating detail [1–5], and some of their near-field behavior will be visited in Chapter 11, when we study small antennas used with portable telecommunications devices. Here, we are interested in array problems, so we model the z-directed half-wave dipole by its far-field elevation directivity pattern, given by E. C. Jordan and K. G. Balmain [6] for arbitrary half-length L wavelengths with sinusoidal current distribution as

$$F(\theta) = \sqrt{G_{dipole}} \frac{\cos[2\pi L \cos(\theta)] - \cos(2\pi L)}{\sin(\theta)[1 - \cos(2\pi L)]} \tag{2.1}$$

and written here in terms of field strength relative to isotropically radiated fields. The wave number is $k = 2\pi/\lambda$, and G_{dipole} is the dipole power gain given by (2.42). On the horizon, at the pattern maximum, the directive gain of a half-wave dipole ($L = 0.25$) is $G_{dipole} = 1.641$, or 2.15 dBi. Spherical coordinate angle θ is measured from the z-axis for a dipole aligned with z. Dipole elements, when arrayed in collinear fashion to form an omnidirectional pattern on the horizon, are highly directional in the vertical plane.

2.3 Pattern Multiplication and Array Factor

The directivity pattern of an array of identically oriented elements is the product of the element directivity pattern with the array factor. The array factor represents the pattern that would be obtained with isotropically radiating elements, but since isotropic elements are not real, it is necessary to multiply the array factor by the element pattern. In an array of N closely spaced dipoles, the $1/r_i$ factors of free space propagation for each i dipole in the far field are very nearly equal, so the combined radiation field will depend on the relative phasing of each dipole at the far-field point. The far-zone directivity pattern, in terms of field strength P_0 in RMS volts per meter, can be written with the element pattern $F(\theta)$ factored out as a multiplier of the array factor:

$$P_0(\theta,\phi) = F(\theta)\sum_{i=1}^{N} V_i e^{jk}\left|\mathbf{r}_i - \mathbf{R}_i\right| \tag{2.2}$$

The voltage excitation at the terminals of each dipole is V_i and the vector location of the dipole is $\mathbf{r}_i$, which has a magnitude r_i:

$$\mathbf{r}_i = r_i\left[\mathbf{z}\cos\theta_i + \left(\mathbf{x}\cos\phi_i + \mathbf{y}\sin\phi_i\right)\sin\theta_i\right]$$

The field point vector **R**, which has a magnitude R_o, is

$$\mathbf{R} = R_o\left[\mathbf{z}\cos\theta + (\mathbf{x}\cos\phi + \mathbf{y}\sin\phi)\sin\theta\right]$$

The *i*th element is pictured in the spherical coordinate system of Figure 2.1.

2.4 Collinear Antennas and Vertical-Plane Pattern Control

We will specialize to the collinear case where $\theta_i = 0$, and all of the dipoles are aligned along the z-axis, so $\mathbf{r}_i$ is purely a function of z, and $|\mathbf{R} - \mathbf{r}_i|$ can be written as

$$|\mathbf{R} - \mathbf{r}_i| = |\mathbf{z}(R_o\cos(\theta) - z_i) + \rho R_o\sin(\theta)| \tag{2.3}$$

The ρ is the unit vector in the x-y plane, and R_o is the magnitude of the vector **R** connecting the origin to the field point. So,

$$|\mathbf{R} - \mathbf{r}_i| = R_o\sqrt{1 + \left(\frac{z_i}{R_o}\right)^2 - \frac{2z_i\cos\theta}{R_o}} \tag{2.4}$$

which reduces in the far zone to

$$|\mathbf{R} - \mathbf{r}_i| \approx |R_o - z_i\cos\theta| \tag{2.5}$$

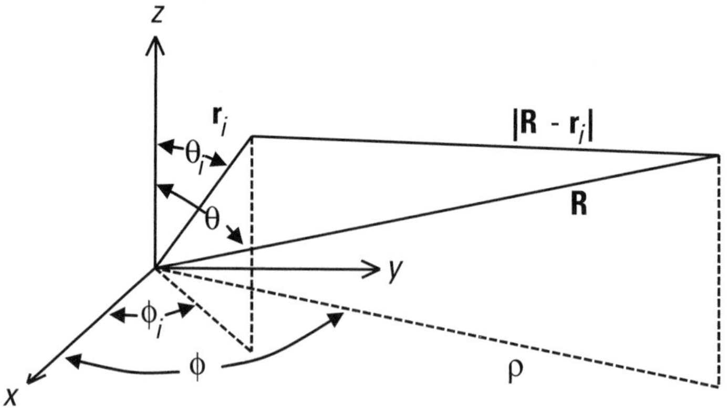

Figure 2.1 Coordinate system for the *i*th radiating element.

where $R_o >> z_i$, and $|\mathbf{R} - \mathbf{r}_i|$ points in the same direction as **R**. Expression (2.2) can now be rewritten for the collinear array as

$$P_c(\theta) = F(\theta)\sum_{i=1}^{N} V_i e^{jkz_i \cos(\theta)} \tag{2.6}$$

The kR_o phase factor common to each element has been suppressed. We retain a complex voltage excitation V_i for each of the i dipole elements. Relation (2.2) and its specialization to (2.6) express a very important principal of pattern multiplication. *The directivity pattern of an array of identically oriented elements is the product of the element directivity pattern with the array factor.*

2.5 Directivity and Beam Width for Omnidirectional Antennas

The elevation plane directivity of a collinear array is proportional to its length. For lengths L much greater than a wavelength λ, the directivity D_d with respect to dipole directivity is approximately

$$D_d = \frac{2L}{\lambda 1.64} = |F|^2 \tag{2.7}$$

where F is the same as in (2.1). It follows that the directivity in decibels relative to a dipole is

$$D_{dBd} = 10\log\left[\frac{2L}{\lambda 1.64}\right] \tag{2.8}$$

A function that can represent the elevation plane pattern of an omnidirectional collinear antenna with all amplitudes equal and all phases identical is

$$f(\theta) = \frac{\sin(b\theta)}{b\theta} \tag{2.9}$$

The parameter b is chosen by relation to the 3-dB beamwidth BW_{3dB}. In fact, from (2.9),

$$b = \frac{159}{BW_{3dB}} \tag{2.10}$$

where BW_{3dB} is specified in degrees. The relationship between beamwidth and the directivity of collinear antennas may be found using the method of A. McDonald [7]. We first define a normalizing factor based on the directivity of a dipole as

$$D_d = \frac{1}{1.64I} \tag{2.11}$$

where

$$I = \int_0^{\frac{\pi}{2}} \left[\frac{\sin(b\theta)}{b\theta} \right]^2 \cos(\theta) d\theta \tag{2.12}$$

Integral I can be evaluated in terms of the sine integral:

$$\mathrm{Si}(x) = \int_0^x \frac{\sin(t)}{t} dt \tag{2.13}$$

Approximate formulas [8] for the sine integral allow us to simplify the expression for I to

$$I = \frac{\pi}{2b} - \frac{1.37}{2b^2} \tag{2.14}$$

Using (2.10) for b allows us to rewrite (2.11) for D_d with respect to dipole directivity:

$$D_d = \frac{62}{BW_{3dB} - 0.0027BW_{3dB}^2} \tag{2.15}$$

The approximation of (2.15) is accurate even for an elementary dipole (BW_{3dB} = 90°). Directivity and the length of omnidirectional antennas are related by (2.7), while beamwidth and directivity are related by (2.15).

2.6 Array Antennas

It is apparent from an examination of the far-field expression of the array pattern (2.2) that the excitation distribution on the array elements and the shape of the far-field radiation patterns resemble a Fourier transform pair. In the case of the collinear antennas, that behavior illustrates the relationship between the length of an omnidirectional antenna and the directive gain in the far field, as will be shown in Section 2.6.1. With more complex structures, two-dimensional (azimuth and elevation) control of the far-field beam can result in null-fill designs, as is shown in Section 2.7 or, additionally, in multiple beams, as explored in Section 2.6.3. Multiple-beam antennas have roots in the defense industry of the 1960s [9–11] and are now emerging in PCS designs, particularly in the new 1.8-GHz PCS services in Europe, to exploit spatial diversity, improve performance under multipath fading conditions, and to improve radio frequency spectrum efficiency through geographic reuse strategies.

2.6.1 Collinear Array and Fourier Transform

The array pattern given by (2.2) is general. Radiating antenna elements may be arranged in any manner in three-dimensional space. In this section, we explore the Fourier transform pair analogy for a collinear (single-dimensional) arrangement of antenna elements and the effect on the far-field pattern along a single dimension of wave number space (elevation angle). The geometry of the antenna in the remaining orthogonal dimensions is essentially a point, so the effect in the corresponding wave number space (azimuth plane) is a uniform amplitude, or an omnidirectional pattern.

The far-zone field-strength pattern of a dipole element was given by (2.1), and a pattern of collinear array of dipoles was given by (2.6). For highly directive arrays, the element pattern of (2.1) may be replaced by unity. In that case, (2.6) may be written as

$$P(k \sin \theta) = \sum_{i=1}^{N} V_i e^{jz_i k \sin(\theta)} \tag{2.16}$$

Replacing the element patterns is the same as having isotropic elements, which means that (2.16) represents only the array factor. This can be compared with the discrete Fourier transform,

$$DF_m = \sum_{i=1}^{N} V_i e^{j(2\pi m/N)i} \tag{2.17}$$

where $(2\pi m/N)i$ corresponds in wave number space to $(k\sin\theta)z_i$. A long antenna on the z-axis will "transform" in the far field to a "short" pattern (narrow beamwidth) in wave number space (k sinθ). We keep this relationship in mind when considering the directivity patterns of high-gain antennas.

2.6.2 Horizontal-Plane Pattern Directivity

Expression (2.2) can be applied to the case of four z-directed (vertical) radiators spaced at 0.25 wavelengths on a square grid. As shown in Figure 2.2, the two elements to the north are fed 90° phase delayed with respect to the two remaining elements. The azimuth (horizontal) plane directivity pattern calculated from (2.2) shows a lobe toward the north and a corresponding null to the south. The directivity at the peak is 20log(2) = 6 dB relative to a single omnidirectional antenna fed with the same power.

[2-6.mcd] The array pattern of Figure 2.2 is generated by summing up four individual omnidirectional radiation patterns that are phased according to a planar array specialization of (2.2). Reproduce the pattern of Figure 2.2. By changing the relative phases of the individual components of the array, the peak of the beam can be directed and other beam shapes can be synthesized.

In the practical case, flat panels of radiators are used at frequencies above 800 MHz to control both horizontal plane and elevation plane radiation patterns. Typical (see [12]) directivity patterns for a practical flat panel antenna often used in the 820–900-MHz cellular telephone service are shown in Figure 2.3. The elevation plane pattern has only 7° beamwidth, while the horizontal plane pattern has 105° beamwidth. The peak directivity is 14 dB referenced to a dipole, or 16.15 dBi. Beam width here is defined, as usual, as the angular width of the main lobe of an antenna far-field radiation pattern as

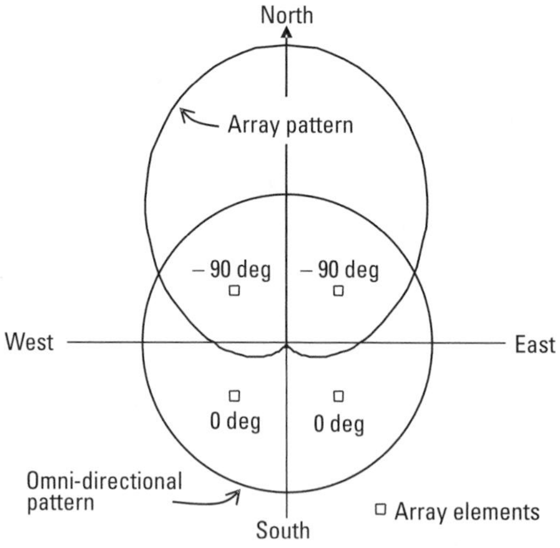

Figure 2.2 Azimuth directivity pattern for a four-element array.

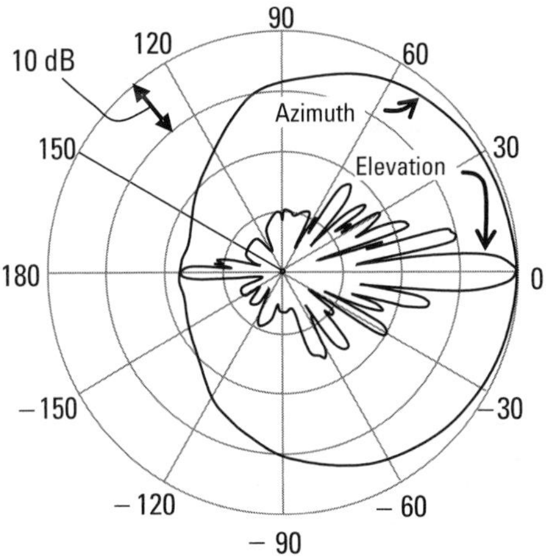

Figure 2.3 Azimuth and elevation plane patterns of a flat panel array.

measured between the amplitude points on the main pattern lobe that are 3 dB below the peak of the main lobe.

2.6.3 Aperture Antennas: Two-Dimensional Transforms

Exploring the general array pattern expression (2.2) further, we can arrange the radiating antenna elements in a two-dimensional plane. A flat panel array antenna was introduced in Section 2.4, and the radiation patterns are shown in Figure 2.3. The z-axis height of that antenna results in the narrow elevation plane beamwidth in the same manner as was described earlier for the collinear antenna. The extension of that antenna along the y-axis results in the narrowing of the azimuth radiation pattern shown in Figure 2.3. The Fourier transform analogy of the two-dimensional array, corresponding to the single-dimensional pair (2.16) and (2.17), consists of the far-field pattern expression

$$P_{2d}\left(k_z, k_y\right) = \sum_{r=1}^{R} \sum_{s=1}^{S} V_{r,s}\, e^{jz_r k \sin\theta}\, e^{jy_s k \cos\phi} \qquad (2.18)$$

and the two-dimensional transform

$$DF_{2d}(p, q) = \sum_{r=1}^{R} \sum_{s=1}^{S} V_{r,s}\, e^{j(2\pi p/P)r}\, e^{j(2\pi q/Q)s} \qquad (2.19)$$

In (2.19), the terms $(2\pi p/P)r$ and $(2\pi q/Q)s$ correspond to $k_z = (k\sin\theta)z_r$ in the azimuth plane and to $k_y = (k\cos\phi)y_s$ in the elevation plane. The function used to shape the antenna-element excitations along the z-axis will affect the elevation plane pattern, while the function used to shape the excitations along the y- (or x-) axis will shape the azimuth pattern. This relationship is used in the design of multiple-beam arrays that are finding their way into PCS designs.

2.7 Pattern Shaping of High-Gain Collinear Antennas

Relation (2.6) can be applied to the study of the elevation plane patterns of collinear antennas. For example, we take a six-element antenna with half-wave dipole elements spaced at intervals of w = 0.82 wavelengths, center to center, along the z-axis. The entire structure is 4.8 wavelengths tall, so we can estimate the directivity from (2.7) as 7.7 dBd. A good estimate of the physical height of a collinear antenna comprising N half-wave dipole elements is to sum up the interelement spacings, add a half-wavelength, and add 0.75m for the mounting hardware:

$$\textit{Physical length} = (N - 1)w + \lambda/2 + 0.75, \text{ m}$$

In Figure 2.4, the field-strength directivity patterns for the dipole and the 4.8 wavelength collinear are compared.

The normalized dipole field-strength peaks at 20log(1.28) = 2.15 dBi (decibels with respect to an isotropic source). The collinear array pattern peak is 20log(3.1) = 9.8 dBi, which is 7.7 dB above the dipole, the same as we get from (2.7). Applying (2.16), we note that the 3-dB beamwidth of the collinear array is about 10.3°. Directivity is achieved at the expense of beamwidth. We further

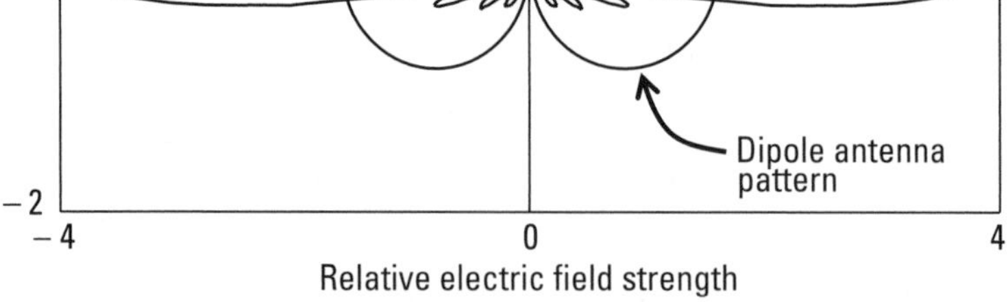

Figure 2.4 Field directivity patterns in the elevation plane of a six-element collinear and a dipole.

observe that pattern directed toward the ground under the main lobe has many nulls. There is a "cone of silence," or at least poor coverage, under a highly directive antenna mounted on a tall building or a tall tower. In the next section, we will tailor the antenna pattern so that the pattern toward the ground is more filled in.

[2-7a.mcd] The collinear array pattern is computed from (2.6). By changing the number of elements on the vertical axis, additional directive gain can be achieved. As an exercise, add vertical elements and observe the effect on the elevation plane pattern shape and beamwidth.

Recall that the antenna far-field pattern and the dipole element excitations are related like a Fourier transform pair. We can now impose element excitations that will tilt the main lobe and fill in the nulls pointing toward the ground. This process is often called *pattern synthesis.* The phase delays and normalized voltage amplitudes for tailoring the six-element patterns are listed in Table 2.1.

Qualitatively, the phase taper of 2.845 rad along the six elements, each spaced 0.86 wavelengths apart, represents a wave tilt of $\tan^{-1}[2.845/(0.86 \times 5 \times 2\pi)] = 6°$. Slightly less element-to-element phase change is chosen for the bottom elements compared with the top elements. A slight amplitude taper for the bottom elements is also chosen. The directivity pattern is computed from (2.2), normalized to a 1V excitation, and shown compared to a dipole pattern in Figure 2.5. The peak occurs nearly 7° below the horizon, as expected from the phase taper given by the approximate determination above.

The down-tilt of 7° was chosen here for illustrative purposes. Typical values are in the 3° to 15° range. Nulls toward the ground have been filled in, and the field strength toward the ground has generally been increased. The peak gain is 9.0 dBi, which is somewhat lower than the 9.8 dBi of the pattern in Figure 2.4. This 0.8-dB reduction in directivity is a consequence of the amplitude and phase tapers. The aperture of a tapered illumination results in reduced peak directivity but gives a measure of control over antenna pattern side-lobe levels.

[2-7b.mcd] The shaping of a collinear array pattern is computed from (2.6) with both amplitudes and phases tailored to shape the radiation

Table 2.1

Normalized Voltage Amplitudes, $|V_i|$, and Phases, $\arg(V_i)$, for a Tapered Illumination on the Six-Element Collinear Antenna

Element i =	**1**	**2**	**3**	**4**	**5**	**6**
$|V_i| =$	0.307V	0.372V	0.438V	0.438V	0.438V	0.438V
$\arg(V_i) =$	0 rad	0.175 rad	0.393 rad	0.655 rad	1.384 rad	2.845 rad

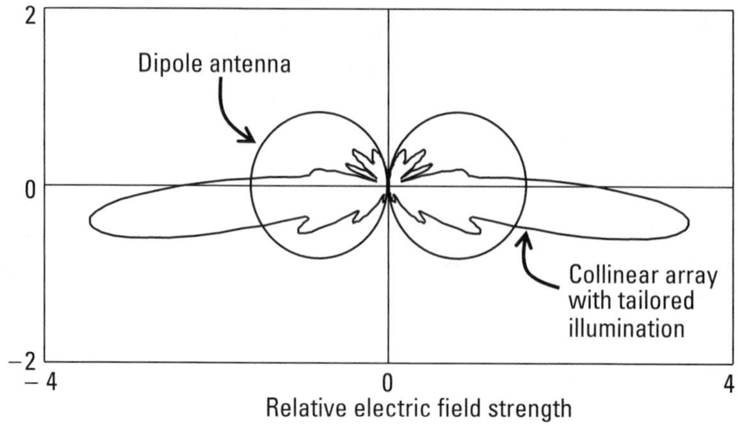

Figure 2.5 Field directivity pattern for a six-element collinear array with a tapered illumination compared with a dipole pattern.

pattern. By changing the phase taper along a collinear array the peak of the pattern can be moved in the elevation plane. Altering the relative amplitudes of the excitations changes the side-lobe levels. A nonlinear phase taper, along with amplitude control, is used to produce a collinear array with filled nulls and a beam-tilted pattern.

The practical impact of shaping the elevation pattern of omnidirectional antennas can be observed by calculating field strengths near the ground for antennas mounted on a tall structure. Field strength E V/m copolarized with the transmitter in terms of power P in watts radiated isotropically is

$$E = \sqrt{\frac{P\eta_o}{4\pi r^2}} \tag{2.20}$$

where $\eta_o = 376.73 \approx 120\pi$ is the intrinsic impedance of free space. We combine (2.20) with (2.6) to get

$$E(d) = \sqrt{\frac{30P}{r^2}}P_c(\theta) \tag{2.21}$$

The term under the square root in (2.21) represents the free space propagation of PW (set to 1W here) at a range of rm from an isotropically radiating antenna. Range $r = (h^2 + d^2)^{1/2}$, where h is the antenna height, and d is the distance along the ground. The remaining term from (2.6) is the collinear array directivity with normalized V_i voltage excitations. Relation (2.21) is applied to the six-element collinear antennas, whose directivities are described in Figures

2.4 and 2.5. The field strengths calculated using (2.21) are shown in Figure 2.6, which shows field strength near the ground for the six-element collinear antennas with and without tailored excitations compared to the field strength from a dipole. Ground contributions are not considered here.

From Figure 2.6, it is clear that the field strength along the ground remains better behaved for the antenna with tailored illumination than for the original collinear antenna. In some regions, the field-strength difference is many tens of decibels and can contribute to severe coverage deficiencies at relatively close distances from the antenna support structure. The dipole provides the highest field strength near the ground for distances closer than about 1,700m, as might be expected from the radiation patterns shown in Figures 2.4 and 2.5.

2.8 Multiple-Beam Antennas

The concept of "intelligent antennas" has been around for many years, especially in military applications, but it is the availability of low-cost semiconductors and new, clever processing algorithms that is beginning to make "smart antennas" a practical reality. This maturing technology provides an important system component for the personal communications systems of the future. Multiple-beam antennas, along with "smart" processing, can generally be divided into to broad categories. In the first group are fixed multiple-beam antenna systems, suitable for transmitting and receiving applications, which use multibeam array techniques, along with infrastructure message traffic control, to exploit frequency

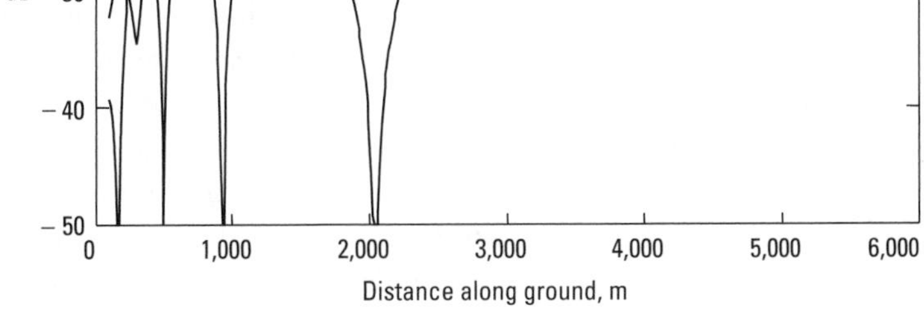

Figure 2.6 Field strengths near the ground for six-element collinear antennas, with and without tailored excitations, compared to field strength from a dipole.

reuse by selective use of narrow high-gain beams. The Butler matrix–fed antenna is an excellent example of this type of multiple-beam antenna. The second group includes a variety of methods and techniques for processing the received signals of multiple antennas so as to separate signals on the basis of angle of arrival. Since the signal characteristics must be known or available at processing time, this second group is most suitable in receive-only conditions.

2.8.1 Matrix-Fed Multiple-Beam Antenna Designs

Arrays using Butler matrix designs for the feeding network are finding application in spatial-diversity PCS designs. The inherent antenna hardware reuse aspect and the relative simplicity with which multiple receiver-transmitter signals can be processed using digital signal processor hardware are opening a wide range of possibilities for innovative system designs.

A very special array antenna is constructed using a Butler matrix [9–11] to feed the antenna elements arranged along a horizontal plane. The matrix has N inputs and N outputs. The N inputs correspond to N distinct $\sin(u)/u$ radiation pattern beams formed by the matrix excitation of N horizontally dispersed antenna elements. Thus, N separate transmitters or receivers may be connected simultaneously to the N separate antenna inputs, each associated with one of the N pattern beams. Each antenna element may itself be a collinear array of some vertical dimension with a fixed elevation plane radiation pattern. The relative antenna-element excitation is given by

$$B_{n,m} = \frac{\exp\left[j\left[n - \dfrac{N+1}{2}\right]\left[m - \dfrac{N+1}{2}\right]\dfrac{2\pi}{N}\right]}{\sqrt{N}} \tag{2.22}$$

where $B_{n,m}$ is the complex voltage excitation of the nth aperture antenna element when the mth port is excited with unit amplitude.

In (2.22), the phase distribution for each beam is linear and has odd symmetry about the center of the array aperture. The amplitude distribution is uniform across the entire array for each beam, and each beam uses the entire array aperture to form a $\sin(u)/u$-type pattern. When designed according to (2.22), the beams intersect at a relative field strength of $2/\pi$, or 3.9 dB down from the beam peaks. The peak of any beam falls on the nulls of all other beams. Examples of Butler matrix power dividing and phasing networks are found in [9–11].

A Butler matrix requires multiple hybrid junctions and fixed phase shifters in far fewer numbers than would be required with common corporate network structures designed for the same task. The number of required hybrids is equal to $(N/2)\log_2(N)$, while the number of fixed phase shifters is equal to

$(N/2)(\log_2 N - 1)$. In general, the hybrids do not necessarily need to provide a 3-dB power split, and the number of beams does not need to be a power of 2.

The radiation patterns of an eight-element Butler matrix–fed array with elements spaced one half-wavelength along the y-axis are shown in Figure 2.7. The direction normal to the array is $\phi = 0°$. The two outer beams, beam 1 (emphasized) and beam 8, begin to show pattern distortions because of the extreme phase taper imposed on the array to achieve those scanning angles. The amplitude of the array-element excitations can be controlled [9] to improve the radiation patterns and to reduce side-lobe levels. Expression (2.22) clearly shows one kind of relationship between the phase-tapered illumination and a complex voltage excitation at multiple antenna ports. Other multiple-beam matrices are possible, such as the Blass matrix described in [11].

[2-8.mcd] An array pattern of a set of radiating elements fed by a Butler matrix will result in as many beams as there are input ports to the matrix. Using (2.22) to represent the antenna-element voltage excitations, compute and plot antenna patterns for each of the possible inputs for arrays with four inputs.

2.8.2 Smart Antennas

Once the exclusive domain of military systems, "smart antennas" are now finding applications in personal communications systems. The term has been

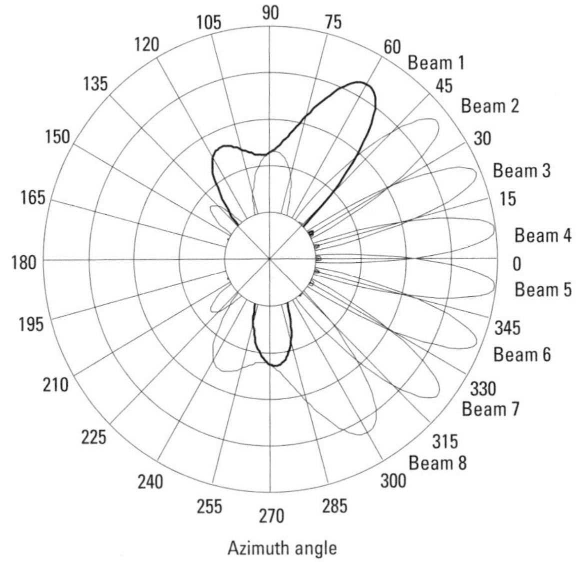

Figure 2.7 Radiation patterns (linear in electric field) of an eight-element Butler matrix array.

applied to a number of signal processing techniques, [13–15], most of which involve proprietary techniques for separating multiple signals on a single frequency channel based on the signal angle of arrival using spatial filtering algorithms. Multiple antennas are required, but the antennas or their actual patterns need not be known a priori in some of the techniques. In general, the output of each antenna drives a linear receiver, whose output is digitally sampled. An algorithm is then applied, for example, to enhance reception of signals in multipath conditions by performing a transformation on the data, which (1) spatially filters out each individual multipath arrival of the same signal, (2) phase-delays and adjusts each multipath arrival, and (3) combines the adjusted signal arrivals in some optimal fashion and presents the result as an output. Alternatively, an algorithm may be applied that (1) identifies multiple signals based on some individual characteristic or "signature" of the signal, (2) separates each individual signal by selectively nulling the others, and (3) presents the separated signal as individual outputs. Figure 2.8 shows a representation of one output of a smart antenna pattern. The algorithm in this case has selectively thrown nulls on signals from "a" and "b" so that the signal from "c" can be received. Simultaneously, additional outputs are available to receive signals from "a" and "b." If enough antenna elements are present, compared to the number of signals to be separated, a pattern "peak" can additionally be steered in the direction of the desired signal.

In a related technique, Multiple Input Multiple Output (MIMO) techniques utilize the randomness imparted to a radio communications channel by

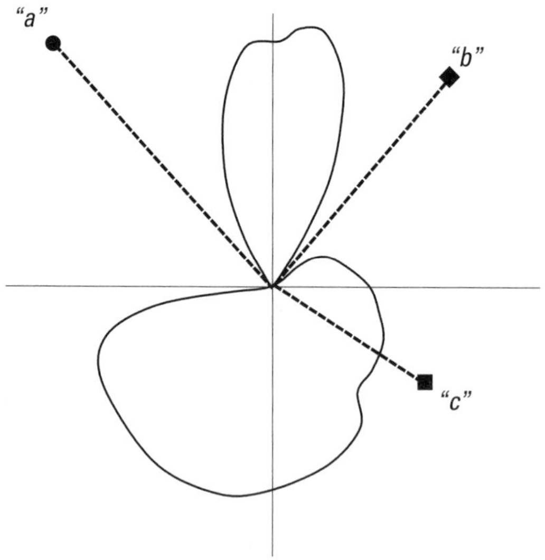

Figure 2.8 A "smart antenna" pattern spatially filtering out "a" and "b" to receive "c."

scattering and multipath to obtain capacity and/or signal-to-noise-ratio (SNR) increases. MIMO systems, as is shown in Chapter 8, utilize multiple antennas at each end of the radio link.

2.9 Proximity Effects in Antennas

The radio links that we are investigating involve fixed-site antennas that are rarely ideally located. The shortage of good, unobstructed sites means that fixed antennas will be located in close proximity to other antennas and, perhaps, towers and masts. Antennas will inevitably interact both with other antennas and with surrounding structures like towers and masts. The approximate analysis presented here is used to estimate that interaction and the effect on radio coverage. An exact analysis of this problem tends to be intractable because the exact geometry and electrical characteristics of the scattering objects are generally unknown. We seek practical approximations in terms of closed-form expressions to canonical electromagnetics problems for which analytical solutions are known. We can then estimate the effect on radio coverage caused by antenna pattern distortions.

2.9.1 Treating Scatterers as Infinitely Long Cylinders

In the following analysis, all antennas other than the driven antenna are treated as passive reflecting cylinders, initially of infinite length. This two-dimensional solution will later be extended to three dimensions. The problem is formulated in two dimensions by assuming a plane wave incident on the "driven" antenna, as well as on the passive reflecting cylinders. Because the geometry is cylindrical, waves are expanded in cylindrical functions. The case of a single two-dimensional scatterer is detailed by R. F. Harrington [16] and others [6, 17, 18], and reflections from the multiple cylinders are treated here in a similar manner. For simplicity, coupling due to multiple interactions is ignored. A correction factor is applied to account approximately for the finite length of the multiple scatterers. Ignoring multiple interactions among the scatterers slightly decreases the accuracy of the result but significantly simplifies the analysis. In any case, limitations exist since treating antennas as passive cylindrical scatterers is itself not entirely correct because these antennas are usually connected to frequency-dependent loads that affect their scattering characteristics.

For simplicity, we restrict ourselves to the case of waves that are normally incident on the cylinders and polarized along the cylinder axis. The far-field pattern of an antenna near m scatterers is given by

$$E_z = E_z^i + \sum_m E_z^m \tag{2.23}$$

Here, waves polarized along the z-axis are assumed. We take the antenna and scatterers to be z directed, as shown in the geometry of a single scatterer in Figure 2.9.

The problem is solved by considering the receiving antenna, illuminated by an incident plane wave field. The incident field E_z^i is written in terms of an outward traveling plane wave written in terms of an expansion of Bessel functions J_n.

$$E_z^i = E_0 \sum_{n=-\infty}^{\infty} j^n J_n(k\rho) e^{-jn(\phi-\pi)} \tag{2.24}$$

The wave number $k = 2\pi/\lambda$, while ρ and ϕ are the coordinates of the field point. Outward-traveling waves scattered by a cylinder are expressed as

$$E_z^S = E_0 \sum_{n=-\infty}^{\infty} j^n b_n H_n^{(2)}(k\rho) e^{-jn(\phi-\pi)} \tag{2.25}$$

where $H_n^{(2)}$ are Hankel functions of the second kind and the unknown scattering coefficients b_n are found from the boundary conditions at the cylinder surface. Hence, the total field is

$$E_z = E_0 \sum_{n=-\infty}^{\infty} j^n \left[J_n(k\rho) + b_n H_n^{(2)}(k\rho)\right] e^{-jn(\phi-\pi)} \tag{2.26}$$

Since E_z is zero at the perfect electric conductor cylinder boundary at $r = a$,

$$b_n = \frac{-J_n(ka)}{H_n^{(2)}(ka)} \tag{2.27}$$

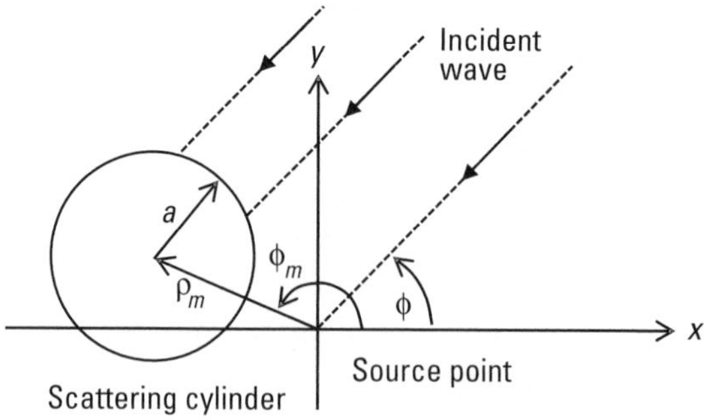

Figure 2.9 Geometry of the scattering problem for the *m*th cylinder.

The m multiple scatterers can be approximated (ignoring secondary reflections) by writing scattered terms of the form of (2.27) for each of the m scattering cylinders. Direct radiation from the line source is combined with an array of m geometrically displaced scatterers:

$$E_z(\phi) = 1 + \sum_m F_m e^{-jk\rho_m \cos(\phi - \phi_m - \pi)} \left[\sum_{n=-\infty}^{\infty} j^n b_{nm} H_n^{(2)}(k\rho_m) e^{-jn(\phi-\pi)} \right] \quad (2.28)$$

Expression (2.28) describes the perturbation to an omnidirectional driven antenna pattern represented by the unity term. The summation on m is the array factor of scatterers located at coordinates (ρ_m, ϕ_m), and the summation on n combines the cylinder function modes for each m scatterer. An amplitude factor F_m, equal to unity for the two-dimensional case, is included to account approximately for the finite length of the m scatterers. The form of (2.28), including the term F_m, assumes negligible coupling between elements, that coordinates are separable, and that F_m depends only on the coordinate ρ. The excitation coefficient for each mode n and at each cylinder m having radius a_m (ignoring multiple interactions among cylinders) is

$$b_{nm} = \frac{-J_n(ka_m)}{H_n^{(2)}(ka_m)} \quad (2.29)$$

We must be satisfied with this first-order scattering approximation in (2.28) because of the approximate nature of the model:

1. Real scatterers are not infinitely long.
2. Scattering antennas exhibit complex impedances because they are connected to loads (transmitters and receivers).
3. The problem is solved for waves normally incident on the cylinder.

Since we are interested in radiation patterns on the horizon, the specialization to normally incident waves is not overly restrictive. This analysis can be readily generalized for arbitrary incidence and polarization (see [17, 19, 20]), but the resulting complexity is not warranted for this study. An approximate accounting for the finite length of the scatterers will be presented next. Finally, measurements will be shown to validate the model.

2.9.2 Modeling the Finite-Length Scatterer

We will restrict the modeling of finite-length cylinders to cylinders that are very long and symmetrically placed with respect to the z-y plane. The scattered field components given by (2.28) decay as the reciprocal square root of distance ρ. This is evident from an inspection of the asymptotic behavior of the Hankel function for large argument $k\rho$.

$$H_n^{(2)}(k\rho) \underset{k\pi \to \infty}{\to} \sqrt{\frac{2j}{\pi k\rho}} j^n e^{-jk\rho} \tag{2.30}$$

Clearly, this is not correct for the three-dimensional case of finite-length scatterers where the far fields must diminish as the reciprocal of distance. The "exact" solution for three-dimensional (that is, finite-length) scatterers is intractable, and numerical methods are often applied. Here, we seek only an approximation that drives the two-dimensional solution to the correct asymptotic form.

We can estimate the behavior of F_m by studying the radiation of a finite-length current filament normalized to the far-field behavior of the Hankel function. When h_m represents the length of the current filament, the field strength due to such a current filament as a function of distance is

$$E(\rho) = \frac{j}{\pi}\sqrt{\frac{\pi k\rho}{2j}} \int_{-h_m/2}^{h_m/2} \frac{e^{-jk\sqrt{z^2+\rho^2}}}{\sqrt{z^2+\rho^2}} dz \tag{2.31}$$

This integral is easily evaluated numerically. For large distances ρ, the phase of $E(\rho)$ is $\pi/4$, while for the close near field, the phase is zero because at close distances, the two-dimensional case is correct. From the numerically derived amplitude of $E(\rho)$ shown by Figure 2.10, we can curve-fit the asymptotic behavior $E(\rho)$ with a function F_m,

$$F_m = \sqrt{\frac{h_m}{h_m - j\rho_m \frac{\lambda}{h_m}}} \tag{2.32}$$

to the magnitude of $E(\rho)$. The curve shown is for a scatterer height of three wavelengths, but the form is general. There is an oscillation around unity (0 dB) with a break point at distance $\rho/\lambda = (h_m/\lambda)^2$, as is evident by the denominator of (2.32).

The function F_m reduces exactly to the form reported by C. A. Balanis [18] for the far field. That three-dimensional solution, E_{3D}, is written in terms of the product of a two-dimensional solution, E_{2D}, and a function depending on scatterer height h_s and distance ρ. The far-field expression reported by Balanis is

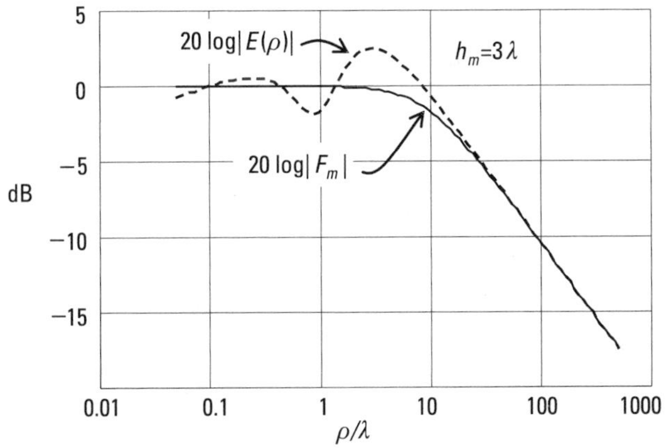

Figure 2.10 Comparison of correction function F_m with the radiation field of a finite-length current filament normalized to the asymptotic behavior of the Hankel function.

$$E_{3d} \underset{\rho \to \infty}{=} E_{2d} \frac{h_s e^{j\frac{\pi}{4}}}{\sqrt{\lambda \rho}} \tag{2.33}$$

which is exactly F_m written for large distances.

2.9.3 Measured and Calculated Patterns Involving Cylindrical Scatterers

The analysis derived in the previous section has been tested experimentally. Six scatterers, each approximately 0.8 wavelengths long and oriented along the z-axis, were arranged along an x-y grid. The relevant dimensions are listed in Table 2.2 relative to the dipole at $(x, y) = (0, 0)$. The calculated and measure fields are shown in Figure 2.11. Again, agreement is satisfactory, despite the use of relatively short scatterers and the fact that, in this case, secondary interactions between the scatterers were ignored. The peak field strength in Figure 2.11 is 5 dB above the omnidirectional dipole pattern.

2.9.4 Application to an Antenna Mounted on the Side of a Tower

The analysis was applied to an omnidirectional antenna side-mounted on a three-corner tower as pictured in Figure 2.12. The tower face was 0.864 wavelengths wide (12 in. at 850 MHz), and the antenna was a half-wavelength away along the $\phi = 0°$ direction. The experimental and calculated patterns for this configuration are seen in Figure 2.13. The pattern peak there is 4.1 dB above the omnidirectional antenna pattern. A pattern ripple of nearly 20 dB is evident in the $\phi = 180°$ direction. The cross-braces on the measured tower were not

Table 2.2
Coordinates and Radii of the Six Scatterers

x (wavelengths)	y (wavelengths)	Radius (wavelengths)
−0.820	0	0.022
−0.137	−0.546	0.022
−0.750	−0.409	0.009
−0.240	0.204	0.009
0.204	0.479	0.005
−0.616	0.409	0.005

modeled, which accounts for the majority of the differences in Figure 2.13 between the measurement and calculation.

Antennas are often placed close to other antennas, as pictured in Figure 2.14. The pattern distortion in this case is very severe, as shown in the field-strength calculations of Figure 2.15. Here, the peak is 4.9 dB above the undisturbed omnidirectional pattern, and nulls appear that are 20 to 30 dB below the peak. This antenna pattern was further analyzed by computing the standard deviation of the pattern values stated in decibels. That standard

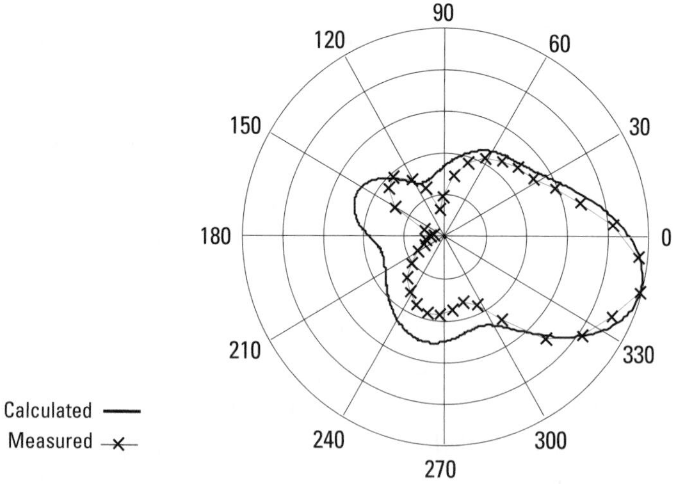

Figure 2.11 Measured and calculated field strengths for a dipole near six scattering cylinders.

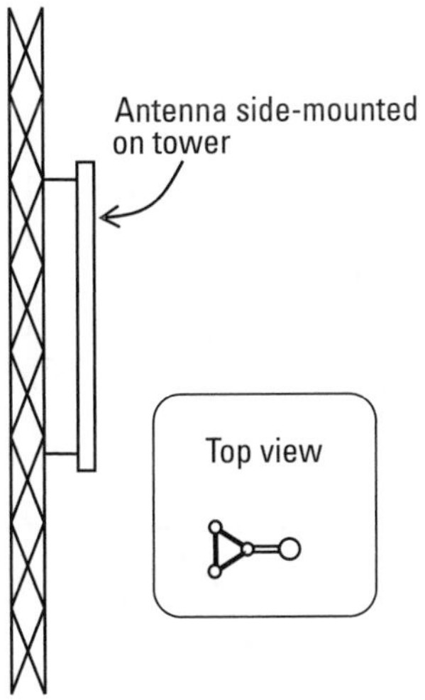

Figure 2.12 An antenna side-mounted on a three-corner tower.

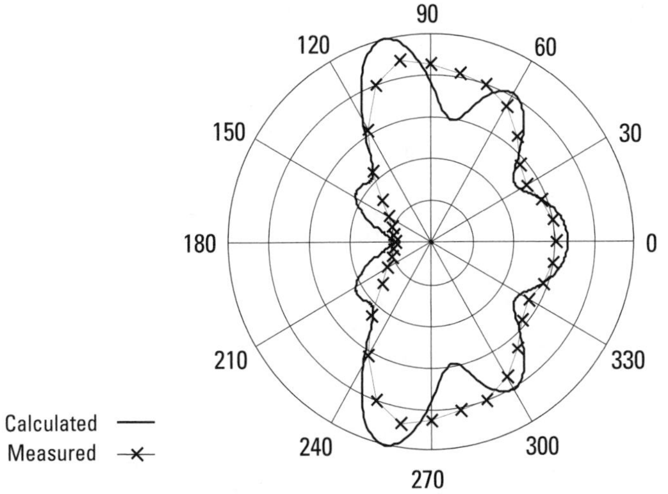

Figure 2.13 Field-strength patterns of an antenna side-mounted on a three-corner tower.

deviation is $\sigma_a = 5.4$ dB, and will be shown in Section 8.4.1 to be a radio coverage design parameter.

[2-9a.mcd] The far-field pattern of a single vertical scatterer near a vertical radiating element can be calculated from (2.28). Try changing the dimensions of and distance to the scatterer and observe the resulting radiation patterns.

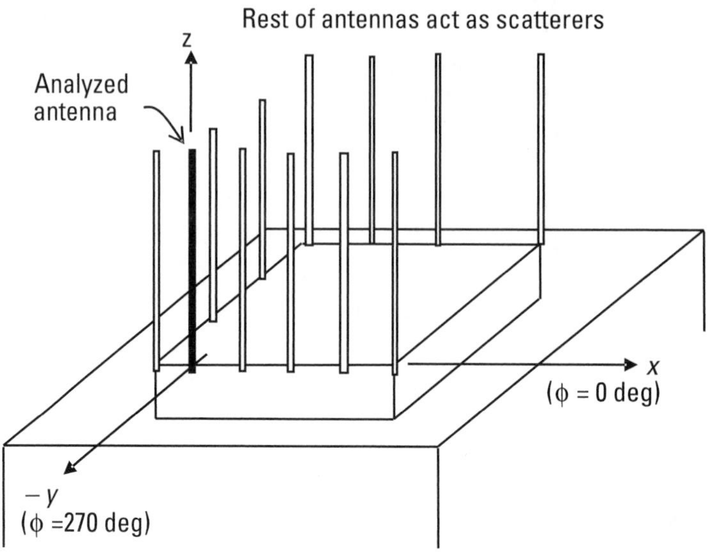

Antenna locations, (*m*)	
x	*y*
4	0
3	0
2	0
1	0
−1	0
−1	2
−1	4
−1	6
0.5	6
2	6
4	6

Figure 2.14 An antenna near other antennas in a typical roof-mounted arrangement. The driven antenna at $(x, y) = (0, 0)$ is analyzed at 930 MHz.

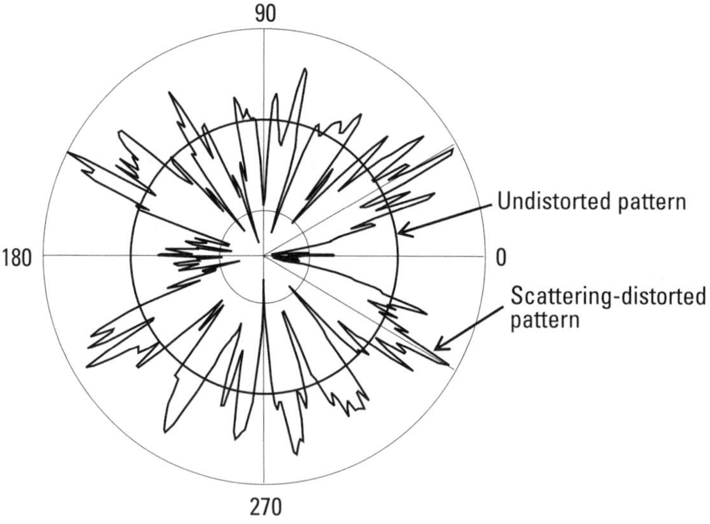

Figure 2.15 Antenna field-strength pattern distorted by scattering from proximate antennas at a common roof site.

2.9.5 Effect of Antenna Distortion on Coverage Range

We will show in Chapter 7 that radiowave propagation can be modeled by a power with distance law. To study the effect of antenna patterns on coverage, we express the power density P at normalized range r,

$$P = \left[\frac{r}{r_o}\right]^{-3.5} \left|E(\phi_z)\right|^2 \tag{2.34}$$

in terms of a power law, 3.5 here, and the magnitude of the normalized antenna pattern E_z given by (2.28). We will show in Chapter 7 that a inverse power law of 3.5 is appropriate for a typical suburban area. The term r_o is a normalizing constant, and we can solve (2.34) for normalized range r in terms of normalized antenna pattern E_z:

$$r(\phi) = \left[\int_0^{2\pi} \left|E_z(u)\right|^2 W(u-\phi)du\right]^{1/3.5} \tag{2.35}$$

In (2.35), the function $W(\phi)$ represents the relative weighting of the power due to local scattering in the vicinity of the mobile or portable telecommunications unit. The local scattering gives rise to multipath fading, which will be discussed further in Chapter 8. Here, we note, this scattering contribution serves to smooth the pattern distortions observed in Figure 2.15.

The effect on range of the antenna pattern distortion can now be estimated by plotting (2.35) in comparison to a circle of unit radius as shown in Figure 2.16. In this example, $W(\phi)$ is a binomial weighting function extending over a 3.6° angular range. In some directions, particularly the regions around the 0° and 180° azimuthes, range is significantly decreased from the nominal range of an omnidirectional, undistorted antenna pattern, shown in the figure as a circle of unit radius. This kind of antenna pattern "ripple effect" was noted by W. C. Y. Lee [21]. The pattern distortion complicates radio system designs and can reduce the effectiveness of some diversity techniques.

🖫 [2-9b.mcd] The far-field directivity (or gain) pattern of an antenna will affect the range according to (2.35). Explore how the range profile varies as a function of the exponent in the path loss model of (2.34).

2.9.6 Parasitically Driven Array Antennas

Sometimes the parasitic coupling of driven antennas to other conductors results in a desirable effect. One case is the Yagi-Uda array [22, 23]; another is the corner reflector antenna patented in 1942 by J. D. Krause (U.S. Patent 2,270,314)

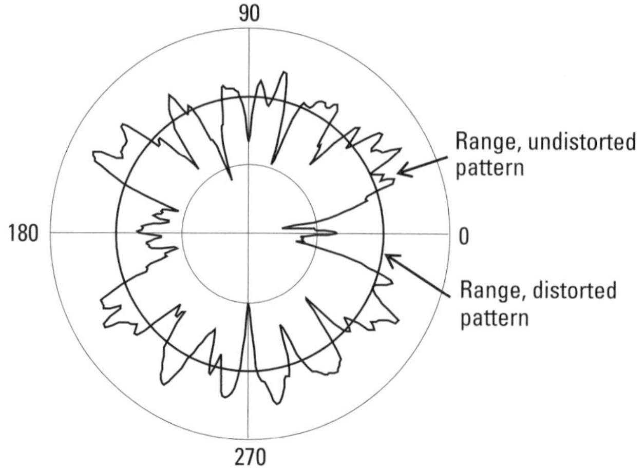

Figure 2.16 Normalized range for undistorted and distorted antenna patterns.

[24]. Both types of antennas find application as source antennas on open-air antenna ranges and as fixed-site directional antennas. The Yagi-Uda array and the corner reflector each rely on mutual coupling to parasitically excite nondriven elements. These elements are actual nearly-half-wave elements on the Yagi-Uda antenna and images of the driven element in the case of the corner reflector antenna.

A moment-method approach such as numerical electromagnetics code (NEC) [25] may be employed for calculating mutual coupling among wire antenna elements of arbitrary lengths, thicknesses, and orientations. The gain, side lobes, and bandwidth characteristics of the array are then optimized by varying the antenna-element spacings' lengths and thicknesses. The dimensions of optimized Yagi-Uda array antennas are given in [26, 27]. Here, the mutual coupling between two elements of half-lengths L_1 and L_2 parallel with the z-axis and separated by (X, Z) is, in simplified form for the case of sinusoidal current distributions on the elements,

$$Z_{12} = \frac{j\eta_0}{4\pi}\int_{-L_2}^{L_2} \frac{G_1 + G_2 - 2G\cos(2\pi L_1)}{\sin(2\pi L_1)\sin(2\pi L_2)} \sin\left(2\pi\left[L_2 - |z|\right]\right)dz \quad (2.36)$$

where

$$G_1 = \frac{e^{-j2\pi\sqrt{[Z+L_1+z]^2+X^2}}}{\sqrt{[Z+L_1+z]^2+X^2}} \quad (2.37)$$

$$G_2 = \frac{e^{-j2\pi\sqrt{[Z-L_1+z]^2+X^2}}}{\sqrt{[Z-L_1+z]^2+X^2}} \tag{2.38}$$

$$G = \frac{e^{-j2\pi\sqrt{(Z+z)^2+X^2}}}{\sqrt{(Z+z)^2+X^2}} \tag{2.39}$$

For the special case of half-wavelength elements, $L_1 = L_2 = 0.25$ wavelengths:

$$Z_{12} = \frac{j\eta_0}{4\pi}\int_{-\frac{1}{4}}^{\frac{1}{4}}[G_1+G_2]\sin\left[2\pi\left(\frac{1}{4}-|z|\right)\right]dz \tag{2.40}$$

Using (2.36) to (2.40), we can construct a matrix of mutual and self-impedances Z_{ij} for antenna elements based on their geometries and write an expression for the currents on the antenna elements with a single driven element having excitation voltage V_0:

$$\begin{bmatrix} V_0 \\ 0 \\ \cdots \\ \cdots \\ 0 \end{bmatrix}^T \begin{bmatrix} Z_{11} & Z_{12} & \cdots & Z_{1N} \\ Z_{21} & Z_{22} & \cdots & \cdots \\ \cdots & \cdots & \cdots & \cdots \\ \cdots & \cdots & \cdots & \cdots \\ Z_{N1} & Z_{N2} & \cdots & Z_{NN} \end{bmatrix} \begin{bmatrix} I_1 \\ I_2 \\ \cdots \\ \cdots \\ I_N \end{bmatrix} \tag{2.41}$$

The currents I_i in (2.41) are the unknowns. The array pattern is now found using (2.1) for the element pattern and (2.2) for the array factor, replacing V_i in (2.2) with I_i from (2.41). For dipoles of arbitrary half-length L, the gain is found by applying (1.34) with the element pattern of (2.1):

$$G_{dipole} = \frac{4\pi\left[\dfrac{\cos\left[2\pi L\cos\dfrac{\pi}{2}\right]-\cos(2\pi L)}{(1-\cos(2\pi L))\sin\dfrac{\pi}{2}}\right]^2}{\displaystyle\int_0^{2\pi}\int_0^{\pi}\left[\frac{\cos(2\pi L\cos(\theta))-\cos(2\pi L)}{(1-\cos(2\pi L))\sin(\theta)}\right]^2\sin(\theta)\,d\theta\,d\phi} \tag{2.42}$$

The expression for gain given in (2.42) is easily found numerically, as shown in the example below [2.9c.mcd].

[2-9c.mcd] Compute the mutual impedances, directive gain, and array pattern of a three-element Yagi-Uda array with elements 0.00254 wavelengths thick and 0.5144, 0.4754, and 0.4554 wavelengths long, spaced 0.09608 and 0.1273 wavelengths apart, respectively. Assume sinusoidal currents.

In the case of the Yagi-Uda array, the parasitic elements are parallel to, and in the plane of, the driven element, as seen in the example of Figure 2.17. A driven element and two parasitic elements are shown. The driven element is a half-wave dipole. The reflector element is slightly longer, and the director element is shorter, than the driven element; hence, they exhibit self-impedances that affect the phase of the currents induced on them in such a way as to form a beam in the direction of the shorter element. The azimuth and elevation plane patterns of a three-element Yagi-Uda antenna are shown in Figure 2.18. The peak gain in the shown case is 7.5 dBi.

The corner reflector antenna consists of a driven element and a reflecting corner that, in the case of a 90° corner, forms three images of the driven element, as seen in Figure 2.19. The array pattern can be found approximately using (2.36) to (2.41), assuming we can neglect the finite size of the reflecting corner. With a half-wavelength driven element and an element-to-corner spacing of a half-wavelength, the directivity of a corner reflector is approximately 12 dBi.

[2-9d.mcd] Compute the directive gain of a 90° corner reflector antenna with a corner-to-element spacing of $S = 0.25$ wavelengths and an element length of 0.4754 wavelengths. Assume sinusoidal element current. What happens to the gain when S approaches zero?

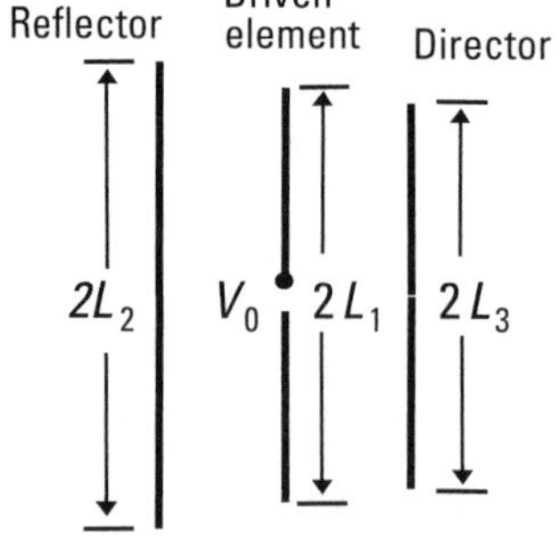

Figure 2.17 A three-element Yagi-Uda array antenna.

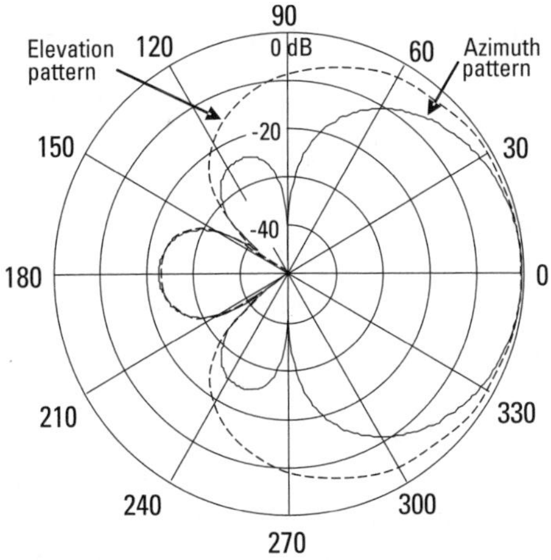

Figure 2.18 Patterns of a three-element Yagi-Uda array in the azimuth (polarization) and elevation planes.

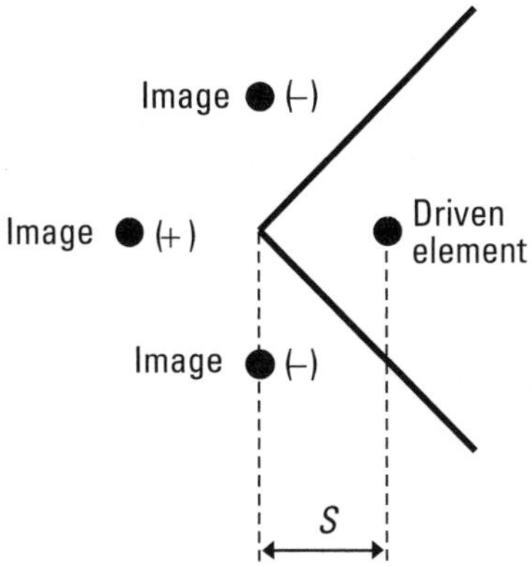

Figure 2.19 Geometry of the corner reflector antenna with corner-to-element spacing *S*.

2.10 Indoor Fixed Sites

Our consideration of fixed sites includes wireless and access points. These usually operate either in wireless local-area networks under the IEEE 802.11 family of standards [28] or personal-area networks operating with IEEE802.15 standards [29] and are located inside buildings. UWB technology is also appearing in the indoor environment with scenarios that can benefit from directional antennas. Increasingly, these networks are using MIMO techniques (see Chapter 8), where performance is sensitive to mutual coupling between antennas.

2.10.1 Wireless Local-Area Network Fixed Sites

Wireless access points are located near walls and other objects, and like building- and tower-mounted antennas, their patterns are distorted by coupling with the environment. The operating frequencies typically include up to 83.5 MHz of bandwidth starting at 2.4 GHz and up to 555 MHz of spectrum in bands between 5.15 and 5.825 GHz (U-NII bands in the United States), although low-data-rate networks can appear at 868 to 868.6 MHz and 902 to 928 MHz, depending on regulations around the globe.

A typical access point antenna is a half-wavelength radiator (dipole) oriented vertically and attached to a small wireless access point. In MIMO applications, multiple antennas are deployed with relatively close separation, and correlation can be estimated by mutual-coupling analysis. The antenna coupling in the horizontal plane is much stronger than the coupling in the vertical plane for vertically polarized antennas. The coupling is found from the ratio of mutual impedance Z_{12} to the self-impedance Z_0 of a dipole using (2.36). Figure 2.20 shows the coupling for broadside and longitudinally aligned dipoles.

The antenna mutual-coupling effect is not an important parameter in vertically separated vertically polarized antennas. The center-to-center spacing starts out at about a half-wavelength and decreases rapidly, while the horizontally spaced vertically polarized dipoles can start out closer together and have significantly more coupling. Mutual coupling is thus largely an issue for horizontally spaced arrays.

2.10.2 Gain Antennas for UWB Pulses

Wireless access points for UWB pulse transmissions will no doubt benefit from gain antennas. These pose some interesting consequences regarding the transmission of pulses. When considered over a wide bandwidth, antennas that are apparently the same when measured using a sine wave may exhibit different behaviors for wideband UWB pulses. Here we will look at two directional antennas that appear to perform "equally" over wide frequency ranges when using narrowband signals. With impulse excitations, however, they perform differently. A pair of parabolic reflector antennas fed by wideband dipoles at their focal points is designed so that their impedance bandwidths match those of a pair of horn aperture antennas.

First, we consider a link between two reflector antennas shown in cross section in Figure 2.21. A pulse sent by the transmitter is propagated by the feed-point dipole toward the reflector and also in the transmission direction. A reflected pulse $2x$ behind it picks up the antenna-area aperture gain, and its polarity is inverted relative to the directly radiated pulse because of boundary condition.

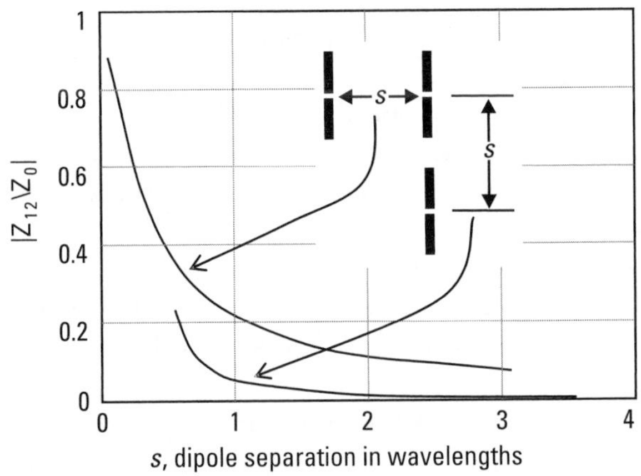

Figure 2.20 Dipole coupling in the horizontal and vertical planes.

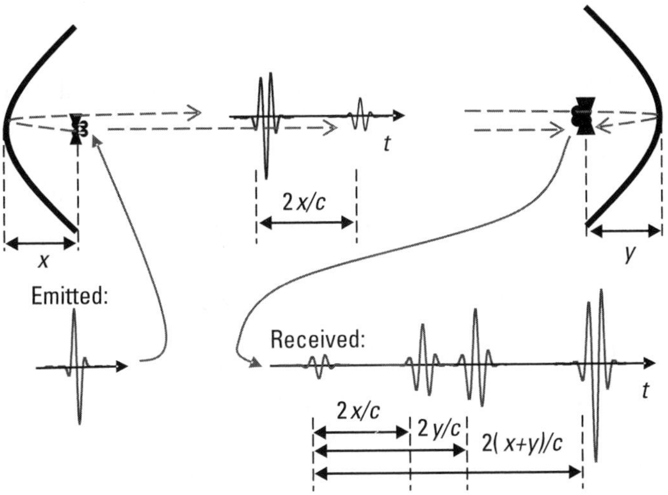

Figure 2.21 A transmission link between two dipole-fed reflector antennas. (After [30].)

This pulse pair is now incident on a receiving parabolic reflector antenna shown in cross section in Figure 2.21. That feed-point dipole receives the pair of directly propagated pulses but also receives the same pair of pulses reflected from the receiving parabola, delayed in time and magnified by the gain aperture. Thus, four pulses are received when one was transmitted. When the same transmission link is used for a sine wave signal, the received signal is a summation of four time-delayed sine waves of four different amplitudes and phases. That summation is a single sine wave.

Next, we consider a horn antenna aperture whose emitted signal comprises just the single pulse (neglecting edge diffraction as we did for the reflector system). If a sine wave had been supplied to the transmitting horn, the propagated field would have been a sine wave and the received signal would have been a sine wave. Compared with the simple horn aperture antenna, the reflector system produces its own "multipath."

Antennas for transmitting UWB pulses must be understood for their behavior in the time domain rather than the frequency domain because the effects of short pulses become observable whenever the retardation time on the antenna is on the order of the pulse duration. For short-pulse signals, the antenna output has a transient response, which includes all of the natural resonant responses of the structure with amplitudes that may be commensurate with the forced impulse response of the system. Resonance effects were not factors in the illustrative examples above but also must be considered.

2.11 Summary

Antennas used on fixed sites have been introduced with particular emphasis on vertically polarized onmidirectional collinear arrangements, which are typical in the telecommunications industry. Directivity is achieved by arraying multiple basic radiators, which results in a narrowing of the elevation plane beamwidth. The discrete array of radiating elements was shown to resemble a discrete Fourier transform. Highly directive antennas having very narrow elevation beamwidths can have coverage problems under a tall antenna support structure. Directivity patterns can be shaped by tailoring the amplitude and phase of each radiating element in an array. In panel-type antennas, feeding networks can result in null fill designs in the elevation plane, as well as multiple beams in the azimuth plane. Pattern shaping results in improved coverage along the ground.

Multiple-beam antennas can provide a system design flexibility that can exploit spatial diversity. Smart antenna technology provides a unique system tool to exploit angle diversity, especially in the uplink path between the portable telecommunications device and the fixed infrastructure antennas.

Since fixed-site antennas are usually located close to other antennas and to towers and masts, severe interactions may occur. The analysis presented above estimates this interaction using practical approximations in terms of closed-form functions for which solutions are calculable. The presented analysis treats all antennas other than the driven antenna as passive reflecting cylinders, initially of infinite length. The solution was extended to three dimensions, and the estimated effect on radio coverage caused by antenna pattern distortions was calculated. The accuracy of this analysis is limited by the following:

1. There are practical difficulties with specifying the exact geometry of a situation (especially on a rooftop).
2. Only vertical scatterers were considered; diagonal cross members, as on towers, were ignored.
3. Actual antennas act as scatterers with complex impedance loads that vary with frequency.
4. Multiple interactions among the scatterers (the mutual-coupling problem) were ignored and would unnecessarily complicate the solution.
5. In the case of roof-mounted antennas, reflections from the roof were neglected.
6. The antennas, along with all scatters, were symmetrically placed with respect to the plane of the incidence.

The Yagi-Uda array and the corner reflector antenna were introduced as examples of directional antennas using parasitically excited elements to form a beam.

Wireless access points make up another class of fixed antenna systems and can include multiple antennas configured for MIMO systems. UWB antennas can also appear in the wireless local and personal-area networks. Some specific behaviors of UWB gain antennas were noted.

Problems

2.1 Find the gain of dipoles of lengths $S = 1, 0.52, 0.5, 0.48, 0.4, 0.2$, and 0.1 wavelengths. Assume a sinusoidal current on the antenna as a function of length. What happens when S becomes infinitesimally small?

Ans:

P2-1.MCD.

2.2 A half-wavelength square patch antenna radiates from the two edges that are $p = 1$ half-wave apart. The in-phase radiating edges look like the magnetic equivalent of a dipole. Using this analogy and assuming a sufficiently large ground plane behind the patch, calculate the directivity of the patch antenna. What happens when the distance between the two edges becomes $p = 0.25$ and $p = 0.01$?

Ans:

P2-2.MCD. $G = 7.15$ dBi for $p = 0.5$, $G = 5.7$ for $p = 0.25$, $G = 5.16$ for $p = 0.01$.

2.3 A patch antenna is formed from a half-wave-wide by quarter-wave-long conductor over a ground plane. One half-wave edge is grounded; the other forms a radiating slot. Compute the directivity.

Ans:

P2-2.MCD. $G = 5.16$ dBi.

2.4 An antenna radiates isotropically with a field pattern given by $E = 6I/r$ V/m. I is the RMS current supplied to the antenna, and r is distance (m). Find the radiation resistance.

Ans:

$P_d = G I^2 R_{radiation}/(4\pi r^2) = E^2/\eta_0$ with $G = 1$, so $R_{radiation} = 1.2\Omega$.

2.5 The z component of the electric field of a resonant half-wavelength dipole located on the z-axis with the feed point at the origin is

$$E_z = \frac{-jI_{rms}\eta_o}{4\pi}\left[\frac{e^{-jkR_1}}{R_1} + \frac{e^{-jkR_2}}{R_2}\right]$$

In the direction of maximum radiation, distances R_1 and R_2 from the dipole tips are equal. Show that the radiation resistance $R_{radiation} = 73.08$. *Hint:* see (1.41) and (1.43) and use $P = I_{rms}^{\ 2} R_{radiation}$.

Ans:

Using the relationship $P_d = G\, P/(4\pi r^2) = E^2/\eta_0$, solve for P and then, with $G = 1.641$ for the half-wave dipole, $R_{radiation} = \eta_0/(\pi G) = 73.08$.

2.6 Two point sources are spaced $\lambda/4$ and radiate with equal amplitude but phase quadrature. What is the pattern in the plane of the sources?

Ans:

$E = \exp(-jx) + j\exp(+jx)$, where $x = k(\lambda/4)\sin(\phi)/2 = (\pi/4)\sin(\phi)$, then expanding by Euler's identity,

$$E = (1 + j)[\cos((\pi/4)\sin(\phi)) - \sin((\pi/4)\sin(\phi))].$$

This is a cardioid pattern with a maximum magnitude of 2 in the $\phi = \pi/2$ direction.

2.7 Two point sources are spaced $\lambda/2$ and radiate with equal amplitude but opposite phase. What is the pattern in the plane of the sources?

Ans:

$E = \exp(-jx) - \exp(+jx)$, where $x = k(\lambda/2)\sin(\phi)/2 = (\pi/2)\sin(\phi)$, which expands to $E = 2j\,\sin[(\pi/2)\sin(\phi)]$. This is a figure eight pattern with maxima of magnitude 2 in the $\phi = \pi/2$ and $3\pi/2$ directions.

2.8 Three isotropically radiating point sources of equal amplitude are arranged at the corners of an equilateral triangle $\lambda/2$ on each edge. What is the pattern when all the sources are in phase? What happens when two of the sources are each the same amplitude and same phase, and the third source is twice the amplitude and in quadrature?

Ans:

Let $y = k(\lambda/4)\sin(\phi)$, and $x = k[\cos(\pi/6)\lambda/4]\sin(\phi)$, then the sources are at $(x, 0)$, $(0, y)$, and $(0, -y)$. The pattern is then given by

$E(\phi) = A\exp(-j\,x\sin\phi) + B\exp(-j\,y\cos\phi) + C\exp(+j\,x\cos\phi)$

When $A = B = C$, the pattern is omnidirectional. When $A = j2$ and $B = C = 1$, the pattern is a cardioid peaking in the $\phi = \pi/2$ (x-axis) direction, the direction of source A. The peak direction reverses when $A = -j2$. The same is true as these amplitudes and phases rotate through the sources.

2.9 Three isotropic sources are in a line and spaced $\lambda/4$ and have amplitudes $+j1$, 3, and $-j1$, respectively. Find an expression for the radiation pattern and plot the field magnitude.

Ans:

$E(\phi) = j\exp(-j\,y\sin\phi) + 3\exp(-j\,y\sin\phi) - j\exp(+j\,y\sin\phi)$, where $y = \pi/2$.

2.10 An omnidirectional vertically polarized antenna is desired for a communications link at 902 MHz. The required directivity is 14 dBi. Find the length of the antenna and 3-dB beamwidth.

Ans:

See (2.7), $L = 10^{14/10}\lambda/2 = 4.17$m (13.7 ft.). From (2.15), $BW_{3dB} = 4.1°$.

2.11 10W is applied to a 14-dBi antenna mounted atop a 100m-tall tower in a 902-MHz communications system. Using free space propagation and ignoring ground reflections—see (1.36)—find the power density at ground level and a distance of 2.3 km.

Ans:

From (1.42), $P_d = 3.8$ mW/m^2. The angle from the top of a 100m tower to a distance on the ground at 2.3 km is 2.49°, the half-power beamwidth of the antenna. The power density is therefore 1.9 mW/m^2.

2.12 A half-wave resonant dipole is placed parallel to and very near ($r = 0.25\lambda$) a very tall and very fat ($2a = 100\lambda$ diameter) metal structure. Give a first-order expression for the far-field pattern. Find the apparent gain of the dipole-structure combination.

Ans:

The structure can be approximated by a perfectly conducting, reflecting, half-space. There is a dipole with a dipole image in the reflector, with current reversed in the image. Using (2.1) and (2.6) and lining up the dipole and its image along the $\phi = \pi/2$, $\theta = \pi/2$ axis,

$$E(\theta) = F(\theta)\,|\exp[-jkd\sin\theta\,\sin\phi] - \exp[+jkd\sin\theta\,\sin\phi]|$$

which reduces to

$$E(\theta) = F(\theta)\,[2\sin(kd\sin\theta\,\sin\phi)]$$

So,

$$kd = 2\pi(0.25) = \pi/2 \text{ so } E(\theta) = 2\,F(\theta)$$

Hence, the apparent field strength is twice that of a dipole in free space, which equals 6 dB greater gain. Note that the solution is valid only for ϕ between 0 and π.

2.13 A Yagi-Uda array has a directive gain $G = 7.5$ dBi. The three-element array has elements 0.00254 wavelengths thick and 0.5144, 0.4754, and 0.4554 wavelengths long, spaced 0.09608 and 0.1273 wavelengths apart, respectively. Find the effective aperture. Comment on the extent of the near field of this antenna.

Ans:

The effective aperture—see (1.43)—is $A_e/\lambda^2 = G/4\pi = 10^{7.5/10}/4\pi =$ 0.447 square wavelengths, or equivalent to a disk 0.377 wavelengths in radius. From Section 1.3.7, we can see that the inner and outer boundaries of the Fresnel region, the "radiating near fields," are between 0.23 and 0.53 wavelengths distance and are commensurate with the equivalent aperture disk radius. This is the effective extent of the reactive near fields around the array as viewed from the direction of maximum gain.

2.14 A pair of dipoles is fed in parallel. The dipoles form a cross aligned along the x- and z-axes, and the feed point is at the origin. One dipole is $H_x =$ 0.496 wavelengths long with self-impedance $Z_x = 71.38 + j34.75\Omega$; the other is $H_z = 0.428$ wavelengths long with self-impedance $Z_z = 47.50 - j84.62\Omega$. Find the feed-point impedance of the pair and the polarization axial ratio AR.

Ans:

P2-14.MCD.

2.15 Find the gain of a lossless corner reflector antenna when the driven element is a resonant half-wave dipole and the element-to-corner spacing approaches zero.

Ans:

P2-15.MCD. Set $H = 0.25$ and $s = 0.00001$. $G = 12.65$ dBi.

2.16 Find the gain of a lossless corner reflector antenna when the driven element is a resonant half-wave dipole and the element-to-corner spacing is a quarter-wavelength.

Ans:

P2-15.MCD. Set $H = 0.25$ and $s = .25$. $G = 12.46$ dBi.

References

[1] Pocklington, H. C., "Electrical Oscillations in Wires," *Proc. of the Cambridge Phys. Soc.*, London, Vol. 9, 1897, pp. 324–333.

[2] Hallén, E., "Theoretical Investigation into Transmitting and Receiving Qualities of Antannae," *Nova Acta Regiae Soc. Ser. Upps.*, Vol. 2, November 4, 1938, pp. 1–44.

[3] Balzano, Q., O. Garay, and K. Siwiak, "The Near Field of Dipole Antennas, Part I: Theory," *IEEE Transactions on Vehicular Technology*, Vol. VT-30, No. 4, November 1981, pp. 161–174.

[4] Balzano, Q., O. Garay, and K. Siwiak, "The Near Field of Dipole Antennas, Part II: Experimental Results," *IEEE Transactions on Vehicular Technology*, Vol. VT-30, No. 4, November 1981, pp. 175–181.

[5] Balzano, Q., O. Garay, and K. Siwiak, "The Near Field of Omnidirectional Helices," *IEEE Transactions on Vehicular Technology*, Vol. VT-31, No. 4, November 1982, pp.173–185.

[6] Jordan, E. C., and K. G. Balmain, *Electromagnetic Waves and Radiating Systems,* 2nd ed., Englewood Cliffs, NJ: Prentice Hall, 1968.

[7] McDonald, A., "Approximate Relationship between Directivity and Beamwidth for Broadside Collinear Arrays," *IEEE Transactions on Antennas and Propagation*, Vol. AP-26, No. 2, March 1978, pp. 341–342.

[8] Abramowitz, M., and I. Stegun, (eds.), *Handbook of Mathematical Functions*, New York: Dover Publications, 1972.

[9] Butler, J., and R. Lowe, "Beam-Forming Matrix Simplifies Design of Electronically Scanned Antennas," *Electronic Design*, April 12, 1961, pp. 170–173.

[10] Moody, H. J., "The Systematic Design of the Butler Matrix," *IEEE Transactions on Antennas and Propagation*, Vol. AP-12, November 1964, pp. 786–788.

[11] Lo, Y. T., and S. W. Lee, (eds.), *Antenna Handbook,* New York: Van Nostrand Reinhold Co., 1988.

[12] Decibel Products, Directional Panel Antennas: 820–960 MHz, Product Literature.

[13] Paulraj, A., R. Roy, and T. Kailath, "Estimation of Signal Parameters via Rotational Invariance Techniques—ESPRIT," *Proc. Nineteenth Asilomar Conf. on Circuits, Systems and Comp.*, Pacific Grove, CA, November 1985, pp. 83–89.

[14] Agee, B. G., A. V. Schell, and W. A. Gardner, "Spectral Self-coherence Restoral: A New Approach to Blind Adaptive Signal Extraction Using Antenna Arrays," *Proceedings of the IEEE,* Vol., 78, April 1990, pp. 753–767.

[15] Talwar, S., M. Vyborg, and A. Paulraj, "Blind Estimation of Multiple Co-Channel Digital Signals Arriving at an Antenna Array," *Proc. Twenty-Seventh Asilomar Conf. on Signals, Systems and Computers,* Pacific Grove, CA, Nov. 1–3, 1993.

[16] Harrington, R. F., *Time Harmonic Electromagnetic Fields,* New York: McGraw-Hill, 1961.

[17] Wait, J. R., *Electromagnetic Radiation from Cylindrical Structures,* London: Peter Peregrinus Ltd., 1988.

[18] Balanis, C. A., *Advanced Engineering Electromagnetics,* New York: John Wiley and Sons, 1989.

[19] Misra, D. K., "Scattering of Electromagnetic Waves by Human Body and Its Applications," Ph.D. Dissertation, Michigan State University, 1984.

[20] Ponce de Leon, L., "Modeling and Measurement of the Response of Small Antennas near Multilayered Two or Three Dimensional Dielectric Bodies," Ph.D. Dissertation, Florida Atlantic University, 1992.

[21] Lee, W. C. Y., *Mobile Communications Engineering,* New York: McGraw-Hill, 1982.

[22] Yagi, H., "Beam Transmission of Ultra-short Waves," *Proceedings of the IRE,* Vol. 16, June 1928, pp. 715–740.

[23] Uda, S., and Y. Mushiake, "Yagi-Uda Antenna," Research Institute of Electrical Communication, Tohoku University, Sendai, Japan, 1954.

[24] Krause, J. D., "The Corner Reflector Antenna," *Proceedings of the IRE,* Vol. 28, November 1940, pp. 513–519.

[25] Burke, G. J., and A. J. Poggio, "Numerical Electromagnetics Code (NEC)—Method of Moments," *NOSC TD 116,* Lawrence Livermore Laboratory, Livermore, CA, January 1981.

[26] Viezbieke, P. P., "Yagi Antenna Design," NBS Technical Note 688, National Bureau of Standards, Washington, D.C., 1976.

[27] Straw, R. D., (ed.), *The ARRL Antenna Book,* Newington CT: The American Radio Relay League, 2006.

[28] O'Hara, B., and A. Petrick A, "IEEE 802.11 Handbook—A Designer's Companion," New York: IEEE Press, 1999.

[29] Gutierréz, J., E. H. Callaway Jr., and R. L. Barrett Jr., "Low-Rate Wireless Personal-Area Network—Enabling Wireless Sensors with IEEE 802.15.4," New York: IEEE Press, 2004.

[30] Siwiak, K., and D. McKeown, *Ultra-wideband Radio Technology*, London: John Wiley and Sons, 2004.

3

Radio Communication Channel

3.1 Introduction

The radio communication channel is introduced here in terms of guided and radiated waves. For our personal communications link involving fixed-site antennas and antennas on body-worn or handheld radio devices, the radio communications channel is a radiated channel. Guided waves are presented in the context of transmission lines whose characteristics are contrasted with those of radiated waves. Transmission lines are important components in communications system designs; they appear as feed lines to fixed-site antennas and may also appear in the increasingly pervasive infrastructures of personal communications systems. Consequently, their behavior and basic characteristics will be presented here in some detail.

The propagation and attenuation functions for transmission lines and basic free space propagation are developed. There are fundamental differences in propagation behaviors in transmission lines as compared with radiated channels. Transmission lines provide a measure of security but are not always the lowest-loss option or the lowest-cost option in system design. Although the economics of system infrastructure will not be considered here in detail, we point out that systems tend to be designed from the market-viability point of view. The choice between radio channels and transmission lines is an economic one based as much on engineering performance as on cost. The performance and characteristics of transmission lines and of the basic free space propagation law are presented in the context of their impact on PCS designs.

Transmission lines and radiated channels have different noise and dispersion characteristics. Basic radiowave propagation will be presented here as the Friis transmission formula. Finally, the polarization characteristics of radiated waves and of antennas will be explored here along with polarization mismatch losses and the coupling of polarizations.

3.2 Guided Waves

Guided waves are signals that are transported by physical guiding structures such as coaxial lines, parallel wire lines ("twisted pair"), waveguides, and optical fibers. Because a medium other than air or a vacuum is involved, there are dissipative losses. Dispersion occurs when the transmission medium has characteristics that vary with frequency and usable bandwidths become dependent of the physical length of the guiding structure.

Wave propagation in a uniform and homogeneous guided wave structure can be described by the ratio of received power P_r to the transmitted power P_t:

$$\frac{P_r}{P_t} = \left| e^{-jk_g d} e^{-\alpha_g d} \right|^2 \tag{3.1}$$

The channel phase propagation constant is k_g, and the first phasor quantity in (3.1) has a unity magnitude. The voltage attenuation constant is α_g measured in nepers/m, so the signal power attenuation is exponential. Attenuation A_g is then

$$A_g = \alpha_g 20 \log(e) = 8.69 \alpha_g \text{ dB / m}$$

There are waveguide structures, such as radial waveguides and spherical waveguides, that have a geometric expansion term multiplying the attenuation term in (3.1). Radial waveguides expand waves in two dimensions; hence, power will propagate with a $1/d$ term multiplying (3.1). Radial waveguides are encountered in certain kinds of power dividers and combiners. Two-dimensional or cylindrical wave propagation was already encountered in Chapter 2, where scattering from infinitely long cylinders was discussed. Spherical waveguides allow waves to expand uniformly in three dimensions; hence, a multiplicative term $1/d^2$ must be applied to (3.1). Radial propagation of spherical waves will be discussed later when the Friis transmission formula is introduced and will be explored again when we study the radiating modes of small antennas in Chapter 11.

3.2.1 Losses in Dielectrics

We will develop the propagation of electromagnetic waves in materials here to obtain the propagation and attenuation factors in dielectric materials and in conductors. Starting with some definitions, the permeability of a material is $\mu = \mu_d\mu_0$, where μ_0 is the permeability of free space, and μ_d for dielectrics is most often near unity. Permittivity is given as $\varepsilon = \varepsilon_d\varepsilon_0$, where ε_0 is the permittivity of free space, and $\varepsilon_d = \varepsilon_r - j\varepsilon_i$ is the complex relative dielectric constant of the material written in terms of the conductivity of the medium σ given in S/m and frequency f given in megahertz or as radian frequency ω. So,

$$\varepsilon_d = \varepsilon_r - j\frac{\sigma}{\omega\varepsilon_0} = \varepsilon_r - j\frac{17{,}975\sigma}{f} \tag{3.2}$$

The propagation constant γ in the material is then

$$\gamma = k\sqrt{\mu_d\left[\varepsilon_r - j\frac{\sigma}{\omega\varepsilon_0}\right]} \tag{3.3}$$

where $k = 2\pi/\lambda$ is the free space wave number. The phase propagation constant k_g in radians per meter is the real part of γ,

$$k_g = \mathrm{Re}\{\gamma\} = k\sqrt{\frac{\mu_d}{2}\left[\sqrt{\varepsilon_r^2 + \left[\frac{\sigma}{\omega\varepsilon_0}\right]^2} + \varepsilon_r\right]} \tag{3.4}$$

and the attenuation constant α_g in nepers per meter is the imaginary part of γ,

$$\alpha_g = \mathrm{Im}\{\gamma\} = k\sqrt{\frac{\mu_d}{2}\left[\sqrt{\varepsilon_r^2 + \left[\frac{\sigma}{\omega\varepsilon_0}\right]^2} - \varepsilon_r\right]} \tag{3.5}$$

The ratio $\sigma/\omega\varepsilon_0\varepsilon_r$ is known as the *dissipation factor* D_g or the *loss tangent* $\tan\delta$ of the dielectric. Often, the losses in a dielectric are specified by the power factor PF of the dielectric. The dissipation factor (loss tangent) is related to the power factor PF of the dielectric by

$$PF = \sin\left(\tan^{-1} D_\varepsilon\right) \tag{3.6}$$

When their values are less than 0.15, the dissipation factor and the power factor differ by less than 1%. The absorption of energy by a dielectric is proportional to the imaginary part of the complex dielectric constant.

Some dielectric materials are dipolar and exhibit resonances as a result of molecular vibrations when subjected to electromagnetic waves. The imaginary part of this orientation polarizability gives rise to absorption of energy by the dielectric material [1]. Materials exhibiting dipolar relaxation have relative permittivity with characteristic resonances of the form

$$\varepsilon_r = \varepsilon_H + \frac{\varepsilon_L - \varepsilon_H}{1 + j\dfrac{f}{f_R}} \tag{3.7}$$

where f_R is the relaxation frequency, and ε_L and ε_H, respectively, are the low- and high-frequency limit values of the relative permittivity. One such material, water, will be studied in detail when we introduce the properties of simulated body devices in Chapter 10.

3.2.2 Losses in Conductors

Transmission lines, like coaxial cables and parallel lines, rely as much on the flow of charges in conductors as on the passage of electromagnetic fields through the dielectric medium. Electromagnetic fields will penetrate conductors, but with a field amplitude that falls off exponentially according to $\exp(-z/\delta_s)$, where z is the distance into the conductor, and δ_s is the skin depth. The general expression for the skin depth is $\delta_s = 1/\alpha_g$, where α_g is given by (3.5), where σ is the conductivity of the conductor, and $\varepsilon_r = 1$.

For a good conductor, the penetration depth in meters simplifies to

$$\delta_s = \sqrt{\frac{2}{\omega \mu_o \sigma}} \tag{3.8}$$

The resistance per unit length of round wire of diameter b and conductivity σ can now be stated by

$$R_s = \frac{1}{\pi b \delta_s \sigma} = \frac{1}{\pi b}\sqrt{\frac{\omega \mu_0}{2\sigma}} \tag{3.9}$$

Notice that the resistance per unit length R_s increases as the square root of frequency ω. The ohmic losses of a round wire depend on the current shape

$$R_{ohmic} = R_s \int_{z_0}^{z_1} I^2(z)\,dz \tag{3.10}$$

where $I(z)$ is the standing wave current normalized to the RMS maximum value.

3.2.3 Coaxial Transmission Lines

Coaxial transmission lines are concentric conductors separated by an insulator. The dimensions, construction, and conductivities of the inner and outer conductors, along with the dielectric properties of the insulator, characterize the performance of the coaxial lines. The capacitance, C_L, and inductance, L_L, per unit length [2] of coaxial cable can be expressed in terms of the outer diameter, d, of the inner conductor, the inner diameter, D, of the outer conductor, as well as the complex permittivity $\varepsilon = \varepsilon_0 \varepsilon_d$ and permeability μ of the insulating material between d and D:

$$C_L = \frac{2\pi\varepsilon}{\ln\left[\frac{D}{d}\right]} \tag{3.11}$$

Since ε is complex, C_L can be seen to be of the form of a lossless capacitance term C_0 in parallel with a conductance term G_0, derived from the imaginary part of (3.2). The inductance per unit length is

$$L_L = \frac{\mu}{2\pi} \ln\left[\frac{D}{d}\right] \tag{3.12}$$

The characteristic impedance of a transmission line is

$$Z_0 = \sqrt{\frac{j\omega L_L + R_c}{j\omega C_0 + G_0}} \tag{3.13}$$

which for low-loss coaxial lines simplified to

$$Z_0 = \frac{\eta_m}{2\pi} \ln\left[\frac{D}{d}\right] \tag{3.14}$$

The constant η_m is the characteristic impedance of the medium between the two conductors. For coaxial lines, C_L and L_L are given by (3.11) and (3.12), while the conductor resistance R_c is derived from (3.9):

$$R_c = \frac{1}{\pi d \delta_{si} \sigma_i} + \frac{1}{\pi D \delta_{so} \sigma_o} \tag{3.15}$$

Expression (3.15) recognizes that the inner conductor (subscript *i*) conductivity σ_i may be different from the outer conductor (subscript *o*) conductivity σ_o and that the respective skin depths are therefore also different. When the dissipation factor D_g is small, and $\mu_d = 1$, then [see (3.1)]

$$k_g = k\sqrt{\varepsilon_r} \tag{3.16}$$

and the approximate attenuation expression, in nepers/m, including conductor losses, is

$$\alpha_g = k\sqrt{\varepsilon_r}\left(\frac{D_\varepsilon}{2}\right) + \frac{R_c}{2Z_0} \tag{3.17}$$

Expression (3.17) can be rewritten in more common engineering terms in decibels per meter as

$$A_g = 0.09102\sqrt{\varepsilon_r}\,D_\varepsilon f + \frac{2.747}{Z_0}\left(\frac{1}{d\sqrt{\sigma_{si}}} + \frac{1}{D\sqrt{\sigma_{so}}}\right)\sqrt{f} \tag{3.18}$$

where f is in megahertz and dimensions d and D are meters. In engineering catalogs for transmission lines, A is sometimes stated in decibels per 100 feet and dimensions d and D are given in inches. Conductivity is likewise customarily given as K, which is defined as a multiplier on $\sigma_{cu} = 5.7 \times 10^7$ S/m, the conductivity of bulk copper. With those customary units [3, 4], (3.18) can be written with A in decibels per 100 feet:

$$A_g = 2.774\sqrt{\varepsilon_r}\,D_\varepsilon f + \frac{0.437}{Z_0}\left(\frac{1}{d\sqrt{K_i}} + \frac{1}{D\sqrt{K_o}}\right)\sqrt{f} \tag{3.19}$$

where dimensions d and D are in inches and frequency f is in megahertz.

In the practical case, the effective conductivity in ratio to bulk copper conductivity K_i and K_o will be about 0.4 to 0.5 for coaxial lines having a stranded inner conductor and a braided shield, even for pure copper conductors. The shielding effectiveness of braided coaxial lines is somewhat limited; the total power external to a 1m length of coaxial transmission line having a braided outer conductor is only about 50 dB below the power transmitted through that line. The effectiveness of solid outer conductors is far greater, but performance is limited by the practical construction of coaxial connectors.

[3-2a.mcd] Expression (3.19) reveals that there are two components responsible for the attenuation of coaxial cable. The dielectric losses are proportional to frequency, while the conductor losses increase with the square root of frequency. Use (3.19) with different parameters for D_ε and for K to see where each loss is dominant as a function of frequency. See also how loss varies with cable dimensions by modeling, for example, RG-58 and RG-213 cables.

Another form for the propagation constant γ for transmission lines can be written in terms of the resistance, capacitance, inductance, and conductance per unit length of the line. In terms of (3.11), (3.12), and (3.15),

$$\gamma = \sqrt{(j\omega C_0 + G_0)(j\omega L_L + R_c)} \tag{3.20}$$

This form of the propagation constant more clearly illustrates the circuit behavior of transmission lines. From (3.20), we can see that a transmission line behaves like a lumped circuit consisting of series inductor elements having series resistive losses and with parallel capacitive elements having shunt resistive losses.

Figure 3.1 shows the attenuation of several types of 50Ω coaxial cables as a function of frequency. The smallest-diameter coaxial line, CA50047, is semirigid solid copper conductor line with a diameter of 0.047 in. and solid polytetrafluoroethylene insulation. The RG-58A and RG-8 (or RG-213) lines

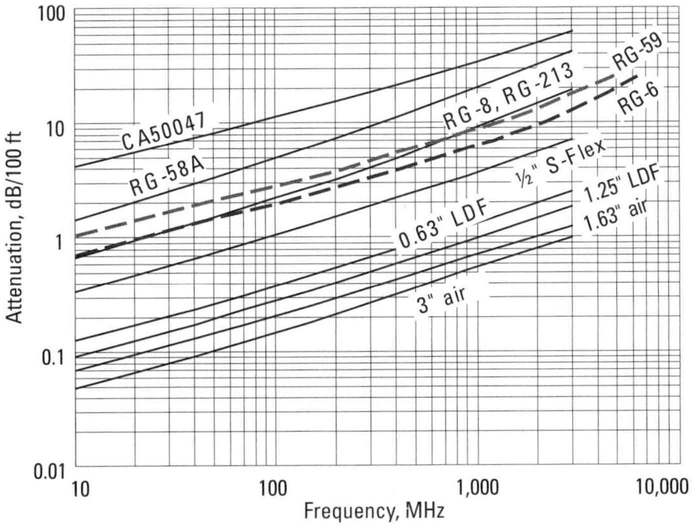

Figure 3.1 The attenuation characteristics of several types of coaxial lines.

have solid polyethylene insulation and a braided outer conductor. RG-58 is approximately 0.25 in. in diameter, and RG-213 is about 0.5 in. in diameter. The rest of the cables in Figure 3.1 are typical of the types used with modern communications system installations. The attenuation tends to vary predominantly with the square root of frequency at the lowest frequencies, where losses are primarily associated with conductors, as seen in (3.18). As frequency increases, the loss asymptotically approaches proportionality with frequency.

Increasingly, community access television (CATV) services are becoming important for the delivery of digital signals and digital communications. The standard coaxial transmission medium is 75Ω cable. Figure 3.1 also shows (in dashes) the attenuation characteristics of RG-59 and RG-6, two popularly used CATV service coaxial cables.

3.2.4 Parallel Transmission Lines

Parallel transmission lines are not often encountered in modern system designs. They are useful primarily at lower frequencies where high impedance antennas, such as end-fed dipoles, are sometimes used and where high mismatch conditions might be encountered by the use of nonresonant wire antennas. Parallel transmission lines are also used when balanced impedances with respect to ground are required. Often, parallel transmission line segments are encountered as matching elements or as components in antenna design. Because parallel transmission lines are often realized as parallel wires separated essentially by air, the attenuation is primarily due to conductor losses. When b is the center-to-center spacing between the two parallel conductors, and d is the conductor diameter, the capacitance per unit length is

$$C_L = \frac{\pi\varepsilon}{\ln\left[\frac{b+\sqrt{b^2-d^2}}{d}\right]} \tag{3.21}$$

The inductance per unit length is

$$L_L = \frac{\mu}{\pi}\ln\left[\frac{b+\sqrt{b^2-d^2}}{d}\right] \tag{3.22}$$

Using (3.13) with the approximation that losses are low, the characteristic impedance of the parallel transmission line is

$$Z_0 = \frac{\eta_m}{\pi} \ln\left[\frac{b + \sqrt{b^2 - d^2}}{d}\right] \tag{3.23}$$

The constant η_m is the characteristic impedance of the medium (usually air), and b is the center-to-center spacing of the two parallel conducts of diameter d. For a parallel transmission line that uses air dielectric, the losses can be modeled by (3.18) by setting D_g to 0 and the conductor diameters $d = D$ with conductivities $\sigma_i = \sigma_o$. For the case of parallel line with air dielectric,

$$A_g = \frac{5.494}{Z_0 d} \sqrt{\frac{f}{\sigma}} \tag{3.24}$$

Wire diameter d is in meters, f is in megahertz, and attenuation A_g is in decibels per meter.

[3-2b.mcd] The characteristic impedance of a parallel transmission line is given by (3.23). Calculate Z_0 with b/d as a parameter. Estimate the characteristic impedance of zip cord, the common household ac cord used with home appliances. Assume that the dielectric constant of the insulation is nearly 1. Is zip cord suitable as a feed line for a dipole antenna?

3.2.5 Minimum Attenuation in Transmission Lines

For a transmission line with a given characteristic impedance, the largest-diameter conductors result in the lowest losses. However, when the dimensions and dielectric material are fixed, there is an optimum impedance level for which losses are minimum [5]. An inspection of (3.18) reveals that the attenuation of coaxial transmission lines can be written in terms of the ratio $R = b/d$. The attenuation due to the conductor losses can then be cast into the functional form

$$f(R) = \frac{1 + R}{\ln(R)} \tag{3.25}$$

for which a minimum can be found by setting the derivative with respect to R of $f(R)$ to zero and solving for the ratio R. The solution, obtained numerically, is $R = 3.591$, for which the corresponding coaxial cable has a characteristic impedance given by (3.14) as

$$Z_0 = \frac{\eta_0}{2\pi} 3.591 = \frac{76.65}{\sqrt{\varepsilon_r}} \tag{3.26}$$

which is in agreement with the value in [5]. This minimum loss behavior is partially the answer to the question, Why do we use 50Ω as an impedance reference level. We note that many of the solid dielectrics suitable for use in coaxial cables have relative permittivity in the vicinity of 2.2, which results in the optimum value for Z_0 of about 50Ω.

A similar minimum can be found for parallel transmission line, where the losses given by (3.24) are written in terms of the ratio of the conductor spacing to the conductor diameters, $R = b/d$. In the parallel transmission line case, the conductor losses take the form

$$g(R) = \frac{R}{\ln\left[R + \sqrt{R^2 - 1}\right]} \tag{3.27}$$

Here, $g(R)$ has a minimum value when $R = 1.8102$, corresponding to a parallel line characteristic impedance of $Z_0 = 143.9\Omega$.

[3-2c.mcd] Starting with (3.14) and (3.18), show that the minimum loss in coaxial transmission lines occurs when the ratio of the outer to inner conductor diameters is $R = 3.591$. Starting with (3.23) and (3.24), show that the characteristic impedance, in air, for minimum loss in parallel transmission line is $Z_0 = 143.9\Omega$.

3.2.6 Summary of Transmission Line Relationships

It is worth mentioning some general relationships on a transmission line connected to a matched load and carrying a time-varying signal. First, the signal voltage-to-current ratio equals the line intrinsic impedance and the matched load resistance. Second, the transmission line impedance equals the square root of the ratio of the line inductance per length to capacitance per length. Thus, for a given power transferred, a higher impedance line has a higher voltage between the transmission line conductors but a lower current. This means that if the primary loss mechanism in the line is due to the conductor losses (I^2R losses), a higher impedance line is preferred (smaller currents for a given power level).

3.2.7 Optical Fiber Transmission Lines

The transmission characteristics of optical fibers can be described by the two parameters of attenuation and dispersion. Attenuation in optical fibers is dependent on the transparency of the materials, usually silica, used to construct the transmission line. Loss is dependent on wavelength, as shown in Figure 3.2. Optical fibers are typically operated at three different wavelength bands: 850, 1,300, and 1,550 nm [6, 7].

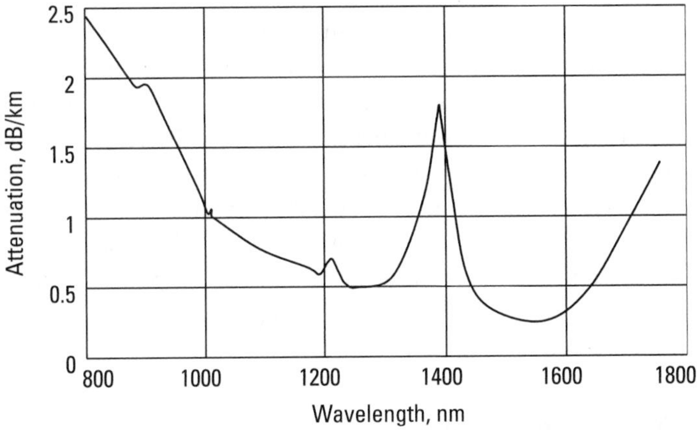

Figure 3.2 The attenuation characteristics of optical fiber as a function of wavelength.

The operation at the 850-nm wavelength has an attenuation of about 2.3 dB/km. Absorption peaks near 1,300 nm are due to OH^- water ions. If they can be avoided in cable manufacture, an attenuation of as little as 0.5 dB/km can be achieved at this wavelength. A third transmission band near 1,550 nm offers the fiber's lowest attenuation of 0.25 dB/km. The 1,300- and 1,550-nm wavelengths are usually chosen for long lines carrying wide-bandwidth signals.

Dispersion, which is a multimode transmission phenomenon, causes energy to travel within an optical fiber by various paths as a result of different internal reflection angles from the fiber wall boundary. Careful design and manufacturing techniques will result in a "single-mode" fiber where the modal dispersion is largely eliminated. There is also chromatic dispersion due to material properties, which vary the dielectric properties of the fiber with wavelength.

The spreading of the light pulses traveling through the fiber because of dispersion limits the permissible optical fiber line length that can be used at a given data rate. The dispersion in fiber optical transmission lines varies with frequency and crosses the zero-dispersion point in the 1,200–1,600-nm wavelength range, depending on the diameter of the cable. With fiber optical transmission lines available today, we can achieve transmission losses equal to a fraction of a decibel per kilometer and distance by bandwidth products of 50 to 100 GHz × km.

3.3 Basic Radiowave Propagation

The primary propagation medium in radiowave propagation is the atmosphere, and wave attenuation is due to geometric spreading, multipath wave interference, and absorptive loss in the medium. As will be shown with the development of the Friis transmission formula, the primary geometric spreading factor in free space power radiation follows an inverse square law with distance. Additional spreading occurs because of multiple reflections due to the multipath wave interference. These multiple reflections can apparently increase the power law of propagation to typically third-power and as much as fifth-power with distance. The SBY propagation model in Section 7.5.1 provides a mechanism for this increase. Urban propagation along with multipath will be considered in detail in Chapter 8. Transmission losses additionally include exponential losses due to building materials, foliage, and atmospheric attenuation (>3 GHz). These, too, will be considered in detail in Chapters 8 and 12.

3.3.1 Friis Transmission Formula

Basic free space propagation attenuation is due to the geometric spherical expansion of waves, so attenuation is inversely proportional to the distance d squared, and the power density at distance d is

$$P_d = \frac{P_e}{4\pi d^2} \tag{3.28}$$

Thus, an effective isotropically radiated power $P_e = G_t P_t$ propagates with a power density P_d W/m^2 at range d according to (3.28). Antenna gain and effective aperture area are related by

$$A = \frac{G\lambda^2}{4\pi} \tag{3.29}$$

So, the received power is $P_r = A_r P_d$. The wave number $k = 2\pi/\lambda$, and the transmission formula in terms of wave number is

$$\frac{P_r}{P_t} = \frac{G_r G_t}{(2kd)^2} \tag{3.30}$$

This form of the propagation law leads to the Friis transmission formula [8]. The Friis transmission formula states the ratio of the received power to transmitted power in terms of the free space propagation law (3.28) and the transmitting and receiving antenna effective apertures A_t and A_r, respectively. So,

$$\frac{P_r}{P_t} = \frac{A_r A_t}{\lambda^2 d^2} \tag{3.31}$$

as seen in Figure 3.3.

A similar transmission formula can be written for UWB pulses as will be shown in Chapter 11.

3.3.2 Comparison of Guided Wave and Radiowave Propagation Attenuation

A comparison of guided wave and radio channel propagation attenuation shows the contrast between exponential and power law behaviors. Using the developed formulas, at a frequency of 4 GHz, we get the attenuation results for free space radiation and for coaxial cable transmission shown in Figure 3.4. Initially, geometric expansion of waves provides much greater attenuation than exponential losses. However, the exponential curve soon overtakes the geometric behavior. Clearly, cable does not always provide the smallest attenuation, and the choice between using cable or RF techniques depends as much on attenuation characteristics as on economics, accessibility, and security.

[3-3.mcd] Compare coaxial cable losses with free space transmission losses given by (3.31) as a function of distance at 4 GHz. What happens if we have urban propagation instead of free space propagation? Use the Siwiak-Bertoni-Yano (SBY) model (see Chapter 7) with a 30m break point to a 3.5 power law.

Figure 3.3 Sergey A. Schelkunoff and Harald T. Friis with Friis' transmission formula. (*Source:* Jean Kennedy Schelkunoff collection, MS-0583-f-3-2-58300001.jpg, Archives and Special Collections, Mount Holyoke College).

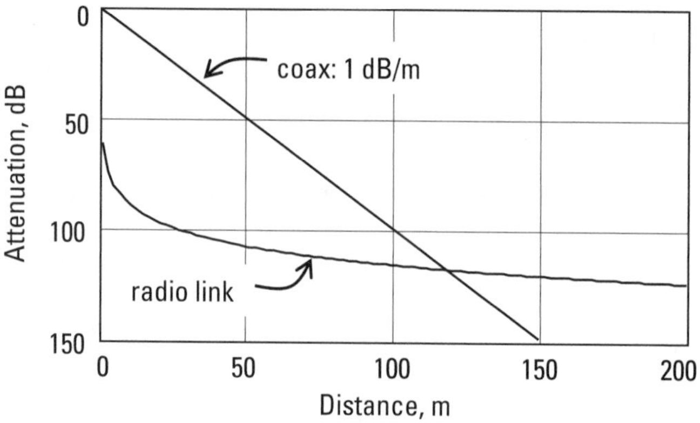

Figure 3.4 Comparison of coaxial cable guided wave and radio channel propagation attenuations.

3.4 Wave Polarization

The polarization of waves is defined in terms of the time-varying vector orientation of the electric field. Antennas and waves each exhibit a polarization characteristic. Losses occur when the antenna and wave polarizations are mismatched. In this section, a method will be developed to describe polarization characteristics and to determine the polarization mismatch loss.

3.4.1 Polarization of Antennas

Many personal communications devices operate with a nominally vertical electric field polarization. Radiowave propagation, especially in urban and suburban regions and within buildings, involves multiple randomly disposed scatterers in the vicinity of the personal communications device. The scattering tends to randomize the polarization of the usually vertically polarized transmitted wave. The radio-link losses due to polarization mismatch will be presented here.

In general, the far-zone field may be written as

$$\mathbf{E} = E_\theta \boldsymbol{\theta} + E_\phi \boldsymbol{\phi} \tau e^{-j\gamma} \tag{3.32}$$

using the notation of [9], where γ is a phase constant. The time domain dependencies in (3.32) are $\cos(\omega t - kd)$ in $\boldsymbol{\theta}$ and $\cos(\omega t - kd + \gamma)$ in $\boldsymbol{\phi}$, where k is the wave number and d is distance. The unit vectors in the $\boldsymbol{\theta}$ and $\boldsymbol{\phi}$ directions are $\boldsymbol{\theta}$ and $\boldsymbol{\phi}$, respectively, orthogonal to the propagation direction. Since the two components $\boldsymbol{\theta}$ and $\boldsymbol{\phi}$ are orthogonal, electric field vector $\mathbf{E}$ traces out an ellipse in space and time with magnitude 1 in $\boldsymbol{\theta}$ and τ in $\boldsymbol{\phi}$. When τ has a magnitude of 1, and $\gamma = \pm\pi/2$, the polarization is said to be circular.

3.4.2 Polarization Characteristics of Antennas

As defined earlier in Section 1.4, we can assign a complex vector $\mathbf{h}_a$ to describe the polarization characteristics of an antenna in terms of the antenna response h_θ in the $\boldsymbol{\theta}$ polarization and h_ϕ in the $\boldsymbol{\phi}$ polarization directions:

$$\mathbf{h}_{\mathbf{a}} = h_\theta \boldsymbol{\theta} + h_\varphi \boldsymbol{\phi} \tag{3.33}$$

For right-hand circular polarization (RHCP),

$$\mathbf{h}_{\mathbf{RHC}} = \frac{\boldsymbol{\theta} - j\boldsymbol{\phi}}{\sqrt{2}} \tag{3.34}$$

and for left-hand circular polarization (LHCP),

$$\mathbf{h}_{\mathbf{LHC}} = \frac{\boldsymbol{\theta} + j\boldsymbol{\phi}}{\sqrt{2}} \tag{3.35}$$

The expressions (3.34) and (3.35) are normalized to unity, but that is neither general nor necessary.

3.4.3 Polarization Mismatch in Antennas

The effect of polarization mismatch loss L_p between a transmitted polarization $\mathbf{h}_{tx}$ and a receiver polarization $\mathbf{h}_{rx}$ is

$$L_p = \left| \frac{\mathbf{h}_{\mathrm{tx}} \cdot \mathbf{h}^*_{\mathrm{rx}}}{|\mathbf{h}_{\mathrm{tx}}| \cdot |\mathbf{h}_{\mathrm{rx}}|} \right|^2 \tag{3.36}$$

Clearly, using (3.36) with definitions (3.34) and (3.35), an RHCP antenna will not receive an LHCP wave. The polarization mismatch loss, for example, between a circularly polarized wave and a linearly polarized antenna is 0.5, or 3 dB:

$$L_p = \left| \frac{\boldsymbol{\theta} \cdot (\boldsymbol{\theta} + j\boldsymbol{\phi})}{|\boldsymbol{\theta}| \cdot |\boldsymbol{\theta} + j\boldsymbol{\phi}|} \right|^2 = \frac{1}{\sqrt{2}}^{2} = \frac{1}{2} \tag{3.37}$$

A linearly polarized antenna used to receive a circularly polarized wave will recover only half the power relative to a circularly polarized antenna of the same directivity. In reciprocal fashion, but perhaps slightly less intuitively, a circularly polarized antenna will receive only half the power from a linearly polarized wave compared with a linearly polarized antenna having the same directivity.

3.4.4 Polarization Filtering: An Experiment in Optics

We can pose an interesting physical experiment that illustrates the behavior of polarized waves. Let's begin, as shown in Figure 3.5(a), with an input plane wave of uniformly, randomly polarized laser light at A traveling in the z-direction. An intervening optical x-polarization filter F1 filters out all but the x-directed electric field components. Since the wave was randomly polarized at A and half of the random polarized signal was filtered out by F1, we might expect the wave power density at B to be half of the wave power density at A. That x-polarized plane wave travels to C and is filtered by an intervening optical y-polarization filter F2, which now filters out all of the remaining wave front. We expect zero power density at D in Figure 3.5(a).

Suppose now that we add an additional filter F3 between F1 and F2, as shown in Figure 3.5(b). Is the power density still zero at D? Not necessarily, especially if filter F3 is polarized in any orientation other than the x- or y-plane.

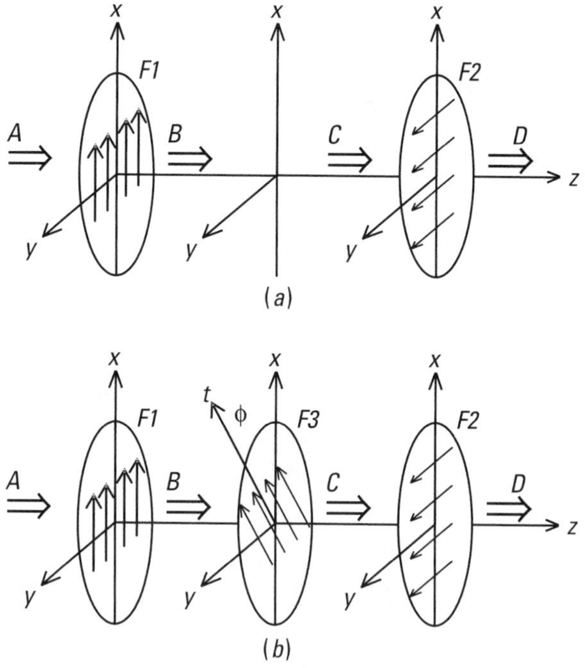

Figure 3.5 Polarization filtering at optical frequencies (a) without, and (b) with an intermediate filter.

In fact, we can show that in (b), the power density at D may be as much as 0.125 times the power density at A, even though more filtering is done in (b) than in (a). The cases in Figure 3.5(a, b) can be analyzed using polarization vectors to describe the signals and the filter. Since the input wave front at A is uniformly random polarizations, the signal can be decomposed into the orthogonally polarized vectors $\mathbf{h}_x + \mathbf{h}_y$, where $|\mathbf{h}_x|$ equals $|\mathbf{h}_y|$. Filter F1 has a polarization characteristic described by the unit vector **x** along the *x*-axis. The power density at B is $|(\mathbf{h}_x + \mathbf{h}_y) \cdot \mathbf{x}|^2$, so the power ratio of B to A is 0.5, and the polarization orientation at B is **x**. In Figure 3.5(a), this *x*-polarized wave front encounters filter F2, which has a polarization characteristic described by the unit vector **y**. Evidently, the vector operation $\mathbf{x} \cdot \mathbf{y}$ is equal to zero, which is the expected result in Figure 3.5(a).

In Figure 3.5(b), however, the wave front at B encounters a filter F3 with a polarization characteristic given by the unit vector $\mathbf{t} = \mathbf{x} \cos \phi + \mathbf{y} \sin \phi$. The power density of the wave front at C is now $|[(\mathbf{h}_x + \mathbf{h}_y) \cdot \mathbf{x}]\mathbf{x} \cdot \mathbf{t}|^2$, which equals $|\mathbf{h}_x\mathbf{x} \cos \phi|^2$ oriented in the **t** polarization direction. This wave front at C encounters filter F2, having polarization characteristic **y** as described above for Figure 3.5(a). The power density at D is now the square of the magnitude of the vector dot product $\mathbf{t} \cdot \mathbf{y}$, or $\sin \phi$, applied to the wave front at C. In fact, the wave front power density at D in terms of the power density at A is

$$D / A = \left|\left(\mathbf{h}_x + \mathbf{h}_y\right) \cdot \mathbf{x}\right|^2 |\mathbf{x} \cdot \mathbf{t}|^2 |\mathbf{t} \cdot \mathbf{y}|^2 / \left|\left(\mathbf{h}_x + \mathbf{h}_y\right)\right|^2 = |\cos\phi \sin\phi|^2 / 2 \quad (3.38)$$

The maximum value of D/A is 0.125 when $\phi = 45°$.

The polarization filters are not simply analog filters but scattering devices that apply vector operations to vector fields. These filters perform operations that are not commutative. We can intuitively understand the operations in Figure 3.5(b) when we interpret the actions of F1, F3, and F2 as radiating elements. F1 can receive energy and reradiate it only in the **x**-polarization, F3 can receive and reradiate only in the **t**-polarization, and similarly F2 can only receive and reradiate in the **y**-polarization. When viewed as receiving and reradiating antennas with (3.31) applied, it is evident that adding the filter F3 in Figure 3.5(b) actually increase the power density at D. This experiment clearly demonstrates that very few scatterers are needed to couple one polarization to its orthogonal polarization.

3.4.5 Polarization Scattering and the Radar Equation

Filters F1 and F2 in Figure 3.5 could just as easily be replaced with a vertically polarized transmitting dipole D1 and a horizontally polarized receiving dipole D2. A third dipole, or half-wave-long wire element D3, could be substituted for

F3 to repeat the experiment with the same effect at radio frequencies. That scenario is exploited in a slight variation in Figure 3.6.

Dipoles D1 and D2 are collocated and orthogonal, so the coupling between D1 and D2 is zero. We use the Friis equation to find the coupling between transmit dipole D1 and D3, then between D3 and D2, and where $\mathbf{x} \cdot \mathbf{t} = \cos(\phi)$. All of the apertures are the same, $A_1 = A_2 = A_3 = A_e$, and the scattering element D3 reradiates all of the received power P_3, so the received power at dipole D2 is

$$\begin{aligned}\frac{P_2}{P_1} &= \left(\frac{A_3 A_1}{\lambda^2 d^2}\left|\hat{\mathbf{x}} \cdot \hat{\mathbf{t}}\right|^2\right)\left(\frac{A_2 A_3}{\lambda^2 d^2}\left|\hat{\mathbf{y}} \cdot \hat{\mathbf{t}}\right|^2\right) \\ &= \left(\frac{A_e}{\lambda d}\right)^4 \left(\frac{\sin(2\varphi)}{2}\right)^2\end{aligned} \tag{3.39}$$

where the substitution $\cos(\phi)\sin(\phi) = \sin(2\phi)/2$ has been made. The result demonstrates the "radar equation"; that is, we see an inverse fourth-power relationship between the radar antennas and the scatterer. It also illustrates a remote "rotational-motion" detector, where the received signal depends on the scatterer-wire rotational orientation.

3.5 Summary

The communication channel was introduced in terms of guided and radiated waves. Guided waves tend to attenuate exponentially because of losses in the medium, while radiated waves tend to attenuate geometrically because of wave expansion and multipath interference. The characteristics of coaxial, parallel line, and optical fiber transmission lines were introduced and the transmission and attenuation characteristics of transmission lines were detailed. The Friis

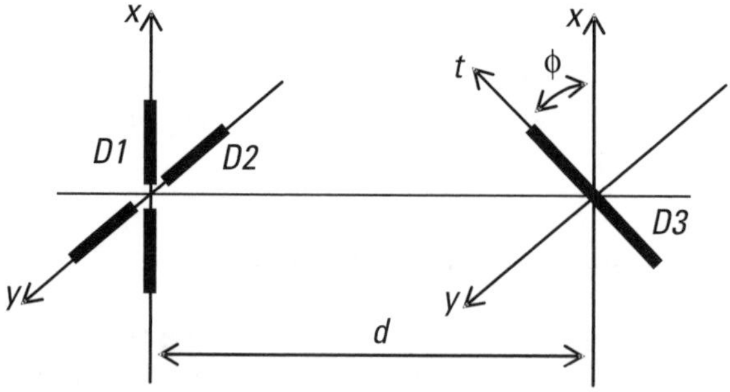

Figure 3.6 Scattering from a rotated half-wavelength element.

transmission formula was introduced as a basic free space propagation law. The polarization characteristics of antennas and waves must be matched for maximum power transfer.

Problems

3.1 Given a dielectric with relative permittivity $\varepsilon_r = 4 - j3$, find the loss tangent and power factor. Are the material properties constant with frequency?

3.2 A dielectric has the properties $\dfrac{\varepsilon}{\varepsilon_0} = 4.9 + \dfrac{73.4}{1 + jf/f_0}$

where f_0 = 19.65 GHz. Find (a) the loss tangent and (b) the frequency, where the imaginary part of $\varepsilon/\varepsilon_0$ is maximum.

3.3 Derive an expression for the ohmic loss of a dipole constructed from a length L of round wire for (a) the ideal short dipole, (b) a short dipole, and (c) a dipole of length $L = \lambda/2$ with sinusoidal current distribution.

Ans:

(a) $R_{ohmic} = R_s/(\pi b)L$ (b) $R_{ohmic} = R_s/(\pi b)(L/3)$ (c) $R_{ohmic} = R_s/(\pi b)(L/2)$.

3.4 A resonant half-wave dipole is constructed from wire with 1,000-S/m conductivity. Find the radiation efficiency.

3.5 An electrically short dipole of total length $h = 0.01$ wavelengths has radiation resistance $R = 200h^2$ and is made from copper ($\sigma = 5.7 \times 10^7$ S/m). Find the radiation efficiency.

3.6 A radio engineer uses a piece of unterminated coaxial cable as a shunt capacitor of 50 pF in a circuit operating at 1 MHz. (a) What length of RG-58 coaxial line ($\varepsilon_r = 2.2$) can she use? What are the properties of this "capacitor" at (b) 100 MHz and (c) 200 MHz?

Ans:

From (3.11), with $\varepsilon_r = 2.2$, $\ln(D/d) = 1.237$, and from (3.8), C = 98.96 pF/m, so the required length is 50.5 cm. At (a), 100 MHz, this is a quarter-wavelength length of unterminated line and appears like a short circuit; at (b), 200 MHz, this is a half-wavelength cable, which reflects the open circuit to its input.

3.7 A 902-MHz-band, 14-dBi omnidirectional antenna costing \$3,000 is used to receive signals at the top of a 100-ft. tower. It connects to a receiver through 200 ft. of coaxial cable. The tower operator wishes to save costs and suggests that RG-58/U cable at \$0.30/ft. be used in this low-power application rather than lower-loss RG-213 at \$0.80. Calculate the losses of the two cables, speculate on the issues involved with respect to the type of antenna selected, and suggest a maximum-performance design option for this installation.

3.8 Find the dimensions of parallel transmission line suitable for feeding a dipole matched to 50Ω terminal impedance.

3.9 Shielded balanced transmission line is constructed from equal lengths of identical coax lines with 100Ω characteristic impedance by connecting their outer conductors together at each end. The two inner conductors form the balanced transmission line pair. Find the characteristic impedance of the combination.

3.10 Design a 25Ω-characteristic-impedance balanced transmission line using (a) parallel coaxial lines and (b) parallel conductors. Comment on the practicality of each.

3.11 An aircraft at an altitude of 57,000 ft. transmits 10W isotropically to a ground station at the nadir. Ignoring ground effects, find the (a) electric field at the ground station, and (b) the power received by a dipole at frequencies 121 MHz and 890 MHz.

3.12 The sun delivers approximately $7f \times 10^{-21}$ W/m^2 to Earth in the frequency range $f = 0.1$ to 100 GHz. How much power is collected by a 2 m^2 solar collector band limited to the above range if (a) the solar collector has a constant aperture and (b) constant directivity as a function of frequency.

3.13 Where does the apparent frequency dependence of free space propagation come from?

3.14 A 1W microwave link uses 20-dBi gain antennas at each end of the 3-km link. Find the received power. What is the received power if the operating frequency is doubled but antennas of the same aperture area are employed?

3.15 A microwave link transmits 100 EIRP (effective isotropically radiated power) at 5 GHz over a 3-km link. What is the power received (a) by a 20-dBi gain antenna at end of the link, and (b) by the same receiving aperture area if the operating frequency is moved to 10 GHz with the same EIRP?

3.16 A CP (circularly polarized) dipole and LP (linearly polarized) dipole are on a common axis and in parallel planes 3m apart. Find the power coupling ratio if (a) the CP antenna transmits, and (b) if the LP antenna transmits.

3.17 The 5-km circularly polarized microwave link in Figure 3.P1 uses the arrangement (a) using 50% efficient mirrors at A and B and a CP receiving antenna at C. Maintenance is required on mirror B, so the system operator temporarily repositions mirror A to use the arrangement in (b). Estimate the signal strength received in (b) compared to A. *Hint:* What is the CP sense after a single specular reflection?

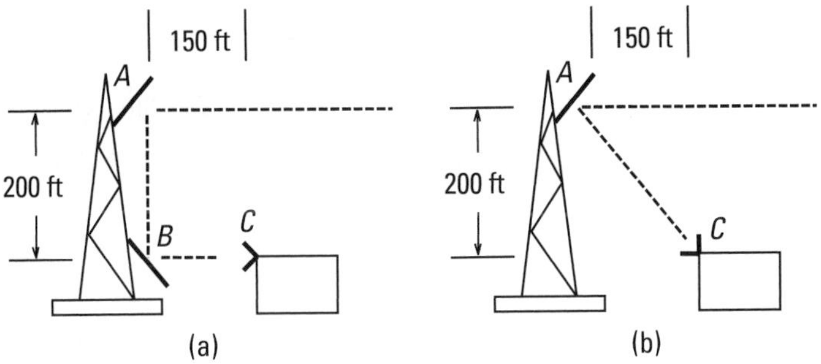

Figure 3.P1 A link in (a) normal and (b) temporary operation.

3.18 Two 915-MHz dipoles in parallel planes and on a common axis are 3m apart. If 1W is supplied to one dipole, what is the power received by the second dipole if (a) the dipoles are copolarized, (b) at 45° with respect to each other, and (c) at 90° with respect to each other?

3.19 Two 915-MHz dipoles in parallel planes and on a common axis are 3m apart and cross-polarized. A large grid of perfectly conducting, finely spaced, thin parallel wires is placed in a parallel plane equidistant between the dipoles. The wires and either dipole form a 45° angle, much like F3 in Figure 3.5. If 1W is supplied to one dipole, (a) find the power received by the second dipole, and (b) find the power received if the second dipole is circularly polarized.

3.20 A lossless coaxial cable delivers a 1W, 10-MHz transmitter signal to a matched load. (a) What is the ratio of the voltage to current in the coaxial cable if its intrinsic impedance is 50Ω? (b) What if its intrinsic impedance is 75Ω? (c) Can you express the signal level in dBm for (a) and (b)?

Ans:

(a) $V/I = 50$; (b) $V/I = 75$; for (c) in (a), 1W is 10log(1,000mW) = 30 dBm. "dBm" is generally used only in 50Ω systems.

3.21 A lossless 300Ω transmission line delivers 1,000W to a matched antenna. What is the RMS voltage between the transmission line conductors?

Ans:

V/I must equal 300Ω, and $P = VI = 1{,}000 = V^2/300$, so $V = 547.7$V.

References

[1] Dekker, A. J., *Electrical Engineering Materials*, Englewood Cliffs, NJ: Prentice Hall, 1959.

[2] Jordan, E. C., and K. G. Balmain. *Electromagnetic Waves and Radiating Systems,* 2nd ed., Englewood Cliffs, NJ: Prentice Hall, 1968.

[3] *Coaxitube Semi-rigid Coaxial Cable*, Precision Tube Company, Inc., Cat. 752-80, North Wales, PA.

[4] *RF Transmission Line Catalog and Handbook*, Times Wire and Cable Company, Cat. No. TL-3, 1970.

[5] Mathieu, G., L. Young, and E. M. Jones. *Microwave Filters, Impedance-Matching Networks, and Coupling Structures*, Norwood, MA: Artech House, 1980.

[6] Nellist, J. G., *Understanding Telecommunications and Lightwave Systems*, Piscataway, NJ: IEEE Press, 1992.

[7] Andonovic, I., and D. Uttamchandani. *Principles of Modern Optical Systems*, Norwood, MA: Artech House, 1989.

[8] Friis, H. T.,"A Note on a Simple Transmission Formula," *Proc. of the IRE*, Vol. 34, 1946, p. 254.

[9] Collin, R. E., *Antennas and Radiowave Propagation,* New York: McGraw-Hill, 1985.

4

Radio Frequency Spectrum

4.1 Introduction

Radio communications involve the use of electromagnetic waves at frequency segments in the radio frequency spectrum. From the PCS point of view, not all segments of the radio frequency spectrum are equally preferred for desired services. This chapter presents an overview of the electromagnetic spectrum and surveys some of the behavior of electromagnetic waves in certain radio frequency bands in view of their potential application to personal communications. Table 4.1 details the radio frequency spectrum and lists some of its uses.

Our survey of the radio frequency spectrum is intended to provide a simple and brief understanding of some of the characteristics of transmissions at various frequencies. We see from Table 4.1 that the frequencies below about 30 MHz are associated with propagation involving the Earth's ionosphere. Radiowave interactions with the ionosphere are quite complex, and only some of the effects will be presented here. The ionospheric phenomena are covered elsewhere [1–4] in great detail, and an HF propagation software tool described in Appendix D is used to illustrate properties of ionospheric propagation.

It is this lower-frequency region that was first exploited commercially, although experimenters like Heinrich Hertz were already experimenting [5] in the VHF band in the 1880s and 1890s. We will show that the primary range of interest in personal communications lies at frequencies above 100 MHz. The particular characteristics of wave propagation at those frequencies will be detailed in Chapters 5 to 7. We will also introduce multiple access techniques in view of their impact on radiowave propagation.

Table 4.1
The Radio Frequency Spectrum and Its Uses

Frequencies	Band	Characteristics	Services
3 Hz–30 kHz	ELF, VLF	High atmospheric noise, Earth-ionosphere waveguide modes, very inefficient antennas	Submarine, navigation, sonar, long-range navigation
30–300 kHz	LF	High atmospheric noise, Earth-ionosphere waveguide modes, absorption in the ionosphere	Long-range navigational beacons
0.3–3 MHz	MF	High atmospheric noise, good ground wave propagation, Earth magnetic field cyclotron noise	Navigation, maritime communication, AM broadcasting
3–30 MHz	HF	Moderate atmospheric noise, ionosphere reflections that provide long-distance links, affected by solar flux density	International shortwave broadcasting, ship-to-shore, telephone, telegraphy, long-range aircraft communication, amateur radio
30–300 MHz	VHF	Some ionosphere reflections at the lower range, meteor scatter possible, normal propagation basically line of sight	Mobile, television, FM broadcasting, air-traffic control, radio navigation aids, land-mobile communications
0.3–3 GHz	UHF	Basically line-of-sight propagation, efficient portable antennas	Television, radar, land-mobile communications, satellite communications, Global Positioning Satellite (GPS), PCS, wireless local-area networking, mobile phones
3–30 GHz	SHF	Line-of-sight propagation, atmospheric absorption at upper frequencies	Radar, microwave links, land-mobile communication, satellite communication, UWB, wireless LANs and PANs, fixed broadband, 3G PCS
30–300 GHz	EHF	Line-of-sight propagation, very subject to atmospheric absorption	Radar, secure and military communication, satellite links, mm-wave personal-area networking
300–10^7 GHz	IR—optics	Line-of-sight propagation, very subject to atmospheric absorption	Optical communications, fiber optical links

4.2 Extremely Low and Very Low Frequencies (<30 kHz)

Frequencies below 3 kHz are generally useful only for communications through seawater because of the extremely severe bandwidth limitation and because it is very difficult to launch any appreciable energy from any reasonably sized antenna. Once launched, however, field strength tends to remain fairly constant around the globe; the Earth is only 133 wavelengths in circumference at 1 kHz. The Earth and the ionosphere form a pair of concentric shells that resonate elecromagnetically at a frequency near 7 Hz. The possibility that the Earth-ionosphere cavity resonates electromagnetically was first reported by W. O. Schumann [6] in 1952 on a theoretical basis. The Schumann resonances have been measured at 7.9, 14, 20, and 26 Hz. The Q factors are in the 2.5 to 7 range for the first three resonances [7]. Recent studies indicate that extremely-low-frequency (ELF) noise, especially in the polar regions, is the superposition of fields of atmospheric (lightning) and magnetospheric origins.

Propagation in the ELF and very-low-frequency (VLF) range is by surface wave and by the Earth-ionosphere waveguide. The effective height of the ionosphere, particularly in the ELF band, is on the order of 90 km. This height is much smaller than a wavelength at ELF, so propagation is by the lowest-order transverse electromagnetic (TE or TM) mode. The Earth-ionosphere waveguide at ELF propagates signals with very low attenuation [8] on the order of 0.1 to 0.5 dB per 1,000 km, in addition to the power with inverse distance law [3] from the Earth-ionosphere TEM waveguide mode of propagation.

The ELF band is primarily useful for communications links through ocean water to submerged submarines. Huge land-based transmitting antennas are required, and relatively low data rates are possible. Basic propagation through water is by evanescent wave because ocean water is highly conductive. The attenuation in one skin depth of seawater, or any other dielectric, is 8.86 dB, so the longest wavelengths are preferred if attenuation is to be kept to usable values. The propagation constant k_w in seawater is complex and involves significant conductivity losses:

$$k_w = k\sqrt{\varepsilon_w - j\frac{\sigma_w}{\omega\varepsilon_o}} \tag{4.1}$$

The relative dielectric constant of water depends on both temperature and salinity. The detailed dielectric properties of saline water will be presented in Chapter 10. For frequencies below about 2 GHz, the approximate value of the dielectric constant is $\varepsilon_w = 80$, and the conductivity is nearly $\sigma_w = 4$ S/m. The attenuation by seawater is severe even at very low frequencies, as seen in Figure 4.1 and reported in Table 4.2. At the low frequencies, there is obviously

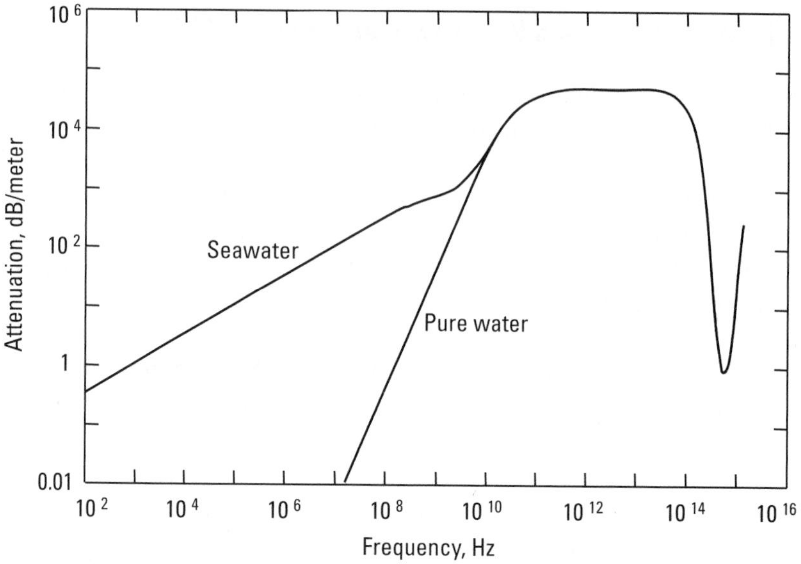

Figure 4.1 Attenuation in seawater and in pure water.

Table 4.2
Attenuation, α, of Radio Waves by Seawater

Frequency	α **(nepers/m)**	**Decibels per Meter**
3 MHz	6.9	61
300 kHz	2.2	19
30 kHz	0.69	6
3 kHz	0.22	1.9

not much spectrum available, and antennas tend to be very high *Q*, limiting the data bandwidth to extremely small values.

Figure 4.1 gives the attenuation in decibels per meter of seawater in comparison with pure water. Clearly, even at the lowest frequency, attenuation in saline water is quite severe. The conductivity due to the saline concentration is the major source of attenuation up to several gigahertz. Above that, the water molecular dipole resonance attenuation predominates to frequencies beyond several hundred gigahertz. There is a window of transparency near 6×10^{14} Hz (green light), which is useful for light wave communications to submerged submarines.

4.3 Low and Medium Frequencies (30 kHz to 3 MHz)

The lower portion of the low frequency (LF) band, from 30 kHz to about 500 kHz, is primarily used for marine and aeronautical radio navigation beacons. This range is very much affected by atmospheric noise. Because of the long wavelengths, reasonably sized antennas in the lower portions of this band are usually very inefficient and very narrowband. The most common use in the upper portion of the band segment is the amplitude modulation (AM) broadcasting service in channels between 535 kHz and 1,705 kHz. The medium frequency (MF) band, from 300 kHz to 3 MHz, is characterized by excellent ground wave propagation. An interesting phenomenon involving electrons moving in the ionosphere under the influence of Earth's magnetic field produces cyclotron noise in this frequency range, but this is not a factor in terrestrial transmissions. Free electrons in the ionosphere spiral under the influence of Earth's magnetic field to generate noise. The Earth's magnetic flux density near the ground at mid-latitudes is $B_0 = 5 \times 10^{-5}$ tesla, corresponding to a magnetic field of 40 A/m, and the electron charge to mass ratio $e_e/m_e = 1.76 \times 10^{11}$ C/kg, so the cyclotron frequency is $\omega_c = 2\pi f_c$, where

$$f_c = \frac{e_e}{m_e}\frac{B_0}{2\pi} = 1.4\,\text{MHz} \tag{4.2}$$

Ionospheric absorption during the daytime is high in this frequency range.

The MF range was once thought to be the upper range of frequencies useful for radio communication. Wavelengths shorter than 200m were left to experimenters and radio amateurs. The experience of the early experimenters and radio amateurs [9], of course, has shown that the interesting radio phenomena involving ionospheric reflections occur above the MF range. Most of the useful frequency ranges for portable communications are well above the MF band.

4.4 High Frequencies (3 to 30 MHz)

Signals at these frequencies are subject to ionospheric reflections, hence are the basis of traditional worldwide radio communications. Recent advances in Earth-orbiting satellites are slowly replacing high frequency (HF) services with comparable satellite-based services. Ionospheric transmission provides a channel of comparatively small attenuation that makes around-the-world propagation possible. The applications in this band are generally narrowband (<10 kHz) with information bandwidths of less than 3 kHz. The ionosphere is the product of solar and cosmic radiation acting on the atmosphere to dissociate free electrons. There is absorption at the lower useful range, and there is a maximum usable frequency (MUF) above which the ionosphere is largely transparent and influences only polarization by the phenomenon of Faraday rotation. The effect of Faraday rotation of polarization will be visited in Chapter 5, when propagation from Earth-orbiting satellites is detailed. Ionospheric physics is a complex and detailed discipline, and only some simple highlights will be illustrated here with the help of readily available software (VOACAP and HamCAP) included on the CD-ROM bound with this text and described in Appendix D.

4.4.1 Ionosphere

The ionosphere is the upper region of the atmosphere (above 50 km), where solar flux and cosmic radiation has ionized the atmospheric gasses. The resulting free electron densities, N_e, are in the range of 10^9 to 10^{12} electrons per cubic meter. The ionosphere is bounded at the highest altitude by the rarity of atmospheric gases and at the lowest altitude by the atmospheric attenuation of the radiation that causes the ionization. The dry atmosphere is well mixed to about the 80-km height. Above 80 km, there are varying densities of ionized gasses with altitude because the composition of the atmosphere varies with altitude; hence, the dissociation of ions varies with altitude. This causes distinct layers, primarily the D, E, F1, and F2 layers.

4.4.2 Layers in the Ionosphere

There are primarily three distinct layers in the ionosphere:

- *D layer:* Prominent in the daytime, this layer vanishes at night. It is characterized by high electron-collision frequency, and absorption is high.
- *E and F layers:* These are most important for communication between 3 and 40 MHz. During the day, the F layer may split into the F1 and F2 layers.

The electron density is shown as a function of height in typical day and night profiles in Figure 4.2. Actual electron densities depend on the solar flux value, which is in turn related to the sunspot number. In the case of the F layer, this relationship is complex as, for example, winter electron densities are generally higher than summer ones. Sunspot numbers have approximately 11-year cyclic behavior. The detailed influence on the ionosphere is further complicated by the Earth's rotation about a tilted axis. This results in day-night and seasonal variations in electron densities around the globe. Relatively good propagation

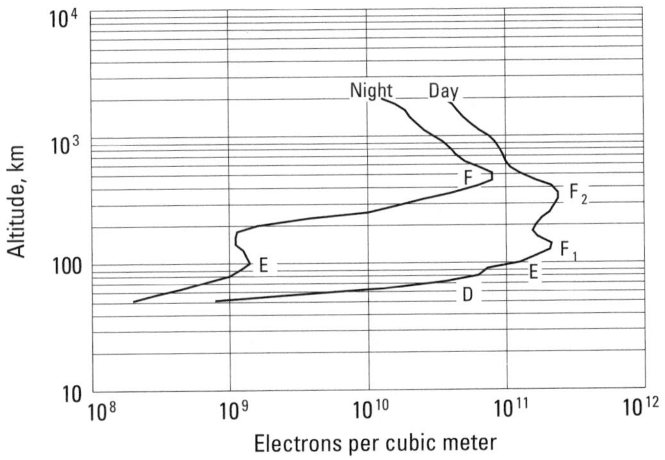

Figure 4.2 Typical electron-density profiles in the ionosphere.

predictions between all regions on Earth can be made on a predicted sunspot number based on observations of the solar flux value.

4.4.3 Ionized Gases

Electrons are about 1,800 times lighter than all other ionized particles in the ionosphere, so the motion of electrons determines the dielectric properties of the ionosphere under the action of electromagnetic fields. We can define the plasma radian frequency for electrons as

$$\omega_p = \sqrt{\frac{N_e e_e^2}{\varepsilon_0 m_e}} \tag{4.3}$$

where N_e is the number of electrons per cubic meter. The dielectric constant of a uniform ionosphere under the influence of the Earth's magnetic field is a tensor. For propagation along the magnetic field lines, and with the electric field perpendicular to the Earth's magnetic field lines, the effective dielectric constant of the plasma having a collision frequency of ν is

$$\varepsilon_d = 1 - \frac{\dfrac{\omega_p^2}{\omega^2}}{1 - \dfrac{j\nu + \omega_c}{\omega}} \tag{4.4}$$

The Earth's cyclotron frequency f_c, hence ω_c, was earlier defined in (4.2). The collision frequency varies between $\nu = 10^3$ per sec at an altitude of 300 km and $\nu = 10^6$ per sec at 90 km altitude.

4.4.4 Ionospheric Reflection

The reflection coefficient for the ionosphere can be approximated for horizontal polarization by

$$\Gamma_h = \frac{\cos(\phi) - \sqrt{\varepsilon_d - \sin^2(\phi)}}{\cos(\phi) + \sqrt{\varepsilon_d - \sin^2(\phi)}} \tag{4.5}$$

and for vertical polarization by

$$\Gamma_v = \frac{\varepsilon_d \cos(\phi) - \sqrt{\varepsilon_d - \sin^2(\phi)}}{\varepsilon_d \cos(\phi) + \sqrt{\varepsilon_d - \sin^2(\phi)}} \tag{4.6}$$

where the relative permittivity ε_d is a tensor but is given by (4.4) for propagation along the magnetic field lines, with the electric field perpendicular to the Earth's magnetic field lines.

Because of the curvature of Earth and the ionosphere, the largest angle of incidence ϕ occurs for a ray leaving at grazing incidence on the Earth as seen in Figure 4.3. That maximum value is $\phi = 74°$ for the case of a thin-layer model of the ionosphere at an assumed height of 250 km.

4.4.5 Maximum Usable Frequency

A first approximation to the refractive index of the ionosphere, ignoring the Earth's magnetic field, is given for frequencies well above the collision frequency:

$$n = \sqrt{\varepsilon_d} = \sqrt{1 - \frac{N_e e_e^2}{\varepsilon_0 m_e \omega^2}} = \sqrt{1 - \frac{81 N_e}{f^2}} = \sin(\phi_i)$$

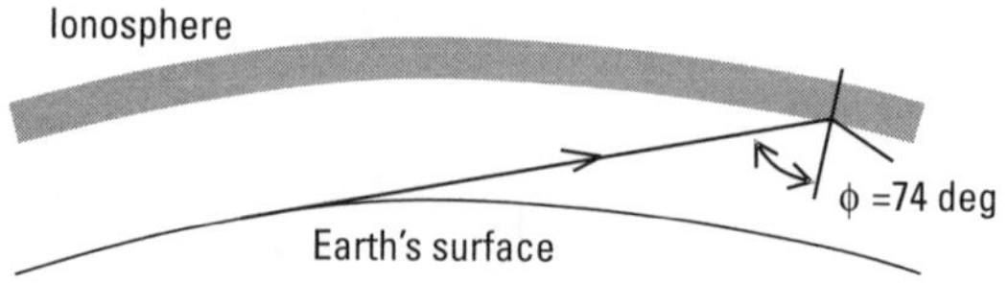

Figure 4.3 A horizontal ray incident on the ionosphere at the largest possible angle.

The highest altitude reached by a nonpenetrating wave is the point at which the highest electron density satisfies the above equation for vertical incidence $\phi_i = 90°$. This defines the critical frequency f_c, in hertz:

$$f_c = \sqrt{81 N_{max}} \tag{4.7}$$

N_{max} is the number of electrons per cubic meter associated with the critical frequency f_c, that is, the value of N_e, which makes the refractive index, n, of the ionosphere equal to zero.

MUF beyond which the ionosphere is largely transparent is then defined in terms of the incidence angle ϕ_i of a ray on the ionosphere:

$$f_{MUF} = f_c \sec \phi_i \tag{4.8}$$

Using a thin-layer model of the ionosphere and assuming a height of 250 km, a tangential ray is incident on the ionosphere at an angle of 74°. When $\phi_i = 74°$, the MUF is

$$f_{MUF} = 32.7\sqrt{N_{max}} \tag{4.9}$$

Because the MUF may show daily variations, it is usual to use a frequency somewhat lower (50% to 85%) than the predicted MUF for communications between two points.

[HamCAP] The MUF can be explored with HamCAP. Preset HamCAP as in Appendix D. Then, on the "Params" tab, enter the end of the link in "DX QTH" (e.g., 22°N and 114°E or use an international prefix VR2 for Hong Kong in the "DX Call" box). Select a month (e.g., May 2006). Select the "short" path and press the "Chart" tab. HamCAP now invokes VOACAP and presents an MUF plot for a 24-hour period in that month, based on the averaged sunspot number in the "Params" tab, which is linked to that month and year. Vary the sunspot number (SSN) to see how it affects the MUF over that path.

4.4.6 Multiple Hops in Shortwave Communications

Long-distance communications using ionospheric reflections usually involve multiple hops. The ionosphere must be reflective at each of the hops; otherwise, a radio link cannot be completed. Paths are along a great circular route between the transmitting and the receiving station. Figure 4.4 shows a rectangular projection map of the Earth with a shortwave radio path at the MUF between Florida, United States, and Hong Kong, China. Straightforward great circular

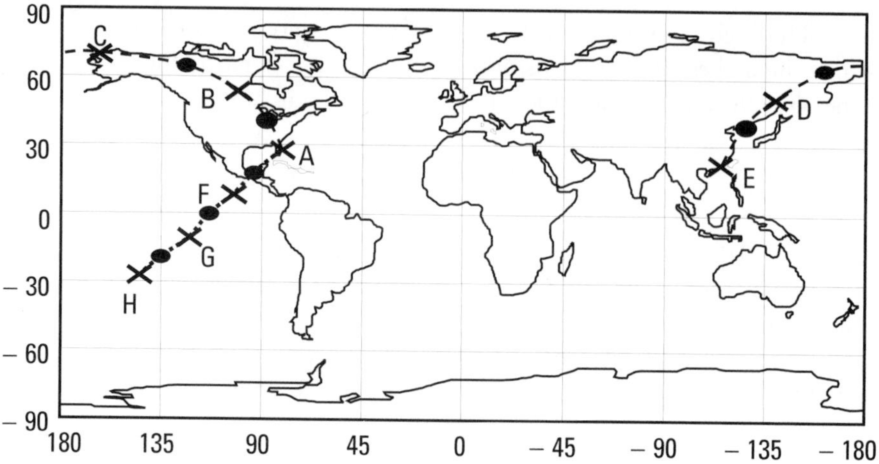

Figure 4.4 Shortwave paths between Florida (A) via multiple ground reflections B, C, and D to Hong Kong (E) and the South Pacific (H) via F and G. The intermediate sky reflections are shown by "•."

route calculations are made between the station at A, north latitude 26.3° and west longitude 80.3°, and the station at E, north latitude 22.4° and east longitude 114.3°, to identify the locations of the sky reflections and ground reflections. There are four hops, and the total distance is 14,400 km. The azimuth bearing from A to E is 342.6°, and the reciprocal direction from E to A is along the azimuth bearing 16.9°. The reflection control point for station A is the sky reflection point located at north latitude 41.5° and west longitude 86.7°, between B between point A and ground reflection B. The reflection control point for station E is the sky reflection point located at north latitude 37.8° and east longitude 114.2°, between point E and ground reflection D. A path from A to H, 45°S by 145°W in the South Pacific, is also shown.

[4-4.mcd] The shortwave communication path between two points on the globe is governed by simple geometric relationships involving (a) the shortest distance along the globe between two points, and (b) the distance between an integer number of reflections from the ionosphere and the ground (e.g., between A and E and between A and H). Try computing the path between widely separated points on the globe. Is the ionosphere necessarily reflective at each of the "sky" points?

Multiple hops in ionospheric propagation are evident in the coverage map shown in Figure 4.5. The propagation path shown between Florida, United States, and the South Pacific Islands (A to H in Figure 4.4) shows signal-to-noise-ratio (SNR) values on the Earth of 7, 58, 2, 56, 2 and finally 49 dB/Hz. The larger values correspond to ground reflection points F, G, and H in Figure

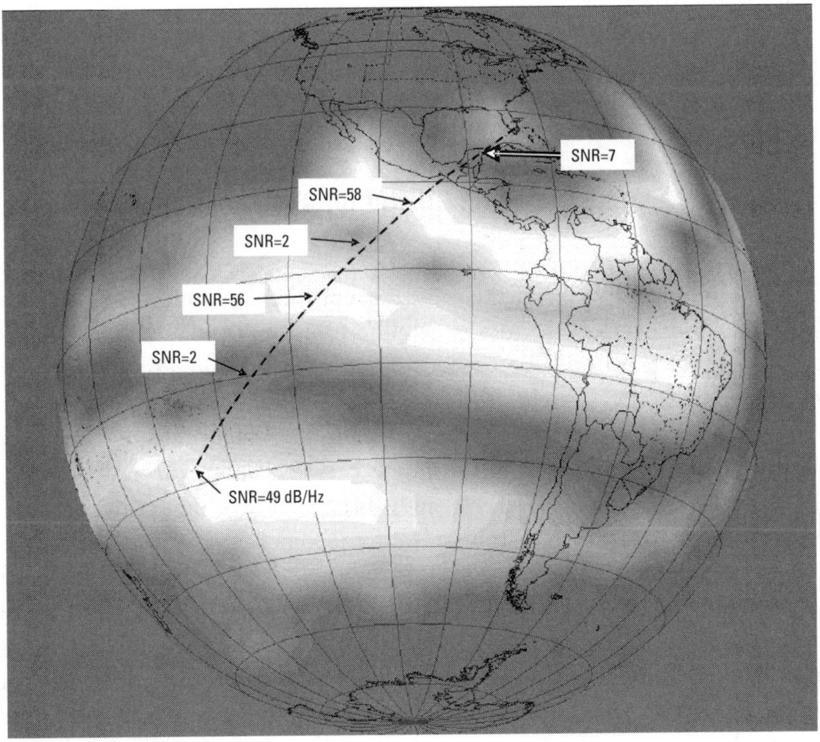

Figure 4.5 A path illustrating ionospheric propagation. (Generated by DXAtlas [www.dxatlas.com]).

4.4, while the lower SNR values correspond to the sky reflection points. That is, the low SNR values are skipped over as the wave is at the ionosphere. These correspond to ionospheric bounces. The propagation map was computed with HamCAP (see Appendix D).

[HamCAP] Multiple hops can be inferred from banding of propagation contours calculated and illustrated with HamCAP. Use the preset values from Appendix D, then on the "Params" tab enter "17" in the SSN window and "May 2006" in the month window. Select the "Map" tab and toggle the 28 MHz button. The SNR/Hz contours for propagation from the "home" location will be calculated and displayed. Moving the pointing device over the map reveals SNR (dB/Hz) and MUF, along with the layers of the ionosphere involved in the path to the cursor location. Try stepping the universal coordinated time (UTC) in the time window to see how propagation varies across the day, or select other frequency buttons to see propagation contours at different frequencies. In particular, the combination of 22:00 UTC and 28 MHz shows distinct banding of propagation to the south, which is identified with ionospheric bouncing. At 10, 7, and 3.5 MHz, the propagation is seen to be increasingly localized, and there is a significant variation between local day and night.

4.5 Very High Frequencies and Ultrahigh Frequencies (30 MHz to 3 GHz)

This is the frequency range of the most current interest in mobile and personal communications services, including satellite-based services. Satellite communications in these bands will be treated in the next chapter. Propagation above 30 MHz is basically straight line, with some refraction by the atmosphere. However, ionospheric reflections are significant up to 50 or 60 MHz due to geomagnetic activity. Additionally, the range between about 30 and 500 MHz, particularly below about 150 MHz, can be used for long-range communications via scatter paths from meteor trails. The frequencies of special interest to personal communications systems (PCS) include cellular telephone services in the 800 and 900 MHz bands, paging and messaging in the 900 MHz band, and the PCS bands from 1,700 to 2,200 MHz as well as personal and local area networks above 2,400 MHz. Propagation in the urban and suburban environment at frequencies in this frequency range will be detailed in Chapters 7 and 8. Here, we consider scattering from meteor trails and some aspects of propagation beyond line of sight.

4.5.1 Communications via Scattering from Meteor Trails

T. L. Eckersley observed in 1932 that scattering from inhomogeneities in the ionospheric ionization could be important in long-distance communications at VHF well above the normal MUF [10]. The possibility of using meteor trails for communications was under discussion by 1950. Actual forward scattering of signals by meteor trails was noticed as early as 1953 by radio amateurs attempting tropospheric contacts on 144 MHz over 300–1,000-km paths [11]. Detailed studies proved the feasibility of burst-mode communications over 1,000-km distances using hundreds-of-watts power and gain antennas in the 30–50-MHz range [12–15].

The Earth's atmosphere is continuously bombarded with an estimated 10^{10} particles a day, representing on the order of 1,000 kg of total mass. The number of particles entering the Earth's atmosphere decreases with increasing size roughly according to

$$N_p \propto 1 / M_p$$

where N_p represents the number of particles of mass greater than M_p. The meteor rate varies hourly and seasonally because of Earth's daily rotation, together with its tilt and motion about the sun. Particles entering the atmosphere become heated and ionize at altitudes between 80 and 120 km.

Meteor scatter communication relies on the sporadic meteors that move around the sun in random orbits rather than the relatively rare spectacular meteors that enter the Earth's atmosphere in groups called meteor showers. The sporadic meteors enter the Earth's atmosphere at velocities approaching 75 km/s and leave ionized trails as long as 25 km. The ionization trail is initially about 1m in radius and can have electron line densities in the range of 10^{10} to 10^{18} electrons per meter.

Meteor trails with fewer than 10^{14} electrons per meter are referred to as *underdense.* In underdense paths, the signals drop off exponentially with a time constant τ_1 in seconds that varies with the square of wavelength λ meters

$$\tau_1 = \frac{1}{2D_d}\left[\frac{\lambda \sec(\theta)}{4\pi}\right]^2 \tag{4.10}$$

The angle θ is formed by the rays incident on and reflected from the meteor trail, and D_d is the ambipolar diffusion coefficient, which depends on height and is of the order of 1 to 10 m^2/sec. Typically, time constants in the range of tens to hundreds of milliseconds are relevant in the 30–50-MHz band. The path attenuation L_{under} for the underdense case takes the form [16]

$$L_{under} = g\left[\frac{\mu_0 e_e^2}{4m_e}\right]^2 2n_e e^{-\frac{t}{\tau_1}} \tag{4.11}$$

where

$$g = \left[\frac{1}{k(d_1 + d_2)}\right]^2 \left[\frac{1}{2\pi}\frac{d_1 + d_2}{d_1 d_2}\frac{\lambda \sin(\alpha)^2}{\left[1 - \cos^2(\beta)\sin^2(\theta)\right]}\right]$$

and

d_1 and d_2	ray distances to reflection point from transmitter and receiver (m)
β	angle between meteor path and plane of rays d_1 and d_2
θ	half the angle between rays d_1 and d_2
α	angle between the field vector and the scattering direction
e_e	electron charge

m_e	electron rest mass
μ_o	free space permeability
n_e	linear electron density in electrons per meter
D_d	ambipolar diffusion coefficient (m^2/sec)
t	time from "instantaneous" trail formation
k	wave number $2\pi/\lambda$

The values for constants are in the list of symbols in the back of this book.

Overdense trails are those with more than 10^{14} electrons per meter. Since at these densities the coupling between individual electrons is significant, the scattering process tends to resemble reflection from a cylindrical surface that encloses the axis of the meteor trail. The electrons initially form a line of small radius where the electron volume density is such that the MUF is sufficiently high to provide ionospheric reflection from the trail. As the electrons diffuse outward, the trail density falls so that the MUF falls below the critical value, and scattering reverts to the underdense type for a short time. For overdense paths, the signal strength initially rises sharply, then drops off smoothly in a shape resembling a smooth pulse of duration τ_2,

$$\tau_2 = \frac{n_e e_e^2}{D_d m_e}\left[\frac{\lambda \sec(\theta)}{2\pi}\right]^2 \cdot 10^{-7}$$

in seconds, and the overdense trail path attenuation, L_{over}, is

$$L_{over} = g\sqrt{n_e \frac{\mu_0 e_e}{4m_e}}\left[\frac{\pi}{e}\right]^2 \tag{4.12}$$

where e is the base of the natural logarithms. Pulse durations are on the order of several seconds for electron line densities in the 10^{15} range and frequencies in the 30–50-MHz band.

Scattering from meteor trails can be described as a specular phenomenon where the incident and reflected rays make equal angles with the axis of the meteor trail. The communication path is essentially from the principal Fresnel zone. Because of the geometry involved, communication paths are relatively directive and hence provide a measure of privacy or security. Path attenuations for meteor trail scattering vary with the inverse third power of frequency, so the lower frequencies, 30 to 50 MHz, are preferred. Near 40 MHz, the path attenuations for the underdense scattering case are on the order of 170 to 220 dB for path distances in the 500–1,000-km range. High power levels (more than 500W) and directive antennas (10 to 15 dBd) are typically required for communications using data rates of several hundred bits per second. The paths are sporadic, very directional, and bursty. They are thus primarily suitable for digital packet communication triggered automatically by beacon signals.

4.5.2 Propagation by Tropospheric Bending

The troposphere is the portion of the atmosphere beginning at the Earth's surface and extending in altitude about 10 km. It is the layer in the atmosphere containing meteorological activity. The dielectric constant of the troposphere is not constant and normally decreases linearly with altitude at a normal rate of about 6×10^{-4}/m. The very slight dielectric gradient causes radio waves to bend as they travel. A common way to work propagation problems is to consider the waves as traveling in straight paths and to compensate for the ray path bending by using an "effective" Earth radius, which is different from the actual radius by a factor K. One standard way of defining the refracting properties of the atmosphere is with the N unit defined as

$$N = \left[\sqrt{\varepsilon_d} - 1\right]10^6 \qquad (4.13)$$

where ε_a is the dielectric constant of the atmosphere. Near the surface, ε_a = 1.0006. An empirical relationship between K and N is

$$K = \left[1 - 0.04665e^{0.005577N}\right]^{-1} \qquad (4.14)$$

and for the most commonly used value, N = 301, the corresponding Earth radius factor is 4/3. The values of the minimum monthly mean values of N throughout the world are published.

When N exceeds approximately 800, the dielectric constant decreases more rapidly than 3.3×10^{-7} per meter, and radio waves traveling parallel to, or at a slight angle above, the Earth's surface may be bent downward sufficiently to be reflected from the Earth. This refraction and reflection continues, and radio energy appears to be trapped or ducted between the maximum height of the bending and the Earth's surface. Ducting can result in long-range transmissions, which can result in interference. The phenomenon does not occur often enough, nor can it be predicted accurately enough, to result in reliable radio communication.

4.5.3 Tropospheric Scattering

Measurements have shown that fairly sharp variations in the atmospheric dielectric constant can occur in regions of the troposphere. Scattering from these irregularities in the troposphere can account for beyond-the-horizon transmissions at frequencies from 30 MHz up to 10 GHz. Experimental results for over-the-horizon transmissions [4] have shown that beyond the horizon, signals decrease with between the seventh and eighth power of distance. An empirical model for beyond-the-horizon propagation is given in Section 7.4. There are seasonal variations of ±10 dB, which correlate with, and are proportional to, seasonal variations in the effective Earth's radius factor *K*. Tropospheric scattering is characterized by fast fading, which is essentially random and follows Rayleigh statistics.

4.6 Above Ultrahigh Frequencies (Above 3 GHz)

Frequencies from a few gigahertz on up find application in satellite-based communications systems. Propagation is essentially line of sight with occasional anomalous tropospheric scattering. In satellite systems, propagation is through the ionosphere, and signal polarization will be rotated because of the combined effects of the Earth's magnetic field and the free ion concentration. The radio waves are also attenuated because of atmospheric absorption, particularly above 10 GHz. There are two primary absorption bands. One, due to the water vapor resonance, is a single line near 21 GHz. The second major absorption band is centered at 60 GHz and is due to the oxygen molecule resonance.

4.7 Picking an Optimum Operating Frequency

The performance of a radio system depends on many frequency-dependent parameters, such as antenna losses, system noise figure, and man-made and atmospheric noise. In a mobile phone or a paging receiver, for example, the receiver antenna's performance is fundamentally limited by the design choice of using a miniature antenna within the radio case. Such an antenna, as will be shown in detail in a later chapter, tends to have increasing losses with wavelength. Atmospheric noise, on the other hand, is lowest in a window roughly between 0.3 and 3 GHz. An actual choice of operating frequency band is usually made on the basis of regulatory considerations and on channel or spectrum availability. Nevertheless, an optimum operating frequency can be found based on technical requirements and parameters, such as antenna losses and equivalent system noise figure.

The equivalent system noise figure is the apparent noise figure, including antenna losses and atmospheric noise. For example here, the total equivalent system noise temperature for a source at T kelvins in the field of view of an antenna can be approximated by

$$T_e = \frac{hf}{k_b}\left[\frac{1}{\exp\left[\frac{hf}{k_bT}\right]-2}+1\right]+\frac{7\times10^{26}}{f^3}+T_b+\frac{f^{2.5}}{10^{25}} \quad (4.15)$$

where h J·sec is the Planck constant, k_b J/K is the Boltzmann constant, T is the source temperature in kelvins, and frequency f is in hertz. The first term becomes recognizable as kT when the approximation for low frequencies ($f < 10^{10}$) is made. The second term is an approximation to cosmic noise contribution, which dominates below 1 GHz. The third term, T_b = 2.726K, is the Big-Bang residual noise temperature of the universe, and the last term approximates atmospheric water vapor, rain and fog. Man-made noise terms are not included in (4.15) and are discussed in more detail in Section 12.2 especially in reference to Figure 12.1. Figure 4.6 shows a wide-frequency view of T_e from (4.15).

[4-7a.mcd] Calculate the equivalent noise temperature (4.15) for sources at 1, 100, and 1,000K. Are there frequencies for which the source temperature is not an important contributor?

The antenna average gain in a body-mounted personal communications device, such as a pager receiver, can be described empirically as a function of frequency f in hertz by

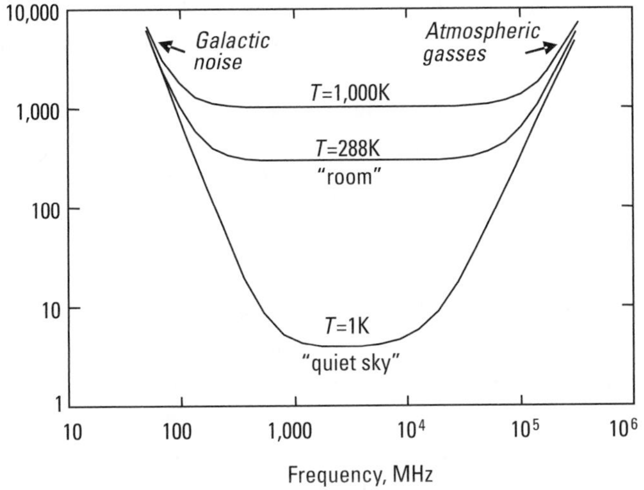

Figure 4.6 Noise temperature, including galactic noise and atmospheric gases.

$$G_r = 20\log(f) - 20\log\left[f + 6\times10^8\right] - 5.5 \tag{4.16}$$

Receiver noise figure *NF* can be empirically described by

$$NF = 4 + f \times 10^{-9} \tag{4.17}$$

The effective noise figure degradation N_e of a receiver having a front-end noise figure *NF* is then

$$N_e = -10\log\left[\frac{T_e}{T_0} + \left[10^{0.1NF} - 1\right]10^{-0.1G_r}\right] \tag{4.18}$$

where T_e is given by (4.15), G_r is from (4.16), *NF* is from (4.17), and T_0 = 290K. Note that noise figure *NF* and receiver equivalent temperature T_{rx} are related by

$$NF = 10\log\left(1 + T_{rx}/290\right) \tag{4.19}$$

referenced to a room temperature of 290K.

An optimum operating range based on just receiver antenna gain and on atmospheric noise can be found by plotting (4.18) as in Figure 4.7. For comparison, the receiver antenna gain G_r and the ratio of equivalent noise temperature to standard room temperature are also shown. The lowest receiver system losses occur between about 0.3 and 3 GHz. This, then, is the optimum band for body-mounted radio operation based on antenna gain and atmospheric noise. A more complete analysis would include various system costs, and a real system design would be based on a minimum-cost approach. Figure 4.7 and analyses similar to the one shown here begin to lead us to an appreciation of why the radio spectrum between 0.3 and 3 GHz is the prime frequency spectrum for personal communications.

[4-7b.mcd] An optimum operating frequency based on noise temperature can be calculated from (4.16) to (4.18), using an equivalent noise temperature that includes all noise sources in the field of view of an antenna as given by (4.15). Calculate the equivalent noise figure, and compare it with the antenna gain model. Calculate the loss due to noise as $10\log(T_e/290)$, and compare it with the antenna loss model.

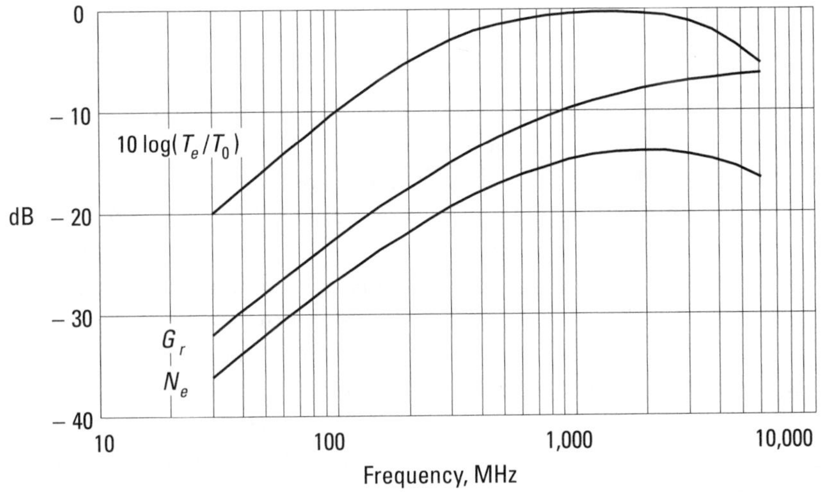

Figure 4.7 An optimization based on the parameters of atmospheric noise and receiver antenna gain.

4.8 Multiuser Communications Systems

A simple communications system linking just two isolated end points is not particularly interesting from the personal communications point of view. The systems of interest link *one-to-many* access, or provide for multiple instances of *one-to-one* access, or facilitate *many-to-one* access. In other words, we are interested in systems that can be accessed efficiently by multiple users. Several broad categories of multiuser techniques will be briefly examined: (1) paging, (2) digital broadcasting, (3) packet radio access techniques, (4) voice and data mobile and cellular systems, and (5) wireless local and personal-area networks, which also include UWB technologies. Since radio spectrum is limited, the number of users per unit bandwidth becomes an important parameter. We will introduce several examples of signaling and modulation methods for multiuser communications in view of their radio propagation channel characteristics. These systems are impacted by propagation impairments in different ways. System capacities are also traditionally specified differently for different services. In paging, for example, capacity is typically stated in busy-hour subscribers per simulcasting region, while telephone capacity is stated in Erlangs. A modern, high-capacity, single-frequency paging system might service 125,000 subscribers, which equates to $1/125{,}000 = 8 \times 10^{-6}$ Erlangs per subscriber. A telephony system typically assumes 0.026 Erlangs per subscriber, or an average of $1/0.026 = 38$ subscribers per radio channel in a cell cluster.

4.8.1 Paging Systems

Paging, traditionally a one-way notification service, is characterized by typically short messages selectively addressed to uniquely identified subscribers and is an example of a one-to-one form of multiple access. Paging also provides information services to multiple subscribers in a one-to-many mode of transmission. Access is by sequential queuing of messages, often with simultaneous synchronized transmissions from multiple sites. A small latency in message delivery allows for very efficient queuing and therefore efficient radio channel utilization. Over the last several decades, paging has evolved into a digital communications method that now includes two-way transmissions and digitally compressed voice. Wide-area paging systems employ multiple transmitters often configured to transmit the digital signals simultaneously to enhance radio coverage in a service area. Propagation delay differences among multiple transmitters, in comparison with digital code bit lengths, therefore play a limiting role in simulcasting systems because of potential intersymbol interference. System capacity, particularly for very-wide-area and nationwide systems, can be increased spectacularly with the addition of a simple reply capability. The reply capability is a form of many-to-one multiple access.

The earliest digital paging codes include GSC, introduced by Motorola in 1973, the NTT code introduced in Japan in 1978, and the POCSAG code introduced in 1981. The most popular signaling format for paging has been POCSAG [17]. The fastest-growing modern high-speed signaling format is the FLEX code [18], which was introduced by Motorola in 1993 and has since gained tremendous worldwide acceptance. The FLEX family of codes further includes ReFLEX, the FLEX protocol for two-way paging, and InFLEXion, the FLEX protocol for voice paging and high-speed data, using advanced digital processing and spectrally efficient modulation techniques (FLEX, ReFLEX, and InFLEXion are trademarks of Motorola). Table 4.3 summarizes some attributes of POCSAG and the FLEX family. The table further shows that two kinds of data, 4-bit numbers and 7-bit characters, are used with paging messages in an effort to keep airtime to a minimum. The details of 4- and 7-bit codes are shown in Appendix C.

POCSAG is a binary FM system whose primary characteristics are shown in Table 4.3. POCSAG is an example of a simple narrowband FM signaling format, which is designed to be used in the VHF and ultra-high-frequency bands (30–932 MHz) with typical channel spacings of 25 or 30 kHz. POCSAG was originally defined with a bit rate of 512 bps and has been "speeded up" to 1,200 and 2,400 bps in an attempt to meet the increasing paging capacity demands of the service providers. POCSAG is a simple protocol (see Figure 4.8), consisting of a synchronization preamble of at least 576 bit reversals, which equals 1 batch, plus 1 code word, followed by a batch of 17 code words, including 1

Table 4.3
Paging Signal Formats

Protocol	Applications	Frequencies	Roaming	Outbound Signaling Speed	Inbound Signaling Speed
POCSAG	4-bit digits, 7-bit characters	Any paging channel	None	512, 1,200, or 2,400 bps	Not applicable
FLEX	4-bit digits, 7-bit characters	Any paging channel	Yes	1,600, 3,200, and 6,400 bps (pagers support all rates)	Not applicable
ReFLEX	4-bit digits, 7-bit characters and two-way data	*Out: 929–932, 940–941 MHz; In: 901–902 MHz	Yes	1,600, 3,200, and 6,400 bps (pagers support all rates)	800; 1,600; 6,400; or 9,600 bps 4-FSK
InFLEXion (for voice paging)	Voice and two-way data	*Out: 929–932, 940–941 MHz; In: 901–902 MHz	Yes	Digitally processed voice	800; 1,600; 6,400; or 9,600 bps 4-FSK
InFLEXion (for high-speed data)	High-speed data, digital voice, and two-way data	*Out: 929–932, 940–941 MHz; In: 901–902 MHz	Yes	Up to 112 kbps multi-channel QAM	800; 1,600; 6,400; or 9,600 bps 4-FSK

* ReFLEX and InFLEXion first appeared in 900-MHz NBPCS channels in the United States but are not otherwise frequency-band specific.

synchronization code word. The first bit in the code word is the message flag, a "0" identifies an address word, and a "1" identifies a data word. The POCSAG code format comprises BCH(31,21) (Bose-Chaudhuri- Hocquenhem cyclic error correcting code) code words with an appended parity bit. The address words identify unique selective call receivers, and subsequently appended data words form the message. A typical message may consist of 10 numeric digits or 40 alphanumeric characters. Because a single bit differentiates address and data code words, only one bit-error per code word may be safely corrected to keep "false address" detection at acceptable levels. Similarly, the reception of two uncorrectable code words in a row terminates message reception to prevent the false appending of a subsequent user's message. From the propagation point of view, POCSAG paging transmissions are susceptible to channel impairments (due to signal fading or simulcast transmission interference) that affect more than 1 bit in a 32-bit code word. Not including synchronization, a typical POCSAG 10-digit numeric message is 3 code words long; a 40-character alphanumeric message is 15 code words long. POCSAG signaling can tolerate

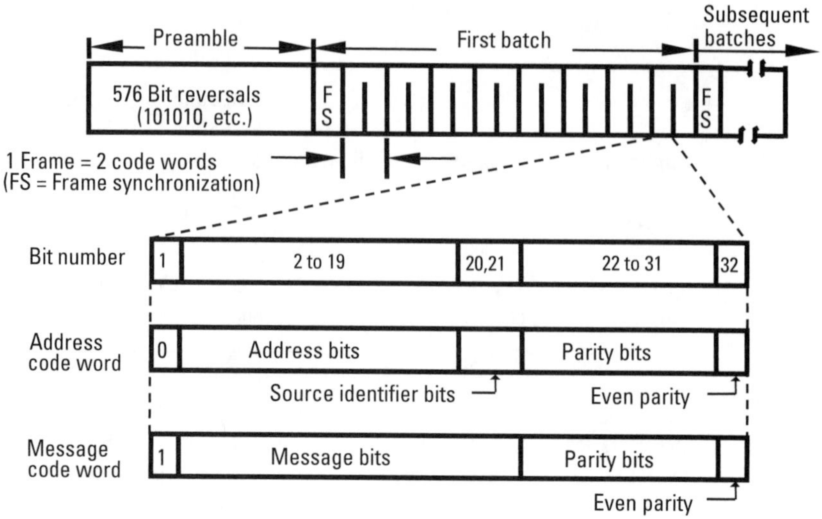

Figure 4.8 POCSAG paging code.

channel impairments no greater in length than 2 ms at 512 bps, 0.8 ms at 1,200 bps, and 0.4 ms at 2,400 bps for correct decoding of transmissions.

The FLEX paging code is a modern multirate signaling format designed to operate efficiently at two-baud rates, 1,600 and 3,200 symbols per second, supporting 1,600-, 3,200-, and 6,400-bps data rates. Two modulation levels are used, binary FSK (Frequency Shift Keying) at ±4,800-Hz deviation and 4-FSK having deviation levels separated by 3,200 Hz. The primary characteristics are shown in Table 4.3. The FLEX code is a synchronous code (see Figure 4.9), comprising four minute cycles of 128 frames. Each 1.875-sec frame consists of a 115-ms synchronization portion followed by 11 blocks, each 160-ms long. The synchronization portion contains "Sync 1" of 144 bits sent at 1,600 bps in binary FSK, followed by "Sync 2," which contains bits sent at rates of 1,600-bps binary FSK and may also contain 4-level FSK symbols at up to 3,200 symbols per second (6,400 bps). Code words embedded in "Sync 1" identify frame information, as well as the modulations in "Sync 2," and define the modulation in the following 11 blocks as either 1,600 bps 2-FSK, 3,200 bps 2-FSK, 3,200 bps 4-FSK, or 6,400 bps 4-FSK. As will be shown in Chapter 8, the multiple data rates provide a system design capability, which can match data throughput with system capability on a regional and subregional basis to optimize network operator investment costs.

The 160-ms FLEX blocks contain 8, 16, or 32 interleaved code words, depending on the signaling code rate. Interleaving is a technique of arranging code words in blocks, here 8, 16, or 32 words to a block, and sending sequentially all of the first bits, all the second bits, and so on. Code words are

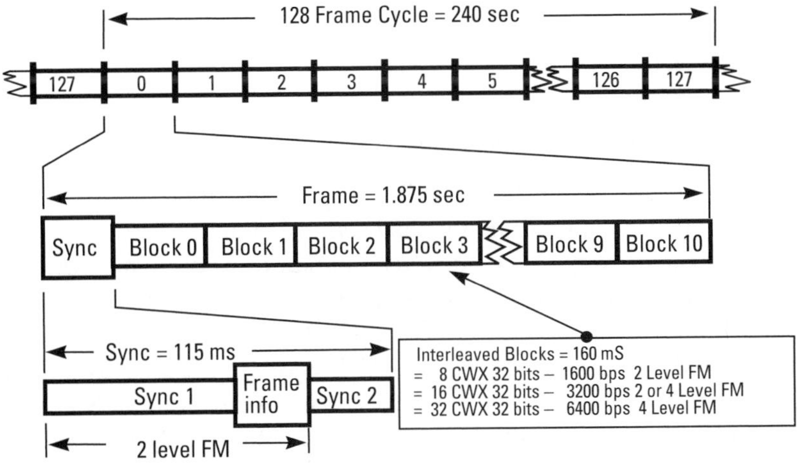

Figure 4.9 FLEX paging code.

BCH(31,21) plus a parity bit. Up to two errors per code word are corrected in the FLEX code, so error bursts of up to 10 ms per block can be tolerated. Because the data is multiplexed at the 3,200- and 6,400-bps rates, the error burst tolerance remains 10 ms, independent of the channel bit rate. The first block contains the necessary block information words, which define the starting positions for address words and vector words, which associate address words one to one with message words. The address, vector, and message words may cross block boundaries as needed. A FLEX frame comprises 4 block information words per 11 blocks. Each block is composed of 8 interleaved code words, and hence there are 384 useful code words per frame. A 10-digit message requires 4 code words, and a 40-character message requires 17 code words. From the propagation point of view, the FLEX code can tolerate propagation impairments up to 10 ms in length.

The FLEX family includes ReFLEX, the FLEX protocol for two-way paging, and high-speed modulations in the form of InFLEXion, the FLEX protocol for voice paging and high-speed data. They currently appear in the U.S. NBPCS spectrum (929–932, 940–941 MHz channels outbound or downlink, paired with 901–902 MHz channels for inbound or uplink transmissions). ReFLEX can appear in 25-kHz or 50-kHz outbound channels paired with 12.5-kHz inbound channels. InFLEXion currently appears in 50-kHz outbound channels paired with 12.5-kHz inbound channels. InFLEXion is an advanced protocol supporting seven subchannels of a spectrally efficient QAM (Quadrature Amplitude Modulation) modulation comprising digitally compressed analog voice at up to 28 equivalent voices or up to 112-kbps data per 50-kHz channel. The uplink channel of InFLEXion uses an ALOHA access technique and can support 800–9,600-bps data rates using 4-FSK modulation having deviation levels separated by 1,600 kHz.

4.8.2 Digital Voice Broadcasting Systems

Digital voice broadcasting (DVB), called digital audio broadcasting (DAB) in Europe, is an example of a one-to-many form of communication and is concerned with the delivery of high-data-rate sound broadcasting for reception by mobile, portable, and fixed receivers using simple nondirectional antennas. The example described here is known as EUREKA 147 (EU-147) and is detailed in an ETSI standard [19]. In essence, the EU-147 system has a bandwidth of 1.5 MHz providing a total transport bit-rate capacity of just over 2.4 Mbps shared by multiple program sources. The system uses differentially encoded QPSK (Quadrature Phase Shift Keying) modulation, coupled with a multicarrier scheme known as orthogonal frequency division multiplexing (OFDM). The basic principle consists of dividing the information to be transmitted into a large number of bit streams having low bit rates individually, which are then used to modulate individual orthogonal carriers such that the corresponding symbol duration becomes larger than the delay spread of the transmission channels. Temporal guard intervals between successive symbols, channel selectivity, and mutipath propagation will not cause intersymbol interference. Three transmission modes, as summarized in Table 4.4, are defined for EU-147 supporting nominal simulcasting transmitter separations of 96, 24, and 12 km. Three transmission modes are supported, providing characteristics ranging from those suitable for terrestrial single frequency (simulcasting) networks to those suitable for satellite delivery.

Table 4.4
EUREKA-147 System Parameters

System Parameter	Mode I	Mode II	Mode III
Frame duration	96 ms	24 ms	12 ms
Null symbol duration	1,297 μs	324 μs	168 μs
Useful symbol duration	1 ms	250 μs	125 μs
N, radiated carriers	1,536	384	192
Nominal maximum transmitter separation	96 km	24 km	12 km
Nominal frequency range for mobile reception	Up to 375 MHz	Up to 1.5 GHz	Up to 3 GHz

The large number N of individual carriers that can be conveniently generated by an FFT (Fast Fourier Transform) process are collectively known as an *ensemble.* In the presence of multipath propagation, some of the carriers are enhanced by constructive signals, while others suffer from destructive interference in selective fading. The EU-147 system provides for frequency interleaving by a rearrangement of the digital bit stream among the carriers such that successive source samples are not affected by the selective fade. This frequency domain diversity is the prime means to ensure successful reception by stationary receivers. Time interleaving of the bits provides further diversity assistance to mobile and moving receivers. The EU-147 implementation of frequency- and time-interleaved OFDM is an example of a multiple access system designed for both high-bit-rate transmission and large spacing of simulcasting transmitters.

4.8.3 Packet Access Systems

Multiple users of a radio channel often need to access other users or a central user in a radio system with minimum coordination to transfer packets of data. Packet radio architectures can be classified as centralized, like two-way paging, where all nodes communicate only with a centralized set of nodes, or distributed, such as AX.25 [20], where nodes communicate directly with other nodes in the system. Personal communications systems typically involve networks of fixed sites that services multiple users who are often beyond direct radio range of each other and hence resemble centralized packet networks in their access methods.

Packet radio networks began in 1970 with ALOHANET [21], a centralized system with the objective of allowing widely scattered users access to the University of Hawaii computer system. ALOHANET required separate downlink and uplink frequencies and used a multiple-access contention protocol known as ALOHA [22] for the uplink. Two-way paging and cellular telephony are modern examples of centralized systems whereby users need to originate a radio message and hence use a contention technique to access a common radio channel. Overlaps in transmissions from multiple users cause packet-data collisions and reduce the capacity of the radio channel.

The ALOHA protocol uses a method of access whereby equal-length packets of data are transmitted randomly by multiple users vying for access to the channel in a free-for-all manner. The throughput T of an ALOHA system is defined as the product of total offered load times the probability P of a successful transmission. The offered load is the product of mean arrival rate of λ packets per second times the packet duration of τ seconds and is assumed to be Poisson distributed. The total traffic $\lambda\tau$ on the channel is the sum of packets generated per time, that is, the offered load T, and the number of retransmitted packets per unit time. Assuming that traffic is Poisson distributed and that the channel is noise-free,

$$\lambda\tau = T + \lambda\tau\left[1 - e^{-2\lambda\tau}\right] \tag{4.20}$$

from which the throughput is

$$T = \lambda\tau e^{-2\lambda\tau} \tag{4.21}$$

When time is divided into equal time slots of length greater than packet duration t, and users are synchronized with, and transmit at the beginning of, the time slots, partial collisions are prevented, and throughput increases at the expense of delay time for repeated transmissions. This form of access is called slotted ALOHA (S-ALOHA), and the throughput is

$$T = \lambda\tau e^{-\lambda\tau} \tag{4.22}$$

Figure 4.10 shows the trade-off between throughput for ALOHA, described by (4.21), and S-ALOHA throughput given by (4.22). S-ALOHA doubles the throughput on a channel but increases the average delay. The maximum throughput ALOHA is $1/2e = 0.184$, while the maximum for S-ALOHA is $1/e = 0.368$.

In a distributed radio network architecture it is possible to use carrier sense multiple access (CSMA), or a "listen-before-talk" form of contention. The

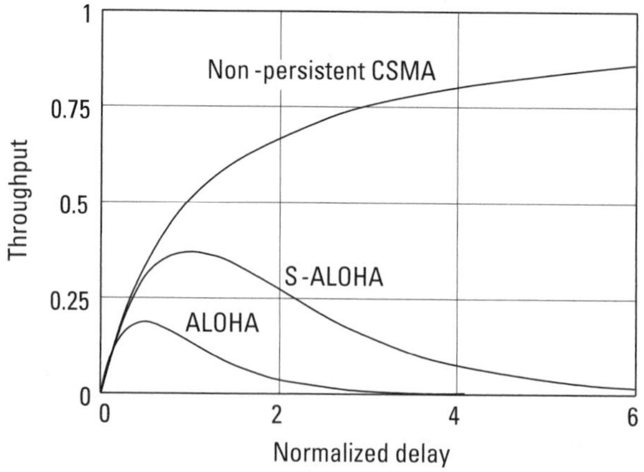

Figure 4.10 Normalized throughput and delay of ALOHA, S-ALOHA, and nonpersistent carrier sense multiple access.

method inherently recognizes that packet lengths are generally much longer than propagation delay times. A station wishing to transmit first listens, then transmits only if a channel is idle. Otherwise, the station waits a small amount of time and listens again with the intent of transmitting. An acknowledgment after transmission certifies a successful packet transfer. Various strategies on the waiting time, retransmission persistence, and acknowledgment lead to varying degrees of channel throughput improvements over ALOHA and S-ALOHA techniques. The normalized throughput T versus delay τ of nonpersistent CSMA access, shown in 4.10, is

$$T = \frac{\tau}{1+\tau} \tag{4.23}$$

Nonpersistent CSMA transmits following a constant or variable delay if the channel is idle; otherwise, it delays and tries again. P-persistent CSMA transmits as soon as the channel goes idle with probability P; if the channel is not idle, it will delay one time unit and try again. The time unit is typically equal to the maximum propagation delay. The 1-presistent protocol transmits as soon as the channel is idle; if there is no acknowledgment, it will wait a random time and try again.

[4-8.mcd] Calculate the normalized throughput versus normalized delay for the ALOHA, S-ALOHA, and nonpersistent CSMA packet radio protocols.

4.8.4 Cellular and Mobile Voice Systems

Cellular telephone is a *real-time voice service* with user characteristics similar to those of landline telephony. It is typically a one-to-one mode of communication. Access to RF channels is via a call setup, followed by assignment of a dedicated pair of logical channels, one for each direction of transmission. Some characteristics of several cellular systems are listed in Table 4.5. Personal communications voice systems can be broadly divided into three categories: (1) analog frequency division multiple access (FDMA) voice systems, (2) time division multiple access (TDMA) systems, and (3) code division multiple access (CDMA) systems. Often, cellular systems use antenna directivity in the design of cell, in effect utilizing a form of space division multiple access (SDMA), although SDMA is usually associated with adaptive antennas and beam steering. Cellular systems approach the coverage problem from a "small cell" point of view. An area is subdivided in small cells served by relatively low-powered transmitters. By reducing the coverage to small cells, it becomes possible to reuse the same frequencies in different cells separated by one ring (7-cell reuse pattern) or

Table 4.5
Cellular and PCS Voice Systems

System	AMPS	GSM	CDMA IS-95	PHS
Introduced	1983	1991	1995	1994
Downlink/ Uplink	869–894 MHz/ 824–849 MHz	935–960 MHz/ 890–915 MHz	869–894 MHz/ 824–849 MHz	1,895–1,918.1 MHz
System bandwidth	25 MHz in each direction	25 MHz in each direction	25 MHz in each direction	23.1 MHz
Channel spacing	30 kHz	200 kHz	1.25 MHz	300 kHz
RF channels	832 pairs	125 pairs	20 pairs	77
Multiple access	FDMA	TDMA/FDMA	DS-CDMA	TDMA/TDD
Slots per RF channel	1 voice per RF channel	8 time slots, each 576.92 μs	64 orthogonal Walsh functions	4 uplink and 4 downlink time slots
Frame duration	Not applicable	4.615 ms	20 ms	5 ms
Modulation	Analog FM voice 10-kbps signaling	$BT = 0.3$ GMSK	Mobile: QPSK Base: OQPSK	$\pi/4$ DQPSK
Modulation/ bit rate	Deviation: 12 kHz voice, 8 kHz control data	270.883333 kbps	1.2288-Mbps chip rate	384 kbps

12-cell clusters or two rings (19-cell reuse pattern) of cells employing different frequencies, as shown in Figure 4.11. Other reuse patterns are also common, including patterns of sectored cells. Since cells are relatively small, mobile subscribers to cellular systems often require handoff to adjacent cells. When subscriber density becomes too large, cells can be split into smaller cells. Cellular systems are thus characterized by (1) relatively low-power transmitters, (2) handoff and central control, (3) radio frequency reuse, and (4) cell splitting to increase subscriber capacity. Four real-time radio telephone systems will be shown as examples of the varied multiple access techniques used by cellular telephone systems. These are AMPS (Advance Mobile Phone System) [23], an analog FM system; Global System for Mobile Communications (GSM, originally Groupe Spécial Mobile) [24], one of the first fully digital systems; CDMA IS-95 [25], a direct-sequence CDMA system (DS-CDMA), and the Personal Handiphone System (PHS) [26], a Japanese low-power system implemented in microcells. These systems, as well as other cellular systems, are described in more

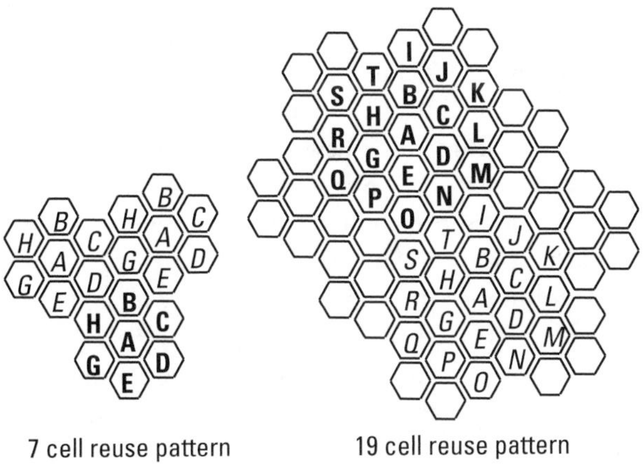

Figure 4.11 Reuse of frequencies A, B, C, and so forth, in mobile phone cellular layouts.

detail in [27]. Each system has unique properties with respect to radio wave propagation impairments.

AMPS, one of the earliest cellular telephone systems, comprises two RF channels separated by 45 MHz in a full duplex mode of two-way analog FM transmission with a peak deviation of 12 kHz. Signaling and channel setup are by FSK modulation, and the voice is analog frequency modulated onto the RF carrier. AMPS signaling uses BCH(40,28) code words on the downlink channel and BCH(48,36) on the uplink channel at 10 kbps. Channel impairments are primarily signal fades within buildings, where signals are weak, and interference from other cells reusing the same frequency. Originally envisioned as a mobile telephone service, AMPS rapidly became a personal cellular system with the introduction of handheld telephone units. Subscriber growth, as seen earlier in Figure 1.1, has been explosive.

One of the most successful digital cellular systems is GSM. The GSM radio-air interface frame structure consists of hyperframes containing 2,048 superframes. The hyperframes are nearly 3.5 hours in length and are important for encryption algorithms that rely on a particular frame number. The 6.12-sec superframes each contain 51 multiframes that are 120 ms long. Multiframes in turn comprise 4.615-ms frames of eight 576.92-μs time slots. The time slots normally contain 148 bits plus an 8.25-bit-long guard period. Of the 148 bits, there are 3 trailing bits, two traffic bursts of 58 bits each, separated by 26 training bits and followed by another 3 trailing bits. Five types of time-slot data bursts are identified in GSM. We are interested here in the normal burst associated with voice data. GSM identifies a 20-ms voice user speech segment of 456 bits. The voice user bits are 50 Type-Ia—the most important bits, which have 3 parity check bits appended, and 132 Type-Ib bits, which with the previous 53 bits are reordered and appended by 4 zero bits. The 189 Type-Ia and Type-Ib bits are then encoded using a rate one-half convolutional code with constraint length $K = 5$, resulting in 378 bits. The least important and unprotected 78 Type-II bits are concatenated to the 378 bits to form a block of 456 bits representing a 20-ms voice segment. These 456 bits are broken up into eight subsegments labeled 0 to 7 of 57 bits each. Interleaving of bits is provided to minimize the effects of radio channels fades. The bits from two consecutive 20-ms (456 bits each) voice segments denoted "a" and "b," 40 ms total, are diagonally interleaved within the same time-slot number over eight frames, F0 to F7, as shown in Figure 4.12. GSM provides interleaving to a depth of 40 ms and provides several levels of error protection, depending on the importance of the bits. Seven additional voices are carried in the same eight frames. GSM systems are typically planned around a seven-cell frequency-reuse pattern, so the frequency-reuse factor is 1/7. Because of the high channel bit rate, GSM provides data bits for channel equalization. Additionally, slow frequency hopping may be implemented on a frame-by-frame basis to help combat certain severe multipath propagation problems. From the perspective of radio propagation serving a single subscriber, GSM is largely a matter of sending bursts of 148 bits at 270,833.333 bps in 576.92-μs intervals using a 0.3 GMSK (Gaussian Minimum Shift Keying) modulation index. The seven-cell cluster layout of GSM systems brings with it the aspect of self-system interference from outlying cells using the same frequency and time-slot combination. An understanding of propagation leads to an understanding of the interference issues.

IS-95 defines a CDMA digital cellular system operating in pairs of uplink and downlink 1.25-MHz-wide channels. Subscriber data is spread to the channel at 1.2288 Mchip/s. The spreading is different for the uplink and for the downlink. On the downlink, all signals in a particular cell are modulated with a 2^{15} length pseudorandom sequence of chips. Orthogonality is preserved because all channels are modulated synchronously. On the uplink channel, signals arrive at the base station by different propagation paths, so a different strategy is implemented. The mobile data stream is first convolutionally encoded using a rate one-third code; then, after interleaving, blocks of six encoded symbols are mapped to one of 64 orthogonal Walsh functions providing 64-ary orthogonal signaling. The resulting 307.2 kChip/s is further spread fourfold by user codes

Frame number	F0	F1	F2	F3	F4	F5	F6	F7
Voice bits	a0, b4	a1, 5b	a2, 6b	a3, b7	a4, b0	a5, b1	a6, b2	a7, b3
	114 bits	114 bits	114 bits	114 bits	114 bits	114 bits	114 bits	114 bits

Figure 4.12 Interleaving of voice bits in GSM.

and base-station-specific codes having, respectively, periods of $2^{42} - 1$ chips and 2^{15} chips. From the propagation point of view, the convolutional encoding, along with the mapping into Walsh functions, results in a better tolerance to interference than could be realized using repetition spreading codes. Both base and mobile receivers employ a rake receiver [28] to resolve and combine some multipath components. The use of at least three rake fingers is specified in IS-95. The average value of energy per bit E_b/N_0 in a single sector of a cell is

$$\frac{E_b}{N_0} = \frac{W / R}{(U - 1)f + N_0 / S} \tag{4.24}$$

for a spreading bandwidth W and bit rate R with U total users in the sector and having an average voice activity factor f. When voice activity detection is used, f is nominally 0.35; otherwise, it is unity. The desired signal power S to noise N_0 ratio is not important when the number of users is large. The frequency-reuse factor on the uplink in multiple-cell CDMA systems is effectively in the range of 0.4 to 0.7 [27] due to out-of-cell interference, depending on cell size and propagation impairments. Both the uplink and downlink data are interleaved in blocks of 20 ms by different methods. Uplink power control, necessary to keep the signal strengths of multiple signals nearly equal at the base receiver, occurs every 1.25 ms in 1-dB steps. CDMA systems are single-frequency, or one-cell-frequency, reuse pattern systems.

The Japanese Personal Handiphone System (PHS) was developed to provide a microcell system involving low-cost base stations that are typically mounted low. As seen in Table 4.5, PHS employs TDMA and time division duplex (TDD) forms of multiple access. Both the transmit and receive functions occur on the same RF frequency in separate time slots. A total of 77 RF channels are provided, 40 for public systems and 37 for private systems. Channels are assigned dynamically based on channel conditions, while cell-to-cell handoffs support only very slow (walking) speeds. The channel data rate is 384 kbps on both the uplink and downlink; only error detection (no correction) is provided. From the channel propagation point of view, the low antennas typical of PHS systems result in small cells. The high channel bit rates without provisions for equalization result in operation only at very slow (walking) speeds within a cell.

4.8.5 Third-Generation Voice and Data Mobile Systems

Compared with second generation (2G) systems, third-generation (3G) technologies provide the higher capacities necessary to transfer simultaneously both voice (a telephone call) and nonvoice data, such as e-mail, instant messaging, music, and video. 3G systems need entirely new networks and new frequency assignments in order to provide support for new services.

The International Telecommunications Union Radiocommunications Sector (ITU-R) initiative, named the Third Generation IMT-2000 Project, has defined 5 MHz of licensed spectrum that enables transmission of 384 kbps for mobile systems and 2 Mbps for stationary systems. The ITU-R originally accepted W-CDMA (Wide-Band Code Division Multiple Access) as part of the IMT-2000 family of 3G standards. Attempts were made to unify W-CDMA (developed by the 3GPP (3rd Generation Partnership Project) Standard body) [29] and CDMA-2000 (developed by the 3GPP2 (3rd Generation Partnership Project 2) standard body) [30] standards in order to provide a single worldwide standard. However, divergent requirements resulted in the two incompatible standards that are currently being deployed and supported globally.

W-CDMA is based on the DS-CDMA technique. It supports two modes of duplex: frequency division duplex (FDD) and TDD. W-CDMA is based on coherent detection of pilot symbols in both the uplink and downlink and provides asynchronous intercell operations. Multiuser detection and smart-antenna technologies are used to enhance both coverage and network capacity. W-CDMA supports multimode transmissions that are based on various traffic types.

Time division-synchronous code division multiple access (TD-SCDMA) is a 3G approach being pursued in the People's Republic of China in an attempt to develop home-grown technology that is not dependent on Western technology [31]. It has been adopted by 3GPP as "UTRA TDD 1.28 Mcps Option."

TD-SCDMA uses TDD, in contrast to the FDD scheme used by W-CDMA. By dynamically adjusting the number of TDD time slots used for downlink and uplink, the system provides support for asymmetric traffic while delivery of various data rate requirements are present. Additionally, since it does not require paired spectrum for downlink and uplink, the resultant spectrum allocation flexibility presents some advantages. TD-SCDMA also uses TDMA in addition to the CDMA used in W-CDMA. The "S" in TD-SCDMA stands for "synchronous," which means that all uplink signals are synchronous.

The CDMA2000 standards CDMA2000 1x, CDMA2000 1xEV-DO, and CDMA2000 1xEV-DV are approved radio interfaces for the ITU-R IMT-2000 standard and a direct successor to the 2G CDMA IS-95 (cdmaOne). CDMA2000 is standardized by 3GPP2. CDMA2000 is an incompatible competitor of UMTS (Universal Mobile Telecommunications System), the other major 3G standard. CDMA2000 is defined to operate at 400, 800, 900, 1,700, 1,800, 1,900, and 2100 MHz.

CDMA2000 1x, the core CDMA2000 wireless air interface standard, is known by many names: 1x, 1xRTT, IS-2000, CDMA2000 1X, 1X, and cdma2000. The designation "1xRTT" (1 times Radio Transmission Technology) is used to identify the version of CDMA2000 radio technology that operates in a pair of 1.25-MHz radio channels. 1xRTT provides almost a double improvement in the system voice capacity compared to IS-95 networks. Although capable of higher data rates, most deployments have limited the peak data rate to 144 kbps. While 1xRTT officially qualifies as a 3G technology, 1xRTT is considered by some to be a 2.5G (or sometimes 2.75G) technology. Compared with IS-95, IS-2000 signaling has 64 more traffic channels on the forward link that are orthogonal to the original set. Some changes were also made to the data link layer to accommodate the greater use of data services—IS-2000 has media and link access control protocols and quality-of-service control.

The initial design of 1xEV-DO, "1x Evolution-Data Optimized," was developed by Qualcomm to meet IMT-2000 requirements. Initially, the standard was called High Data Rate (HDR); it was renamed to 1xEV-DO after it was ratified by the International Telecommunications Union. It was then given the numerical designation IS-856. The title of the 1xEV-DO standard document is "cdma2000 High Rate Packet Data Air Interface Specification." The name "cdma2000" (in lowercase) is another name for the 1x standard, numerically designated IS-2000. When deployed with a voice network, 1xEV-DO requires a separate 1.25-MHz bandwidth radio channel. The successor to the first revision of the standard, 1xEV-DO Rev. 0, is called 1xEV-DO Rev. A. This Rev. A offers fast packet establishment on both the forward and reverse links, along with air-interface enhancements that reduce latency and improve data rates. In addition to the increase in the maximum downlink (forward-link) data rate from 2.4576 Mbps, in Rev. 0, to 3.1 Mbps, Rev. A, which has a 12-fold improvement in the maximum uplink (reverse-link) data rate, from 0.15 Mbps to 1.8 Mbps. EV-DO Rev. A supports low latency services (as low as 50 ms), including Voice over IP (VoIP) and video telephony on the same carrier with traditional Internet packet-data services. EV-DO Rev. CDMA2000 1xEV-DV (1x Evolution-Data/Voice) supports downlink (forward-link) data rates up to 3.1 Mbps and uplink (reverse-link) data rates of up to 1.8 Mbps. 1xEV-DV can also support concurrent operation of legacy 1x voice users, 1x data users, and high-speed 1xEV-DV data users within the same radio channel.

4.8.6 Broadband Wireless Access Systems

The IEEE 802.16 Working Group on Broadband Wireless Access Standards, which was established in 1999, is focused on specifications for the global deployment of broadband wireless metropolitan-area networks (WMANs). A related future technology, mobile broadband wireless access (MBWA), is also under development in the IEEE 802.20 Working Group.

Although the 802.16 family of standards is officially dubbed WMAN, it has been commercially referred to as "WiMAX" (see [32]). The Worldwide Interoperability for Microwave Access, or WiMax, Forum is an industry forum focused on defining and certifying compatibility and interoperability of products compliant with 802.16 standards.

The first 802.16 standard was approved in December 2001 and was followed by two amendments—802.16a and 802.16c—to address issues of radio spectrum and interoperability. The original WiMAX standard (IEEE 802.16) was specified for operation in the 10–66-GHz frequency range. 802.16a was updated in 2004 to add support for the 2–11-GHz frequency range. There is no global spectrum allocation, and allocations vary by region. 802.16d was updated to 802.16e in 2005. Revision 802.16e uses scalable OFDM as opposed to the nonscalable version used in the earlier 802.16d standard [33]. Most interest in recent years has developed around the 802.16d standard since the lower frequencies suffer less from signal attenuation and therefore give improved range and in-building penetration.

IEEE 802.16-2005 (formerly named, and still best known as, 802.16e, or Mobile WiMAX) provides an improvement on the modulation schemes stipulated in the original (fixed) WiMAX standard. It allows for fixed wireless and mobile non-line-of-sight (NLOS) applications primarily by enhancing the orthogonal frequency division multiple access (OFDMA). It provides improved NLOS performance using antenna-diversity techniques (either adaptive antenna algorithms, MIMO, or both), increased system gain by use of denser subchannels in indoor deployments, and improved channel coding schemes such as turbo coding or low density parity check (LDPC) codes.

4.8.7 Wireless Local-Area Network Systems

The IEEE 802.11 family of standards (see Table 4.6) encompasses a set of wireless local-area network (WLAN) standards that operate in unlicensed spectrum in the 2.4- and 5-GHz bands. The 802.11 family of physical layer air interfaces is defined as the "b," "a," and "g" amendments to the original standard. Security was originally included and was later enhanced via the 802.11i amendment. 802.11n incorporates MIMO technology.

The original 1997 release of the standard IEEE 802.11 defines an infrared (IR) and an ISM (Industrial Scientific and Medical) band physical layer using either frequency hopping (FH) or direct sequence spread spectrum (DSSS). The original standard also defines carrier sense multiple access with collision avoidance (CSMA/CA) as the medium access control (MAC) method. All devices that communicate through the access point compete for access point time on a random interruption basis, which can cause distant devices to be repeatedly interrupted by closer devices. This fundamental property of 802.11 MAC makes services such as VoIP that require a predetermined quality of service difficult to achieve for a large number of active devices. Widespread adoption of 802.11 networks occurred only after 802.11b was ratified.

Table 4.6
The IEEE 802.11 Family of Standards

Protocol	Release Date	Operating Frequency Band	Typical Data Rate (Mbps)	Maximum Data Rate (Mbps)	Indoor Range (m)
802.11	1997	2.4-GHz ISM	1	2	—
802.11a	1999	5-GHz U-NII	25	54	~30
802.11b	1999	2.4-GHz ISM	6.5	11	~50
802.11g	2003	2.4-GHz ISM	25	54	~30
802.11n	2007 (draft)	2.4-GHz ISM or 5-GHz U-NII	>100	>200	~30

The 802.11b and 802.11g standards operate in the 2.4-GHz ISM band as license-free devices under Part 15 of the U.S. FCC rules and regulations. Hence, 802.11 and 802.11b/g equipment cannot interfere with licensed equipment in the 2.4-GHz band and must accept interference from other devices in that band, including licensed amateur radio transmitters and other unlicensed devices, such as cordless phones, Bluetooth devices, and ISM devices like microwave ovens that operate within the same band. However, the 802.11a standard operates in the 5-GHz bands, which are not affected by 2.4-GHz-band devices. Different countries have different regulatory support. To date, 802.11a is permitted in the United States and Japan but restricted in other areas, such as the European Union. Use of this higher frequency range also brings the disadvantage of poorer propagation compared with the 2.4-GHz range.

The 802.11b amendment to the original standard has a maximum raw data rate of 11 Mbps and uses the same CSMA/CA media access method defined for 802.11. The 802.11b standard uses a DSSS with complementary code keying (CCK) modulation technique and is usually configured in a point-to-multipoint configuration wherein an access point communicates with one or more clients. With high-gain external antennas, the protocol can also be used in fixed point-to-point arrangements for outdoor applications.

The 802.11a amendment to the standard uses the same MAC protocol as the original standard and defines a 5-GHz-band physical layer based on a 52-subcarrier OFDM modulation technique. The 48 data carriers and 4 pilot subcarriers are separated 0.3125 MHz (20 MHz/64). The subcarriers can be BPSK (Binary Phase Shift Keying), QPSK, 16-QAM or 64-QAM modulated, which in combination with various forward-error-correction rates, provide data rates of 6, 9, 12, 18, 24, 36, 48, and 54 Mbps. Actual throughput is roughly half the data rate. The modulation occupies a 16.6-MHz bandwidth in the 20-MHz channels. A symbol duration of 4 ms with a guard interval of 0.8 ms provides a good measure of protection from multipath effects.

The 802.11g physical layer specification adds the OFDM format for a maximum data rate of 54 Mbps in the 2.4-GHz spectrum and is backwards compatible with 802.11b. The 802.11g standard has found good consumer acceptance but suffers from the same interference issues as 802.11b in the already crowded 2.4-GHz band.

The IEEE 802.11 Task Group continues to develop new amendments to the 802.11 standard [34] to support data rates in excess of 100 Mbps. The new standards build upon previous 802.11 standards by adding MIMO technology, which uses multiple transmit and receive antennas to allow for increased data throughput through spatial multiplexing and increased range by exploiting the spatial diversity.

4.8.8 UWB Systems

Ultrawideband communications systems are permitted in the 3.1–10.6-GHz frequency spectrum on an unlicensed basis by FCC regulations in the United States (see [35]). Other countries will likely base their regulations on the recommendations of the ITU-R [36]. Regulations outside the United States will be similar to those of the United States above 6 GHz but restrictive below 6 GHz. The regulations are designed to provide adequate protection from harmful interference to incumbent services in the spectrum. The regulations do not describe or mandate any particular modulation or signaling method; hence, they encourage multiple and innovative solutions subject to broad guidelines regarding signal bandwidth, effective isotropically radiated power (EIRP), and peak power. Consequently, a rich variety of UWB architectures and signaling formats have emerged in this nascent industry, including a low-data-rate IEEE 802.15.4a standard supplement with localization capabilities [37].

Each UWB signaling format (see [38, 39]) has unique attributes and targets a unique set of applications. Time-modulated UWB (TM-UWB), for example, relies on precision placement of pulse sequences that are coded in time and is uniquely suited for low-probability-of-detection (LPD) applications and those involving distance measurements and imaging. Both pulse-UWB in the form of DS-UWB and multiband OFDM (MB-OFDM) provide similar performance in the high-data-rate wireless personal-area network (WPAN) space. MB-OFDM targets personal computer–centric and wireless USB–like computer platforms, while pulse-UWB, implemented as CWave (from Pulse~Link, Inc., Carlsbad, CA) or DS-UWB, is suitable for digital entertainment applications such as streaming video. The marketplace is realizing that multiple and diverse UWB architectures are emerging to satisfy a variety of applications at various price points, data rates, and energy-consumption appetites.

UWB signals can achieve their UWB bandwidths by virtue of short-duration pulses (where the bandwidth is the inverse of the pulse duration) or by aggregation of narrowband carriers that persist on the channel for relatively long bursts of time. Modulation and coding are applied to convey information on the UWB signal and to "whiten" the spectrum. In the case of UWB pulses, information can be coded in pulse amplitude (similar to PAM (Pulse Amplitude Modulation)), pulse polarity (analogous to BPSK), orthogonal pulse waveform shapes (analogous to QPSK), or pulse pair-polarity (as in DPSK (Differentila Phase Shift Keying)) or by pulse position in time. Furthermore, multiple combinations of these modulations can be applied to exploit clever and efficient coding mechanisms like *M*-ary biorthogonal coding (M-BOK) and time-frequency pulse-hopping modulations.

Pulse-UWB using the direct-spreading method uses a sequences of subnanosecond pulses, each of which occupies the full UWB bandwidth in a lower band between 3.1 to 4.9 GHz and/or a higher frequency band spanning 6 to 10.6 GHz. The systems combine various spreading code sequences with various forward-error-correction rates to achieve data rates from as low as tens of Mbps to well over a Gbps. Because UWB pulses are less than 2 ns in duration, they can be free from the effects of multipath fading.

The MB-OFDM physical layer (PHY) protocol employs symbol-by-symbol frequency hopping of a 500-MHz-wide OFDM signal over channels spaced 528 MHz apart. The MB-OFDM signal persists on the channel for 242.4 ns and has a 937.5-ns repetition interval on each of at least three channels. Hence, each carrier behaves like a narrowband (4-MHz-wide) signal and is subject to full Rayleigh fading. A basic 640-Mbps channel rate, in combination with various forward-error-correction rates, supports between 55 and 480 Mbps if all three hopping channels are employed.

4.9 Summary

The communications channel was introduced in terms of the characteristics of wave propagation in certain bands. The lowest frequencies were seen to operate in an Earth-ionosphere waveguide. Noise is particularly high, and resonant antennas are especially large in the lowest frequency bands. The bands between 3 and 30 MHz are subject to ionospheric reflection, which makes relatively reliable long-distance communications possible. Above about 30 MHz, reflections from meteor trails can lead to some interesting radio systems. The frequencies between 30 MHz and 3 GHz were identified as the prime frequency spectrum for personal communications systems, and now that range extends beyond 10 GHz. We introduced some modern multiple access techniques in view of their radio channel characteristics. We described the many mobile and cellular voice and data systems now deployed around the world and identified wireless local- and personal-area systems, including those using UWB technology.

Problems

4.1 An ELF communication link is operating at 10 kHz. The transmitter supplies 10,000W to an h = 50m tall vertical antenna with unity directivity and having resistive losses of $R_{loss} = 5\Omega$. The receiving antenna is 3m tall with 3Ω loss resistance. Find the efficiencies of the (a) transmitting and (b) receiving antennas.

Ans:

Antenna efficiency is $R_{radiation}/(R_{radiation} + R_{loss})$ = 0.001/(5.001) or –36.5 dB transmit antenna; $4 \times 10^{-6}/3$ or –58.7 dB receive antenna.

4.2 An ELF submarine communications link operates at 3 kHz. The transmitter supplies 2 MW to a vertical antenna h = 100m tall with unity directivity and having resistive losses of $R_{loss} = 1\Omega$. The receiving antenna is 60m long with 5Ω loss resistance. The submarine antenna is at a depth of 10m. Find (a) the power dissipated as loss in the transmitting antenna, (b) the efficiency of the receiving antenna, and (c) the additional path attenuation due to propagation through seawater.

Ans:

Antenna efficiency is $R_{radiation}/(R_{radiation} + R_{loss})$, or (a) 0.0004/(1.0004) or –34.0 dB transmit antenna, power dissipated is 2 (1 – 0.0004/1.0004) MW = 1.9992 MW; (b) Receiver efficiency is $1.4 \times 10^{-4}/10$ or –45.4 dB; (c) seawater attenuation is 19 dB.

4.3 An HF communications link uses dipole antennas. Allowing 30 dB for ionospheric and ground reflections, how much power is received if 100W is transmitted when the distance is 8,000 km and (a) the frequency is 29 MHz, and (b) the frequency is 14 MHz? (c) Using (4.15), what is the received signal-to-noise ratio?

Ans:

Use free space propagation,

$P_r = (1/1{,}000)\, P_t G_t G_r/(4\pi d/\lambda)^2 = (100/1{,}000)(1.641)^2/(4\pi d/\lambda)^2$,

which gives (a) 1.8 pW (–115.5 dBm) at 29 MHz and (b) 0.18 pW (–109.1 dBm) at 14 MHz. (c) The noise power is –154 dBm/Hz at 29

MHz, and –144.5 dBm/Hz at 14 MHz; so SNR is 38.5 dB at 29 MHz and 35.4 dB at 14 MHz.

4.4 If a 7-dB SNR is required for reliable communications, what is the maximum available bandwidth in the HF communications system of problem 4.3?

Ans:

At 29 MHz: 38.5 – 7 = 31.5 dB, $BW = 10^{3.15}$ = 1,413 Hz. At 14 MHz: 35.4 – 7 = 28.4 dB, $BW = 10^{2.84}$ = 692 Hz.

4.5 Find the critical and the maximum usable frequencies for a tangential ray incident on the ionosphere at an angle of 74° when ionospheric electron density is 1.8×10^{12} per cubic meter.

Ans:

$$f_{crit} = \sqrt{81 \cdot 5.8 \times 10^{11}} = 6.85\,\text{MHz}$$

$$MUF = 32.7\sqrt{N_{\max}} = 32.7\sqrt{5.8 \times 10^{11}} = 24.9\,\text{MHz}$$

4.6 A 40-MHz meteor scatter communications system has a bandwidth of 1,000 Hz and requires a 10 dB SNR to function. How much effective radiated power is needed if the receiver antenna has a 15-dBi directivity and path loss is 200 dB?

Ans:

$P_r = P_t G_t G_r / (4\pi d/\lambda)^2$ in decibels: EIRP_{dBm} + 15 dBi – 200 dB = $P_{r,dBm}$ The required receiver power is –158.2 dBm (noise) + 10 dB + 10log(1,000) = –118.2 dBm, so the required EIRP = –118.2 + 200 – 15 = 66.8 dBm or about 4.8 kW.

4.7 Calculate the number of paging subscribers that can be handled by a POCSAG 512-bps system. The average message is 10 digits, and the busy-hour traffic is 0.25 calls per subscriber. Assume 90% system efficiency and ignore the synchronization sequences. What happens when the data rate increases to 1,200 bps?

Ans:

At 512 bps, POCSAG has a throughput of 3,600 × 0.9 × 512/32 = 51,840 code words per hour. A 10-digit message requires three code words. Hence, at a rate of 0.25 messages per subscriber, capacity = 51,840/3/0.25 = 69,120 subscribers. At 1,200 bps, the system capacity increases to 69,120 × 1,200/512 = 162,000 subscribers.

4.8 Calculate the number of paging subscribers that can be handled by a FLEX 6,400-bps system. The average message is 10 digits, and the busy-hour traffic is 0.25 calls per subscriber. Assume 90% system efficiency and ignore the synchronization sequences.

Ans:

At 6,400-bps FLEX has 4 block information words, followed by 11 blocks of 8 interleaved code words, or 348 usable code words per frame and 32 frames per minute at 30 batches per preamble, so there are 668,160 code words per hour; hence, capacity is $668,160 \times 0.9/4/0.25 =$ 601,344 subscribers.

4.9 Voice messaging is a very desirable user function. If a typical voice message is 4 sec long, how many users can be handled at 0.25 calls per busy hour using standard analog voice modulation? On a full system, what is the relative cost impact to a service provider of a voice subscriber compared with a FLEX 6,400-bps subscriber that expects a 40-character message?

Ans:

At 4 sec per call at 100% channel utilization efficiency, there are $3,600/4/0.25 = 3,600$ subscribers per channel. The same channel could support $668,160 \times 0.9/17/0.25 = 141,492$ message subscribers in FLEX 6,400, or about 39 message pagers per voice pager. In a large system, a service provider would be inclined to fill up a system with messaging subscribers rather than standard voice subscribers.

4.10 Assume that an AMPS cellular system can be laid out in clusters of 19 cells within which frequencies cannot be reused. What is the total number of traffic channels per cell if 12.5-MHz downlink and 12.5-MHz uplink bandwidth are allocated to the service provider.

Ans:

The total number of up- and downlink channel pairs, taking into account 10-kHz guard bands and 30-kHz channels, is (12,500 kHz – 20 kHz)/30 = 416, which means 416/19 = 21 are available in a cell.

4.11 Assuming a seven-cell frequency-reuse pattern, find the maximum number of data user bits per second per cell per hertz of bandwidth in an AMPS system if all but one channel per cell are devoted to data at 9,600 bps.

Ans:

There are 832/7 – 1 = 117 channels per cell in 25 MHz. So,

(117)(9,600)/(25 MHz) = 0.0449 bps/Hz/cell.

4.12 Assuming that hexagonal cells are split into three sectors, devise a frequency-reuse strategy for sector clusters of one and two rings. Show that a minimum of 9 frequencies are needed in one ring of sectors and 21 frequencies are needed in two rings of sectors.

4.13 How many traffic channels are available in a GSM system having 25 MHz each of uplink and downlink bandwidth? If cells are clustered in groups of three for frequency reuse, what is the number of available traffic channels per cell? How does this compare with an AMPS system?

Ans:

GSM has eight time slots in each 200-kHz RF channel. So, 25 MHz in each direction has (25,000/200) 8 = 1,000 total channels. With a cell cluster of three, the number of channels per cell is 1,000/3 = 333. An AMPS 25-MHz system would have 833/7 = 119 channels per cell. Everything else being equal, GSM provides 333/119 = 2.8 times the channels of an AMPS system for the same total bandwidth.

4.14 Assuming a nine-sector frequency-reuse pattern, find the maximum number of user bits per second per cell per hertz of bandwidth in a GSM system if all but one RF channel per cell are normal traffic channels.

Ans:

GSM delivers 58 bits per 576.92-μs time slot in normal traffic channels, and there are 125/9 = 13 RF channels per cell. So,

(58 bits)/(576.92 μs) (13 – 1 freq)(7 time slots)/(25 MHz) = 0.338 bps/Hz/cell

4.15 Find frame efficiency for normal GSM time slots.

Ans:

A frame is eight time slots (TS); TS = 156.25 bit intervals, of which (6 + 26 + 8.25) = 40.25 are overhead bits. Frame efficiency is then TS efficiency = (156.25 – 40.25)/156.25 = 0.742.

4.16 Find the total number of simultaneous CDMA users in a single cell if W = 1.2288 MHz, R = 13 kbps, and the minimum acceptable E_b/N_0 is 10 dB and (a) an omnidirectional antenna is used at the fixed site with no voice activity detection, and (b) sector antennas are used in a three-sector cell having voice activity detection.

Ans:

From (4.19): (a) $N = 1 + (1{,}228{,}800/13{,}000)/10 = 1 + 9.5 = 10$.

(b) N per sector = $1 + (1{,}228{,}800/13{,}000)/10/(.35) = 1 + 27 = 28$, and there are three sectors, so there are 84 simultaneous users per cell.

4.17 A 1.2288-Mbps chip rate CDMA system of Table 4.4 carries data at 13,000 bps per user. Find the maximum bits per second per hertz per cell if one control channel per cell is needed, the cell has three sectors, and 6-dB C/I is needed by each user. What other assumptions are implicit?

Ans:

There are 64 – 1 available code channels per cell and 21 codes per sector, and the coding gain is 10log(1,288.8/13) = 19.76 dB. Take away 6 for C/I, which leaves 13.76 dB or up to $10^{1.376} = 23$ codes. All 21 codes per sector plus 1 per cell can be used. The bits per second efficiency is (13,000)(63)/(1.25 MHz) (0.4 to 0.7 reuse efficiency) = 0.262 to 0.459 bps/Hz/cell. Implicitly assumed is power control per user sufficient to justify 6 dB C/I.

4.18 Calculate the total bits per second, the symbol rates, and the frame rates sent on a EUREKA-147 digital voice broadcast system operating in (a) Mode-I, (b) Mode-II, and (c) Mode-III.

4.19 A COFDM (Coded OFDM) modulation format with parameters the same as EUREKA-147 is to be used in the downlink of a cellular data system. Estimate the bits per second per hertz per cell.

Ans:

EU-147 delivers 2.4 Mbps in 1.536 MHz. In a cellular system, a seven-cell reuse strategy would be needed, so (2.4)/(1.536)/7 = 0.223 bps/Hz/cell.

4.20 Show that the maximum throughput for ALOHA and S-ALOHA are $1/2e$ and $1/e$, respectively.

4.21 An ALOHA protocol is operated at 9,600 bps and is shared by a group of N stations. On average, the stations transmit 512 bit packets every 1,000 sec. Find the maximum useful value of N.

Ans:

The offered normalized load is $\lambda\tau = N512/9600$ per 1,000 sec = 5.33×10^{-5}, and "useful" is near the peak, $\lambda\tau = 1/2e$; hence, $N = 1/(2e5.33\times10^{-5})$ = 3449.

Another way: maximum throughput is $1/2e$ of capacity = $9{,}600/2e$ = 1,766 bps divided by (512/1,000); the average data rate per user = 3,449.

References

[1] Davies, K., *Ionospheric Radio*, London: IEE/Peter Peregrinus Press, 1989.

[2] Hall, M. P. M., and L. W. Barclay, *Radiowave Propagation*, London: IEE/Peter Peregrinus Press, 1989.

[3] Collin, R. E., *Antennas and Radiowave Propagation*, New York: McGraw-Hill, 1985.

[4] Jordan, E. C., and K. G. Balmain, *Electromagnetic Waves and Radiating Systems*, 2nd ed., Upper Saddle River, NJ: Prentice Hall, 1968.

[5] Bryant, J. H., *Heinrich Hertz—The Beginning of Microwaves*, New York: IEEE Press, 1988.

[6] Schumann, W. O., "Uber die strahlungslosen Eigenschwingungen einer leitenden Kugel, die von einer Luftschicht und einer Ionenspärenhülle umgeben ist," *Z. Naturforsch*, Vol. 72, 1952, pp. 149–154.

[7] Bashkuev, Y B., et al., "Global Electromagnetic Resonances of Earth-Ionosphere Cavity in Middle Latitudes of Asia," *EMC 90, Proc. of the International Wroclaw Symposium on Electromagnetic Compatibility*, Warsaw, Poland, June 26–29, 1990.

[8] Burrows, M. L., *ELF Communications Antennas*, London: Peter Peregrinus Press, 1978.

[9] DeSoto, C. B., *Two Hundred Meters and Down*, Newington, CT: The American Radio Relay League, 1936.

[10] Eckersley, T. L., "Studies in Radio Transmission," *J. IEE*, Vol.71, September 1932, pp. 405–454.

[11] Bain, W. F., "VHF Meteor Scatter Propagation," *QST* (The American Radio Relay League, Newington, CT), April 1957.

[12] Forsyth, P. A. et al., "The Principles of JANET—A Meteor Burst Communications System," *Proceedings of the IRE*, December 1957, pp. 1642–1657.

[13] Eshlemann, Von R., "On the Wavelength Dependence of the Information Capacity of Meteor-Burst Propagation," *Proceedings of the IRE*, December 1957, pp. 1710–1714.

[14] Vernarec, E., "Digital Voice Rides Micrometeor Trails," *Microwaves & RF*, November 1986, pp. 39–42.

[15] Schilling, D. L., (ed.), *Meteor Burst Communications, Theory and Practice*, New York: John Wiley and Sons, 1993.

[16] Dulukhanov, M., *Propagation of Radio Waves*, trans. B. Kuznetsov, Moscow: Mir Publishers, 1971.

[17] Annex I Radio Paging Code No. 1 [POCSAG], *CCIR Recommendation 584-1,* International Radio Consultative Committee, International Telecommunications Union, Geneva, 1986.

[18] Annex I Radio Paging Code No. 3 [FLEX], *CCIR Recommendation 584-3,* International Radio Consultative Committee, International Telecommunications Union, Geneva, 1997.

[19] ETSI EN 300 401 V1.4.1 (2006-01), European Standard (Telecommunications Series) Radio Broadcasting Systems; "Digital Audio Broadcasting (DAB) to Mobile, Portable and Fixed Receivers," available at http://webapp.etsi.org/action/OP/OP20060526/en_300401v010401o.pdf (last accessed October 25, 2006).

[20] Karn, P., H. Price, and R. Diersing, "Packet Radio in the Amateur Service," *IEEE Journal on Selected Ares in Communications,* Vol. 3, No. 3, pp. 431–439, May 1985.

[21] Stallings, W., *Data and Computer Communications,* 3rd ed., New York: Macmillan Publishing Company, 1991.

[22] Lam, S. S. "Satellite Packet Communications, Multiple Access Protocols and Performance," *IEEE Transactions on Communications,* Vol. COM-27, October 1997, pp. 1456–1466.

[23] Young, W. R., "Advanced Mobile Phone Service: Introduction, Background, and Objectives," *Bell System Technical Journal,* Vol. 58, January 1979, pp. 1–14.

[24] Mehrota, A. *GSM System Engineering,* Norwood, MA: Artech House, 1997.

[25] TIA/EIA Interim Standard 95, *Mobile Station—Base Station Compatibility Standard for Dual-mode Wideband Spread Spectrum Cellular System,* July 1993.

[26] Personal Handiphone System, *Japanese Telecommunications System Standard,* RCR-STD 28, December 1993.

[27] Rappaport, T. S., *Wireless Communications, Principles and Practices,* Upper Saddle River, NJ: Prentice Hall, 1996.

[28] Price, R., and P. E. Green. "A Communication Technique for Multipath Channel," *Proceedings of the IRE,* March 1958, pp. 555–570.

[29] Third Generation Partnership Project, available at www.3gpp.org (last accessed October 25, 2006).

[30] Third Generation Partnership Project 2, available at www.3gpp2.org (last accessed October 25, 2006).

[31] TD-SCDMA Forum, available at www.tdscdma-forum.org/en/resources (last accessed October 25, 2006).

[32] WiMax Forum, available at www.wimaxforum.org/home (last accessed October 25, 2006).

[33] IEEE Standard 802.16e-2005, February 28, 2006.

[34] IEEE 802.11 Project, available at http://grouper.ieee.org/groups/802/11 (last accessed October 25, 2006).

[35] U.S. 47 C.F.R. Part15 Ultra-Wideband Operations FCC Report and Order, April 22, 2002.

[36] Draft New Recommendation ITU-R SM.[UWB.COMP], Impact of Devices Using Ultra-wideband Technology on Systems Operating within Radiocommunication Services, Geneva, October 12–20, 2005, Annex 2 to TG 1/8 Chairman's Report, Document 1-8/443-E, January 17, 2006.

[37] Part 15.4: Wireless Medium Access Control (MAC) and Physical Layer (PHY) Specifications for Low-Rate Wireless Personal-Area Networks (LRWPANs) (Amendment of IEEE Std 802.15.4), IEEE P802.15.4a(tm)/D4, August 2006.

[38] Bahreini, Y., and K. Siwiak, "An Interoperability Mode for Data Intensive UWB Systems," Wireless Networking and Communications Group symposium, Austin, TX, October 21–22, 2004.

[39] Siwiak, K., and D. McKeown, *Ultra-wideband Radio Technology*, London: John Wiley and Sons, 2004.

5

Communications Using Earth-Orbiting Satellites

5.1 Introduction

Earth-orbiting satellites have the potential to provide personal communications in a manner so pervasive that it can change the fabric of society. Satellites, in the simplest viewpoint, are a communications infrastructure in the sky; they are "fixed sites" that transcend economics as well as national borders. Satellites can instantly provide modern communications to regions on Earth that are otherwise unreachable by the terrestrial infrastructure. Communications satellites were at first perceived largely as a means to reach remote and isolated places. Commercial communications satellites began in the early 1960s with the relatively low orbiting Telstar and Relay satellites, followed by Syncom, the first communications satellite in a geosynchronous orbit that was popularized by Arthur C. Clarke [1]. Soon it was realized that satellites could provide long-distance telephone and television circuits more economically than could the established common carriers. Satellites became the "trunked line" infrastructure for the highest-demand circuits. Prices steadily dropped as the capabilities of satellites increased. Throughout the 1960s into the 1970s, most satellite capacity was devoted to telephone and television traffic. The trend, however, was toward digital communications via satellites. Data transmissions, including digitized voice/audio (XM Satellite Radio and SIRIUS Radio services), and, more recently, digital video broadcasting (DVB) and broadband IP traffic are the norm. Satellites are becoming an on-demand multiple access infrastructure capable of linking people and computers on an unprecedented scale. Terrestrial systems, in the meanwhile, have evolved to provide wireless telephone access in the form of cellular telephones in most of the populated regions of the industrialized nations. While the evolution of broadband wireless services is still underway, the future now points to personal communications accessed via Earth-orbiting satellites, and we appear to have come full circle: with the launch of the Iridium satellites beginning on May 5, 1997, Earth-orbiting satellites are poised to provide on-demand personal communications access from rural and maritime areas and from the most remote regions of the planet. Personal communications via Earth-orbiting satellites is the main focus of this chapter.

5.2 Satellite Orbit Fundamentals

Satellites move in orbits that satisfy Kepler's laws. They provide the vantage point of altitude, which results in potential coverage over wide geographic areas. Personal communications using satellites include a few additional effects that become significant in the path link description compared with terrestrial path links. These including large Doppler frequency shifts, atmospheric absorption, and Faraday rotation of polarization. The fixed site is no longer fixed but orbiting the Earth.

This section will introduce the basic laws of Kepler and provide methods of analyzing Earth-orbiting satellites. Readily available satellite orbital elements are described in view of the proliferation of software for accurately calculating satellite orbits. The different types of orbits will be investigated, and their potential use in personal communications will be explored.

5.2.1 Orbital Mechanics

All orbits are essentially elliptical. They are governed by Kepler's laws, but with perturbations. The orbit of a satellite is an ellipse with the center of the Earth at one focus. The line joining the Earth's center of mass and the satellite sweeps equal areas in equal time intervals. The square of the orbital period T, in seconds, of a satellite is proportional to the cube of its mean distance R_a, in kilometers, from the Earth's center. So,

$$T = \frac{2\pi R_a^{1.5}}{\sqrt{\mu_\oplus (1+m_a)}} \tag{5.1}$$

The gravitational parameter for Earth is $\mu_\oplus$ (see List of Symbols), and m_a is the ratio of satellite mass to the Earth's mass. Since the mean distance is the average of the apoapsis and periapsis radii [2], R_a is the semimajor axis (half the length of an ellipse) of the orbit. Thus, the period of an elliptical orbit depends only on the size of the semimajor axis. The velocity, v_a, in kilometers per second, of a satellite at distance r km from the Earth's center and in elliptical orbit around the Earth is

$$v_a = \sqrt{\mu_{\oplus}\left[\frac{2}{r} - \frac{1}{R_a}\right]} \tag{5.2}$$

Expressions (5.1) and (5.2) are sufficient to study simple satellite motion but are not accurate enough to describe real satellites whose orbits are perturbed by Earth's gravitational anomalies, as well as by the masses of the moon and sun. Furthermore, Earth rotation must be taken into account to orient the orbit with respect to the stars.

The velocity V_a, in kilometers per second, relative to the Earth at the equator of a satellite orbiting the Earth at an angle of inclination I degrees with respect to the equator is

$$V_a = v_a\sqrt{\left[\cos(I) - \frac{2\pi R_a}{T_s v_a}\right]^2 + \sin^2(I)} \tag{5.3}$$

Most satellites orbit the Earth from west to east (that is, the various inclination angles are smaller than 90°) because an eastward launch uses the Earth's rotational velocity to obtain some of the energy needed to achieve orbital velocity, hence the negative sign in the brackets of (5.3). From simple geometry, the slant range D_s in kilometers to a satellite in a circular orbit around the Earth is

$$D_s = \sqrt{\left(R_e + H\right)^2 + R_e^2 - 2R_e\left(R_e + H\right)\sin\left[E + a\sin\left[\cos(E)\frac{R_e}{R_e + H}\right]\right]} \tag{5.4}$$

In terms of the elevation angle E to the satellite, the orbital altitude above Earth is H km, where R_e is the Earth's radius. The angle ϕ formed by a line from the satellite to a vertex at the Earth's center and to the observer is

$$\phi = a\sin\left[\frac{D_s}{R_e + H}\cos(E)\right] \tag{5.5}$$

Equations (5.2) to (5.5) are useful in estimating the performance of communications systems using Earth-orbiting satellites, but they are insufficient to predict the orbits of real satellites around the Earth. A detailed description of orbiting satellites requires the consideration of many more effects, including the gravitational influences of the sun and moon. Additionally, frictional effects due to trace atmosphere and perturbations due to gravitational anomalies on Earth perturb the orbit away form the ideal described by Kepler's laws. Detailed treatments [2, 3] are beyond the scope or requirements of this text but are required for accurate prediction of satellite orbital calculations. Orbital data is available for a large number of Earth-orbiting objects based on North American Air Defense Command (NORAD) and National Aeronautics and Space Administration (NASA) measurements that allow accurate orbital calculations using readily available software.

5.2.2 Orbital Predictions

The Earth rotates about its axis with a sidereal (meaning, "with respect to the stars") period of 23 hours, 56 minutes, and 4.09 seconds (T_s = 86,164.09 sec). The orbit of a satellite around Earth is described by its semimajor axis (radius from the Earth's center for circular orbits), hence its orbital period, the inclination of its orbit with respect to the Earth's equator, the orientation with respect to the stars of where the ascending node of the satellite crosses the Earth's equator. The orbit, perturbed by many forces of orbital drag, Earth's gravitational anomalies, and the gravitational influence of all other celestial bodies, especially the moon and the sun, differs from ideal two-body Keplerian orbital behavior, and, as a result, satellites require station-keeping thrusters to maintain required orbits precisely.

Orbits of satellites, particularly those of satellites in low to medium Earth orbit, may be computed using any of several general-purpose orbital-prediction computer programs [4–8] that are currently available. These computer programs compute satellite positions from measured orbital elements available from NASA, NORAD, and other sources. The basic orbital elements are

- *Epoch* is the reference time for the orbital-element parameters and consists of the epoch year, followed by the day of the year, including a fractional part of the day. An orbit number is often associated with the epoch time.
- *Orbital inclination* is the angle between the orbital plane and the equatorial plane.
- *Right ascension of ascending node* (RAAN) is an angle measured in the equatorial plane from a reference point in the sky called the *vernal equinox* to the point where the satellite orbit crosses the equator going from south to north (ascending node).

- *Argument of perigee* is the angle measured at the center of the Earth from the ascending node to the point of perigee.
- *Eccentricity* is the ellipticity of an orbit.
- *Mean anomaly* is the angle that varies uniformly in time during one revolution. It is defined as 0° at perigee; hence, it is 180° at apogee.
- *Mean motion* is the number of orbits per day.
- *Drag* is the perturbation contribution due to residual atmosphere. Drag is one-half the first time derivative of mean motion and is usually given in orbits per day squared.

The basic orbital elements are useful in calculating precisely the orbits of objects for which elements are available. The currently available orbital-tracking computer programs that use the published satellite elements may be used to study a theoretical orbit by entering the appropriate test element set. Additional orbital calculation methods and relationships between different kinds of orbital elements and orbit descriptions can be found in [2, 3, 9, 10].

5.2.3 Types of Orbits

The orbits of satellites that might be used for personal communications systems can be grouped into four categories: low Earth orbit (LEO), medium Earth orbit (MEO), geostationary orbit (GEO), and elliptical Earth orbit (EEO). Each has advantages and disadvantages in PCS designs. Table 5.1 shows sample orbital elements for each of the satellite orbit types in units suitable for use with most of the general-purpose orbital-prediction computer programs.

The Iridium (Iridium is a trademark of Iridium, LLC) system [11, 12] is a LEO constellation of 66 satellites (plus 6 in-orbit spares) approximately 778 km above the Earth in six equally spaced orbital planes at nearly 86.4° inclination to the equator. Every point on Earth is in view of this "infrastructure in the sky," and the system is planned as a complement, or adjunct, to the existing "islands" of terrestrial communications infrastructures.

The MEO characteristics are typified by the Global Positioning Satellite (GPS) system of 24 satellites occupying six orbital planes inclined about 54.8° from the equator and at an altitude of about 20,000 km. The GPS satellites are spaced so that at least four are simultaneously visible from each point on Earth. Although the GPS system is not a communications system, GPS receivers are increasingly being included in personal communications devices. The GPS system is often selected for time keeping and site synchronization in multiple-transmitter terrestrial systems.

Table 5.1
Orbital Elements for Four Types of Satellite Orbits

Orbital Element	LEO	MEO	GEO	EEO
Satellite	Iridium	GPS	INTELSAT	Molniya
Epoch	2006 231.40822226	2006 230.32290049	2006 230.72851175	2006 230.72851175
(revolution of epoch)	(rev 48,635)	(rev 11,658)	(rev 74,158)	(rev 412)
Orbital inclination	86.4009°	54.76780°	0.01080°	62.08880°
RAAN	171.4706°	263.9384°	1.62870°	214.5329°
Argument of perigee	77.8669°	155.0343°	148.7212°	266.2404°
Eccentricity	0.0002548	0.0097258	0.00026580	0.7242485
Mean anomaly	282.2821°	205.50360°	129.1125°	16.2853°
Mean motion	14.34216969	2.00563941	1.00274158	2.00707324
Drag	–0.00000e–000	–0.00000e–000	–0.00000e–000	0.00000e–000

Orbital elements can be found at http://celestrak.com/NORAD/elements.

Satellites that are in orbits having a period of one sidereal day are referred to as having a geosynchronous orbit. If, additionally, the inclination of the orbit with respect to the equator is nearly zero and the orbital path is highly circular, the orbit appears stationary to an observer on the Earth. Such an orbit is called a geostationary orbit, or GEO). Table 5.1 lists the orbital parameters of the INTELSAT satellite as an example of a GEO satellite.

Sometimes, highly elliptical orbits are chosen for communications because of certain desirable geometric relationships between the orbit and the intended coverage area. The EEO of the Molniya satellite is an example. The Molniya satellites were designed to provide television coverage to the high latitudes of Siberia in Russia since those high latitudes are not well covered by the equatorial geostationary satellites. The Molniya orbits are highly eccentric, having approximately one-half sidereal-day period paths, which appear at apogee over Siberia for a useful six-hour operational period. There is a second apogee over Hudson Bay, Canada. When viewed from a fixed point on Earth, the orbital track appears to make a somewhat distorted, open-ended "U" shape with the Earth tucked in just above the curved part of the "U." The two open ends are, respectively, about 41,000 km above Siberia and Hudson Bay. Because at apogee the orbit appears from Earth to be slow moving, very modest tracking capabilities are required. Four satellites can provide essentially continuous coverage over the high latitudes of one hemisphere of the Earth. The Molniya satellites have been in service since 1963, but their future in the Russian television service is uncertain.

5.2.4 Big LEO Systems

The three big LEO systems, Globalstar, Iridium, and Odyssey, are the first to be licensed in the United States, and they have agreed to cooperate in an effort to secure global authorizations for the portions of the radio frequency spectrum to be used by their mobile phones. Globalstar and Odyssey, which employ CDMA, or code division multiple access, share a segment of spectrum for their mobile links. That spectrum segment can accommodate other global systems employing compatible technologies. Iridium is a TDMA, or time division multiple access, system and uses a separate segment for its mobile links. The agreement conforms with the International Telecommunication Union's frequency authorizations for global mobile systems.

Globalstar LP, based in San Jose, California, is a partnership of 12 international telecommunications service providers and equipment manufacturers that are building a global mobile satellite telephone system to be operational in 1998. Globalstar's dual-mode (cellular-satellite) handsets will be compatible with the world's existing cellular and wire-line networks. Globalstar will sell access to the Globalstar system to a worldwide network of regional and local telecommunications service providers, including its strategic partners.

Iridium, LLC, is an international consortium of leading telecommunications and industrial companies funding the development of the Iridium system. The Iridium system is a 66-satellite telecommunications network designed to provide global wireless services to hand-held telephones and pagers virtually anywhere in the world. The first five Iridium satellites were successfully launched from Vandenburg Air Force Base on May 5, 1997, at 1855 UTC on board a Delta II vehicle; this was followed by the launch of seven more satellites on a Proton vehicle from Baikonur, Kazakhstan at 1402 UTC on June 18, 1997. Iridium, LLC, has since gone through a reorganization.

Odyssey Telecommunications International, Inc., (OTI), which has TRW and Teleglobe as initial shareholders, was to develop a constellation of 12 Odyssey satellites orbiting approximately 10,354 km above the globe. The project was cancelled, and the license has been returned.

The reorganization of Iridium, LLC, and the demise of Odyssey underscore the difficulty of the economics model for communications satellites. This is particularly true as terrestrial mobile phone technology becomes more and more pervasive globally.

5.3 Satellite Propagation Path

The radio path link to Earth-orbiting satellites is explored in this section. Basic path loss tends to be due to free space propagation, but with a multipath component that is somewhat different than is usually experienced from terrestrial paths (see Chapter 8). Doppler frequency shift is a significant factor in all types of orbits, including, to a lesser degree, the geosynchronous orbits. The kinds of coverage available from satellites in various orbits will be explored here. That communications between handheld terminals via satellites is technologically feasible is not in dispute. The question for personal communications systems is one of economic feasibility in situations that are demanded by subscribers for such services. People are generally located in buildings and cars or on urban or suburban streets with significant shadowing to the free space propagation path. Such systems need additional margin in the tens of decibels over the free space path for reliable and desirable communications. One characteristic of satellites is that the weight of a satellite, hence its design and launch cost, is roughly proportional to the total RF power available for communications. Furthermore, RF power from satellites is greatly more expensive than RF power from terrestrial transmitters wherever terrestrial transmitters are available. That general rule of thumb means that *x* dB of additional satellite link margin results in *x*-dB higher costs to the system provider, hence to the subscriber. The net historical trend in the cost of RF from a satellite has been downward. Pervasive personal communications from satellites will occur when the economics become favorable for the type of coverage desired.

5.3.1 Path Loss in a Satellite Link

Radio path attenuation, F dB, is essentially free space loss, given earlier by (Chapter 3) and written here in terms of distance D_s in kilometers and frequency f in megahertz:

$$F = 32.4479 + 20\log(fD_s) \tag{5.6}$$

Expression (5.6) accounts only for the direct line-of-sight paths and neglects atmospheric absorption. The paths of interest in personal communications are generally from low- to medium-Earth-orbital heights, and the typical maximum elevation of the satellite is not far above the lowest design elevation angle. Figure 5.1 shows the free space path loss at 1,625 MHz for the relatively rare case of a direct overhead pass of the LEO satellite example shown in Table 5.1. Typically, the satellite arcs above the horizon to a median elevation of less than 50°, as will be shown later. Figure 5.1 plots the path loss as a function of time, and it is interesting to note that the satellite is at large path losses for a large percentage of time. Additionally, because the personal communications device will typically have small, broad antenna beams, scattering and multipath require additional signal margin than is shown in Figure 5.1.

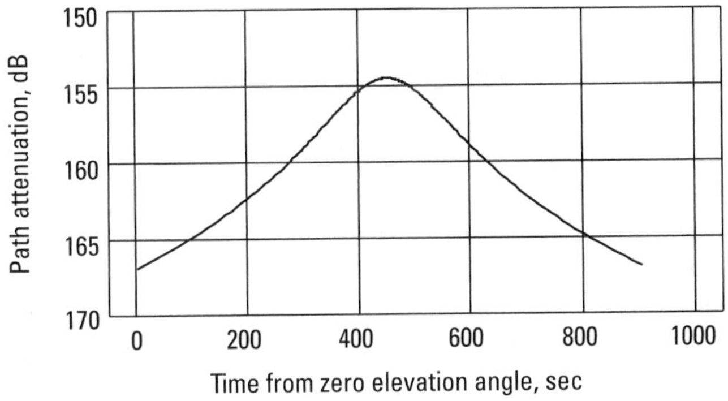

Figure 5.1 Path loss at 1,625 MHz for an overhead LEO satellite pass.

[5-3a.mcd] Compute the path attenuation at 1,650 MHz for orbital heights between 500 and 50,000 km and elevation angles of 15° and 90°.

Satellite line-of-sight links are typically described in terms of the following:

- Transmitted power P_{TX} dBW;
- Antenna gain G dBi;
- Illumination level W dBW/m^2 at the receiver;
- Free space path loss F;
- System noise temperature T_{sys} K;
- Carrier-to-thermal noise temperature C/T dBW/K;
- Carrier-to-noise ratio C/N dB;
- Carrier-to-noise-density ratio C/N_0 dBHz;
- Figure of merit G/T_{sys} dBi/K.

The system noise N is $k_b T_{sys} B$, where k_b J/K is Boltzmann's constant, and B Hz is the noise bandwidth. The object of a link margin calculation is to provide a value of C/N that is sufficient for reliable system performance under the worst-case design conditions, including margin for rain fall, polarization mismatch, angular alignment of antennas and, so, is represented by L_{total}, which can be in the range of 3 to 8 dB.

$$C/N = P_{TX} + G - F - L_{total} + G/T_{sys} - 10\log(k_b) - 10\log(B) \quad (5.7)$$

The link margin M is the carrier to noise ratio in dB relative to the required carrier to noise ratio in dB C_r/N:

$$M = C / N - C_r / N \tag{5.8}$$

In systems like television receive-only (TVRO) Earth stations, the system noise temperatures can often be significantly lower than a room temperature of 290K, and antennas of substantial gain are oriented line-of-sight (LOS) to the satellite. G/T of tens of decibels is not unusual. Furthermore, the additional losses represented by L_{total} characterize a path that is direct LOS between high-gain antennas. In contrast, personal communications systems, especially those involving LEO satellites, operate with handheld terminals located within buildings. In a building or urban environment without a line-of-sight path to the satellite, the antenna gain can be considered, at best, antenna efficiency since directivity is nearly irrelevant in multipath, and the system noise temperature must include the noise figure at room temperature. For a handheld terminal with a 2-dB noise figure, the receiver noise temperature from (4.19) is about 170K, which adds to the antenna noise temperature of 290K. Assuming an antenna efficiency of perhaps 90%, the figure of merit G/T is –25 dB/K. The personal communications link must additionally provide margin for signal degradations due to multipath propagation. In this regard, satellite-based and terrestrially based personal communications systems are similar.

5.3.2 Doppler Shift

The Doppler shift of a signal as observed on Earth for a satellite in Earth orbit is found from (5.2) to (5.5) as

$$f_{dop} = f \frac{V_a}{c} \cos(E + \phi) \tag{5.9}$$

where c is the velocity of electromagnetic waves in kilometers per second, and V_a is from (5.3). Note that special care must be used in assigning the Doppler shift direction when applying (5.9). As shown in Figure 5.2, a positive Doppler shift is observed on the downlink from an approaching and transmitting satellite. In transmitting to the approaching satellite, Doppler shift must be compensated for at the Earth uplink by transmitting in a frequency lowered by the Doppler amount. The Doppler effect will shift the ground-transmitted signal up to the required nominal frequency. The ground station receives an approaching satellite "high" and transmits to it "low." Once the satellite has passed its closest point to the observer, the downlink transmission exhibits a negative Doppler shift, and the uplink frequency is compensated for by transmitting higher [AQ:

in a frequency increased by by the Doppler amount. Now the ground station receives the receiving satellite "low" and transmits to it "high." The Doppler correction in two-way communications systems is performed on the ground because there are multiple users accessing a single satellite from various points on the ground, each instantaneously with an individual geometry and hence a unique Doppler shift. Figure 5.2 shows that when plotted versus time, the Doppler shift for the LEO case in Table 5.1 at a nominal frequency of 1,625 MHz is near its extreme values for a significant percentage of the time. Furthermore, the Doppler rate of change is dramatic as the satellite passes through its zenith. Typical satellite passes will have lower Doppler shifts than the extreme case shown, and the corresponding Doppler rate will also be smaller. Although the specific case examined was for a LEO satellite, the same holds true for MEO satellites. GEO satellites also exhibit Doppler frequency shifts because the satellites can drift to slightly inclined orbits which trace a figure eight ground track about the equator.

[5-3b.mcd] Compute the Doppler frequency at 1,650 MHz for orbital heights between 500 and 50,000 km and orbital inclinations of 0° and 90°. Find the orbital height for which Doppler shift vanishes.

5.3.3 Coverage from Satellites

Coverage from satellite systems tends to be from relatively low elevation angles as viewed from the Earth. World coverage in satellite systems that are intended to cover the entire globe can be viewed as repeated quadrilaterals ABCDA that form triangles whose vertices are the orbiting satellites, as shown in Figure 5.3. For systems like Iridium, all orbital planes converge at the poles, and the triangle areas vanish there. The probability density function (PDF), $p(E)$, of an elevation angle E is related through (5.5) to angle ϕ, which is proportional to the radius in the coverage footprint of the satellite. Clearly, there is much more area at the

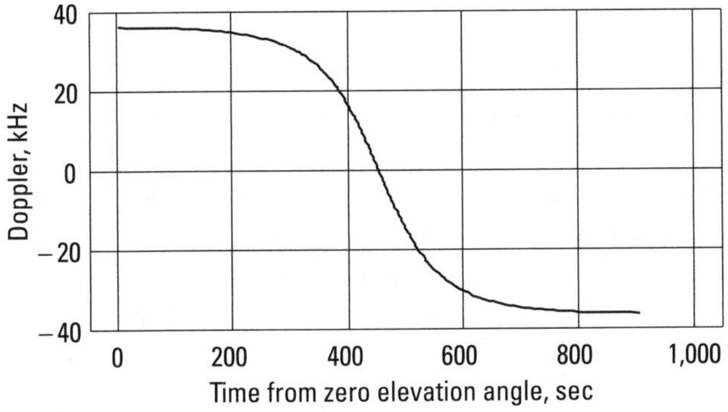

Figure 5.2 Doppler frequency shift at 1,625 MHz for a LEO satellite.

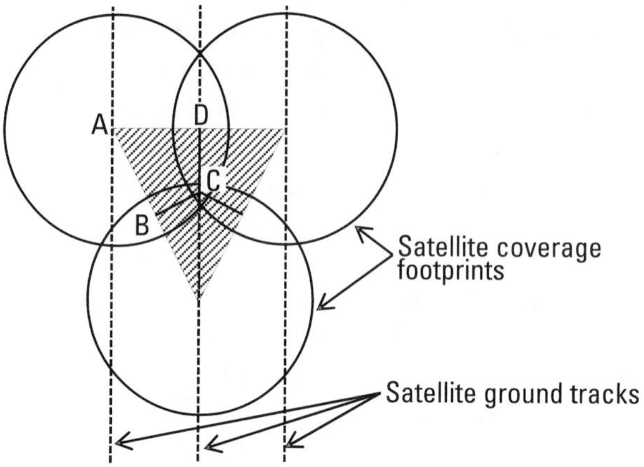

Figure 5.3 Coverage from a LEO satellite system.

limits of the footprint than at the center, so lower elevation angles are more probable than higher ones. *Coverage from satellites is therefore typically from nearly the lowest design elevation angle.*

For the LEO satellite parameters shown in Table 5.1, the minimum elevation angle is essentially along the line CD in Figure 5.3 and equals about $E_{min} = 4^o$ at the equator. The PDF is approximately

$$p(E) = \frac{\phi(E)}{\int_{E_{\min}}^{\pi/2} \phi(\alpha) d\alpha} \quad (5.10)$$

where $\phi(E)$ is from (5.5). The corresponding cumulative distribution function (CDF) is

$$P(E) = \int_{E_{\min}}^{E} p(\alpha) d\alpha \quad (5.11)$$

Figure 5.4 shows the PDF and CDF calculated for a latitude of 45°. The median elevation angle at that latitude is 31°, and the minimum elevation angle is 13°.

Figure 5.5 shows the median and minimum elevation angles as a functions of latitude calculated for a system of LEO satellites with orbits of the type shown in Table 5.1 and arranged in a triangular fashion shown in Figure 5.3. At the higher latitudes, many satellites begin to converge, and the curves of Figure 5.5 become irrelevant. The point of the figure is that for personal communications systems, typical coverage from the satellite is from the lower elevation angles. Coverage from nearly overhead is a rare event. Low angle coverage is also typical

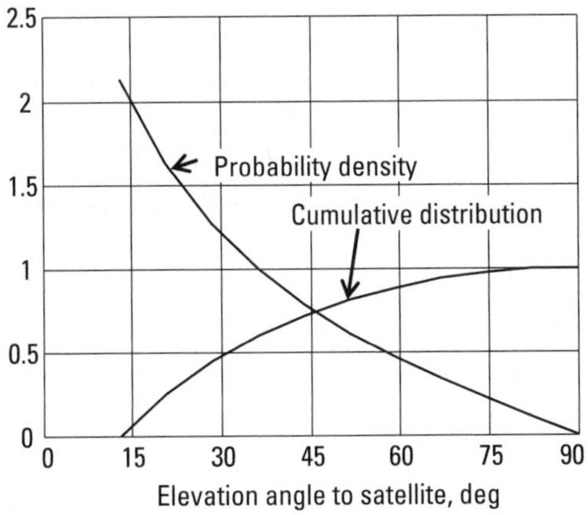

Figure 5.4 Distribution and density functions of elevation angle at 45° latitude.

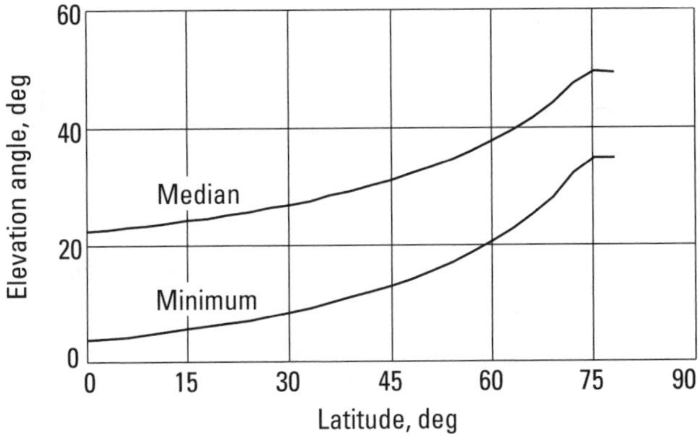

Figure 5.5 Median and minimum elevation angles as a function of latitude.

of GEO satellites, a fact that motivated the use of EEO satellites such as the Molniya series of satellites to deliver television signals to Siberia.

5.3.4 Link Characteristics from Earth-Orbiting Satellites

The satellite to Earth-bound subscriber path link is essentially a free space path with an excess path loss due to the multipath effects of ground and building reflections. By comparison, the more common satellite-to-Earth-station antenna link is typified by TVRO service and is well covered elsewhere [9, 13–15]. The TVRO link generally avoids excess scattering loss with very highly directional antennas in a line-of-sight path to the satellite and benefits from a generally lower system noise figure because the TVRO antenna field of view is limited to a sky with a noise temperature significantly lower than "room temperature." These lower noise temperatures result in smaller satellite transmitter power requirements and, in the case of sky radio surveys [16], higher sensitivities. The noise temperature associated with personal communications systems using satellites is not different from that of the terrestrially based systems because personal communications devices use antennas that have broad beamwidths.

In personal communications using LEO and MEO satellites, there is rarely a line-of-sight path between the satellite and the subscriber personal communications device. The statistical descriptions of signals subjected to multipath and shadowing, as well as the methods of system design using statistical signal descriptions, are covered in Chapter 8. Here, the specific statistical characteristics of signals from Earth-orbiting satellites are described. The satellite path link typically involves a diffraction from a rooftop followed by a single or multiple reflections from the ground and nearby buildings. The probability distribution function describing this behavior has yet to be determined, but measurements [17, 18] using GPS satellites have shown that path loss in excess of free space loss has a median value of between about –2 dB (that is, 2 dB of enhancement over free space loss) for the suburban residential environment, about 13 dB for the urban mid-rise environment, and 16 dB for the downtown urban "canyons." Furthermore, the statistics for the large-scale variation (after averaging multipath fades) are not the same as for the case of terrestrially based systems. After averaging the fast-fading components, the signal statistics for the measured [17] GPS satellite signal strengths that are weaker than the median tend to appear lognormal with a standard deviation in the 4–6-dB range. This case can be understood as a signal diffracted by a single rooftop and arriving at the personal communications device by several subsequently reflected paths. Signal strengths above the median level appear to have at least a partial line-of-sight path to the satellite and are characterized by a standard deviation on the order of 2 dB.

Table 5.2 shows some typical path link characteristics for the orbits listed and described in Table 5.1. LEO satellites are characterized by coverage from a median elevation angle of between 25° and 45°, depending on the latitude. Fairly high Doppler frequency shifts are typical, and signals are characterized by fading, which takes on different characteristics, depending whether a direct-line-of-sight path exists to the satellite. The propagation delay is relatively small; hence, LEO orbits are particularly attractive for near-real-time applications, such as two-way voice. Because the LEO satellite has a restricted view of the Earth (about 1,700 km from a 15° elevation angle), as many as 66 are needed to cover the globe simultaneously. The MEOs cover a radius of about 6,900 km on Earth from a 15° elevation angle, so fewer are needed. The GPS system requires 21 satellites to provide global coverage from up to four satellites simultaneously. Orbits at heights between about 1,500 and 20,000 km must

Table 5.2
Link Characteristics for Several Satellite Orbits

Parameter	LEO	MEO	GEO	EEO
Satellite	Iridium	GPS	INTELSAT	Molniya
Path loss 1.6 GHz:	−163 dB	−184 dB	−189 dB	−189 dB
2.5 GHz:	−167 dB	−188 dB	−193 dB	−193 dB
Doppler 1.6 GHz:	36 kHz	9.5 kHz	<1 kHz	<1.5 kHz
2.5 GHz:	55 kHz	13 kHz	<1.5 kHz	<2.4 kHz
Round-trip propagation delay at (maximum distance)	15.3 ms (2,200 km)	160 ms (24,000 km)	267 ms (40,000 km)	273 ms (41,000 km)
Number of satellites	66, plus 6 in-orbit spares	24, including 3 in-orbit spares	Multiple, but high latitudes are not covered	4 per hemisphere

contend with Van Allen radiation [9] due to both protons and electrons trapped in the Earth's magnetic field. Satellites designed for orbits in this range must use radiation-hardened electronics. The Doppler characteristics of MEO satellites are significantly less severe than those of LEO satellites, but propagation delays are much greater. A GEO system is characterized by satellites in equatorial orbits having orbital periods and directions equal to Earth's rotational period and direction.

Communications using GEO satellites are characterized by relatively high propagation delays, very high path loss, and a constant position relative to the observer. The constant geometry makes it difficult to exploit path link improvements using repeated transmissions as a strategy because the repeats will be highly correlated. GEO satellites can cover a large footprint on Earth (one-fifth Earth's surface), but there is no coverage at high and polar latitudes.

EEOs, as typified by the Molniya satellites, provide coverage at high latitudes from high elevation angles. There is a trade-off between LEO coverage with its additional shadowing losses, and EEO coverage from nearly overhead but at larger distances than LEOs. From Table 5.2, the basic path loss differential is seen to be about 26 dB, but shadowing losses at the typical low elevation angles of LEO coverage cut that difference to perhaps 10 dB for similar coverage. That, together with the requirement for fewer satellites (8 versus 66) to cover the globe, presents an interesting economic trade-off for satellite communications systems, particularly for data systems where propagation delay is not a significant factor.

System designs must have path link margins to provide a useful calling success rate in the shadow-fading and mutipath environment of a satellite system. Coverage into buildings needs additional margin. Since coverage is very much an economics issue in satellite systems, economics and customer expectations will determine the actual design coverage criteria. If the system has good coverage at an acceptable cost, then it can be commercially successful.

5.4 Polarization Effects in Signals from an Orbiting Satellite

Propagation between satellites and the ground are influenced by several factors, which depend on the polarization of the wave. The relative geometry is time varying, except perhaps for the GEO-TVRO type of operation, so a linear polarization is difficult to orient, especially for a personal communications device that may be operated in a random orientation. Furthermore, waves passing through the ionosphere are subjected to varying degrees of Faraday rotation of polarization. Both of these factors are effectively nullified by the use of circular polarization in personal communications systems. Some effects, however, especially diffraction by roof edges and the corners of building structures, as well as reflections from various planar surfaces, are polarization sensitive, and the orthogonal polarization components of circular polarization will be affected differently.

5.4.1 Effects of Reflections and Diffractions

The reflection coefficient for arbitrary incidence angle is generally different for orthogonal polarizations, as will be shown in Chapter 6. In fact, a complete specular reflection of a right-hand circular (RHC) polarized wave at normal incidence will result in a left-hand circular (LHC) wave, which, of course, is cross-polarized to the original wave. Diffractions are also polarization sensitive; for the case of a conducting wedge [19], the diffraction coefficient $G^{\pm}$ from the geometric theory of diffraction (GTD) is

$$G^{\pm} = \frac{e^{-j\pi/[4\sin(\pi^2/\phi_n)]}}{\frac{\phi_n}{\pi}\sqrt{2\pi k}}\left[\frac{1}{\cos[\pi^2/\phi_n] - \cos[\pi(\phi-\phi')/\phi_n]} \pm \frac{1}{\cos[\pi^2/\phi_n] - \cos[\pi(\phi+\phi')/\phi_n]}\right] \quad (5.12)$$

where k is the wave number, ϕ_n is the exterior wedge angle, and ϕ and ϕ' are the angles of incidence and diffraction, as picture in Figure 5.6. In (5.12), the upper

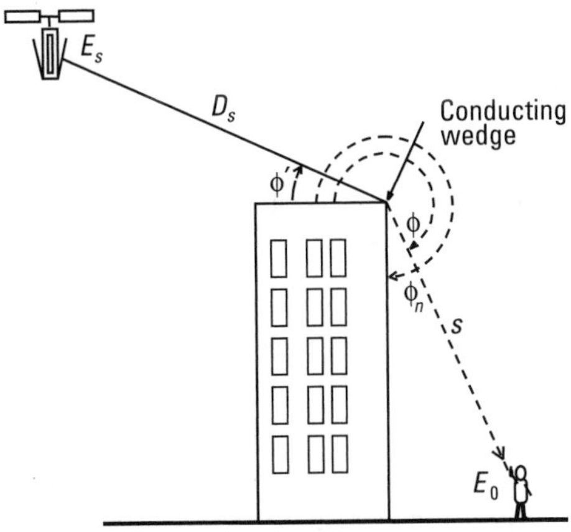

Figure 5.6 Geometry for diffraction by a conducting wedge.

(+) sign is used for polarization parallel to the plane of incidence, and the lower (–) sign is used for polarization perpendicular to the plane of incidence. The wedge vertex is perpendicular to the plane of incidence. The electric field at the field point that is not close to a shadow or a reflection boundary is then

$$\frac{E_o}{E_s} = \frac{e^{-jk(D_s+s)}}{D_s} \sqrt{\frac{D_s}{s(D_s+s)}} G^{\pm} \tag{5.13}$$

where E_s is the spherical wave source amplitude and E_o is the field point amplitude, both in volts per meter. In satellite communications applications, the distance D_s is given by (5.4), and distance s is from the wedge to the field point. Since G is polarization sensitive, the diffracted field E_o at the field point given by (5.11) is also polarization sensitive.

5.4.2 Faraday Rotation of Polarization

Linearly polarized systems are subject to the effects of polarization rotation when waves propagate through the Earth's ionosphere in the presence of the Earth's magnetic field. The effect is complex and involves the solution of Maxwell's equations for an anisotropic medium, which is described by a tensor. A wave incident on the ionosphere will split into two modes (called *ordinary* and *extraordinary*), and modes propagate differently through the ionosphere. Upon emerging from the ionosphere, the waves will recombine into a plane wave, but the polarization will usually have changed. The effect is named after Michael Faraday; more detailed analysis is available in [20, 21] and will not be repeated here. The number of turns, however, for the special case of propagation along the Earth's magnetic field vector and with the electric field perpendicular to magnetic field vector is given here by

$$Turns = \frac{e_e^3 B_0}{2c\varepsilon_0 \left(m_e 2\pi f\right)^2} \frac{\int N(z)dz}{2\pi} \tag{5.14}$$

where B_0 is the Earth's magnetic field strength ($\approx 5 \times 10^{-5}$ tesla at the Earth's surface), e_e is the electron charge in coulomb, m_e is the electron mass in kilograms, ε_0 is the free space permittivity, f is frequency in hertz, c km/sec is the wave velocity in free space, and $N(z)$ is the free-electron-density profile per cubic meter along the propagation path. Most Faraday rotation occurs in the 90–1,000-km height range. From (5.14), it can be seen that the effect is inversely proportional to the square of frequency, hence is most noticeable at lower frequencies.

A recent measurement [22] of field strength at VHF from a space shuttle orbiting at a height of 300 km presented an opportunity to observe the phenomenon of Faraday rotation of polarization in propagation through the Earth's ionosphere. The measurement involved comparing the gain characteristics of two different VHF antennas aboard the orbiting vehicle Columbia during mission STS-55. Signal strengths were sampled every half-second for one second in one linear polarization followed by one second in the orthogonal polarization for the duration of an orbital pass using two 13-dBi-gain Yagi antennas configured as pictured in Figure 5.7. The calibrated signal strengths were recorded and subsequently analyzed to report antenna gains [22].

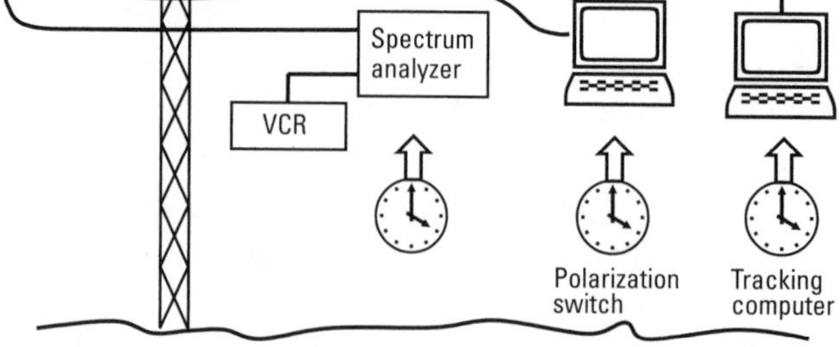

Figure 5.7 Configuration for signal strength test.

Since linear polarization was transmitted and two orthogonal polarizations were measured, the angle of the polarization vector could be computed from the measured data. Figure 5.8 presents the magnitude of the measured polarization angle for orbit number 61 of the test and compares the result with a calculation based on (5.14) using an electron-density profile in electrons per cubic meter coarsely approximated by

$$N(h) = \left[\frac{h}{320}\right]^2 4 \times 10^{12} \tag{5.15}$$

where h is height in kilometers. Only a change in the polarization angle can be detected, not the total number of turns. The maximum elevation angle for the orbit 61 pass was 12°, so the geometry did not provide for a great change in the ionospheric path. Figure 5.9 extends the comparison to the next orbit, 62, where the maximum elevation angle was 60°, using the same electron density. The change in the ionospheric path length was much greater on this orbital path, and a correspondingly greater polarization rotation can be seen in Figure 5.9 compared with Figure 5.8. The measurement and the approximate calculated model substantiate the necessity of using circular polarization in satellite communications applications involving relative motion between the space and ground segments.

5.5 Summary

Orbital mechanics relative to radio wave propagation from Earth-orbiting satellites was presented. Several orbits that are interesting for personal communications applications were explored. The LEO had characteristics most suitable for near-real-time communications, like two-way voice, because the low Earth

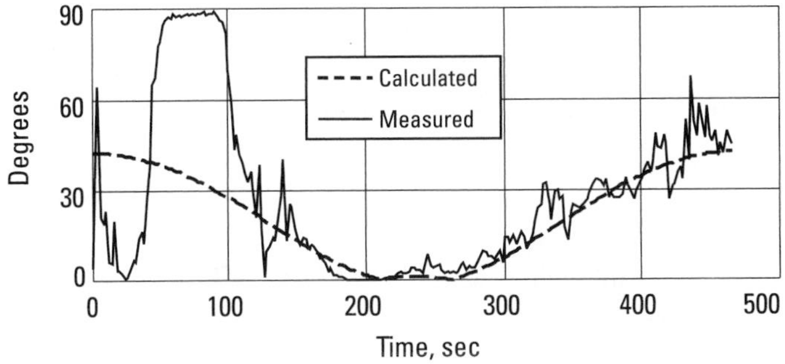

Figure 5.8 Faraday rotated polarization angle for measurements during orbit 61.

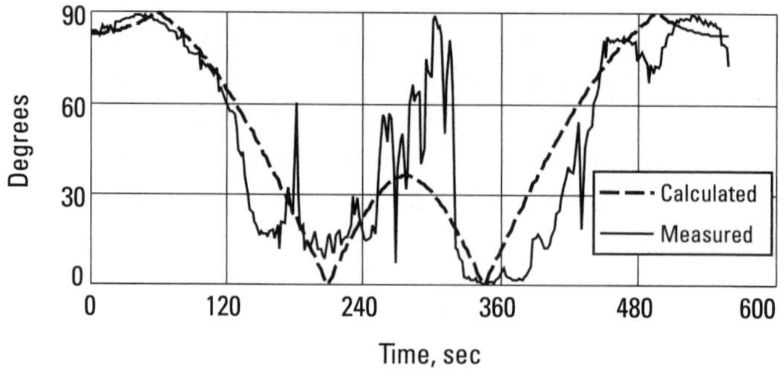

Figure 5.9 Faraday rotated polarization angle for measurements during orbit 62.

orbits result in the smallest path delays, and the lower path attenuation tends to result in lower subscriber costs for a given path link margin. The MEO is similar to the LEO but may need to contend with Van Allen radiation, depending on the exact altitude. The GEO has the unique property of presenting a fixed geometry between the space and ground segments, but this property may be of limited utility in personal communications systems. The EEO presents an interesting design trade-off problem because the orbit can be engineered to present an apogee at a high elevation angle at high latitudes on Earth. The high altitude of the EEO at apogee increases path loss, but the high elevation angle can mean lower diffraction and shadow losses compared with the GEO case.

Polarization effects include diffraction and reflection coefficients that are polarization sensitive, as well as Faraday polarization rotation, which is a factor at VHF. Personal communications using satellites differs significantly from fixed-point satellite services, such as TVRO service. The fixed services usually rely on a line-of-sight path to a GEO satellite and can control the cost of RF from the satellite by using high-gain antennas. Personal communications services will be used by subscribers who have small handheld or body-worn devices with broadbeam antennas that most often are not in a line-of-sight path to the satellite. Furthermore, additional signal margin is desired for coverage into buildings. The engineering problem is then one of economic optimization, given the enormous costs of the space segment.

Problems

5.1 Identify the following as either LEO, MEO, geosynchronous, or geostationary:

Orbital Period (sidereal days)	Eccentricity	Orbital Inclination (°)
1.0	0.00001	0
1.0	0.70	63
0.5	0.00001	56
0.07	0.00001	0
0.07	0.00001	86

5.2 Calculate the circular orbital radius and orbital height for a satellite orbiting the Earth at the equator with a period of 6,030 sec.

5.3 Find the orbital radius, orbital height, and inclination for a geostationary satellite.

5.4 Show that the sun is 333,432 Earth masses, given that the sun is 1.496×10^8 km from Earth and Earth's orbital period is 365 days.

5.5 The Earth's moon is 384,400 km away and has 1/83.6 of Earth's mass. Show that the lunar orbital period is 27.3 days.

5.6 Approximately how many times a day is a satellite located at high elevation angles from a point near the equator if the satellite orbital height is 305 km, at an inclination of (a) 3°, and (b) 60°?

5.7 The space shuttle orbits in a 28.3° inclined orbit at a height of 305 km. Find (a) the area of the zero elevation angle footprint on Earth, and (b) given that an observer is within that footprint, find the probability that the elevation angle is greater than 60°.

Ans:

(a) Insert (5.4) with $E = 0$ into (5.5) then, $Area_1 = (\phi R_e)^2 \pi$

(b) Find $Area_2$ for $Elev = 60°$; $Prob = Area_2 / Area_1$.

5.8 A satellite transmits 20W to an Earth station 40,000 km away. Find the received power when the transmitter antenna gain and the receiver effective aperture are (a) isotropic and 0.0026 m^2, (b) 30 dBi and 0.0026 m^2, (c) isotropic and 1 m^2, and (d) 30 dBi and 1 m^2.

5.9 A LEO satellite transmits 1W at 1.6 GHz to an Earth station 2,000 km away. Find (a) the power received by a unity-gain antenna, and (b) assuming a required carrier-to-noise ratio of 12 dB at the antenna terminals, find the excess link margin.

5.10 In a system of LEO satellites, each carries a total RF transmitter capability of 450W, orbits the Earth in a circular orbit at 5,000 km, and has high inclination with respect to the equator. Sequential scanning of 48 subbeams from the satellite is used to cover the footprint. Voice subscribers access the satellite with an average of 0.25W of RF power. Find (a) the number of satellites to essentially cover the globe from elevation angles of at least 15°, and (b) the maximum number of simultaneous users who can be served.

5.11 A LEO satellite transmits an average RF power of 400W at 60% efficiency and uses 15% efficient solar cells. Assuming total dc-to-dc conversion efficiency of 90% and that 3.85×10^{20} MW are radiated isotropically from the 1.5×10^{8}–km distant sun, find the minimum solar cell area. Comment on the requirement for, and capacity of, storage batteries.

Ans:

$2 \times 400/0.6/0.15/0.9 = 9{,}877$W needed while in the sun. Solar constant is 3.85×10^{20} MW/$(4\pi[1.5 \times 10^{11}\text{ m}]^2) = 1{,}362$ W/m^2, so Area = $9{,}877/1{,}362 = 7.3$ m^2.

5.12 Calculate the figure of merit G/T dBi/K for a handheld terminal Earth station with a 73% efficient antenna and 1-dB noise figure operated in a suburban home.

Ans:

System noise temperature (4.19) is 75K + 290K = 365K, G/T = 10log(0.73/365) = –27 dB/K.

5.13 A satellite transmits 40 dBW EIRP at 10 GHz from 40,000 km. Find the power received by a 40-dBi gain antenna.

5.14 A TVRO Earth station uses a 10-ft. in diameter, 55% efficient dish operating at 3.9 GHz with a receiver system noise temperature of 75K. Find the figure of merit.

5.15 A 1.6-GHz satellite system delivers a field strength of 30 dBμV/m on Earth. Find the power receiver by a unity-gain antenna. Assuming a fully reciprocal 2,000-km path, find the satellite antenna gain if the Earth station transmits 1W.

5.16 The International Space Station orbiting at a height of 354 km and a 51.7° inclination transmits a 430-MHz signal. Find the Doppler frequency shift extremes observed on Earth.

5.17 A communications satellite orbits Earth in an 86° inclined orbit at a height of 780 km and operates at 1.62 GHz. Find the transmit and receive Doppler frequency shift extremes for an approaching and receding satellite.

5.18 The space shuttle orbiter *Atlantis* is to rendezvous with the International Space Station; it is in the same orbital plane at a height of 354 km but is 3,000 km distant. A ground observer, expecting a near overhead pass, wishes to monitor a 432-MHz communication between the two spacecrafts. Find the minimum receiver bandwidth needed.

5.19 A LEO satellite operating a 100-kHz bandwidth telemetry link at 3.95 GHz with 6-dBW EiRP, polarization and pointing error losses of 4.4 dB, and an Earth station figure of merit $G/T = 4$ dBi/K needs $C/N = 8$ dB. Find (a) the link margin at a range of 704 km, (b) the link margin if the system bandwidth drops to 10 kHz, and (c) the suitability of a 10-kHz bandwidth system for a personal communications system.

Ans:

$C/N = P_{TX} + G - F - L_{total} + G/T_{sys} - 10\log(k_b) - 10\log(B)$, $M = C/N - C_r/N$

(a) $M = 6 - 161.3 - 4.4 + 4 + 228.6 - 50 - 8 = 14.9$ dB

(b) $M = 14.9 + 10\log(100/10) = 24.9$ dB

(c) PCS implies unity-gain antenna (90% efficient at best) operating with a 2-dB noise figure at room temperature, so $G/T = -25$ dB/K. Hence, net margin is $24.9 + (-25 - 4) = -4.1$ dB. The system has insufficient margin.

5.20 A GEO satellite is operating a 36-MHz bandwidth link at 3.72 GHz with 36-dBW EiRP, polarization and pointing error and atmospheric losses of 4 dB, and an Earth station figure of merit $G/T = 25.2$ dBi/K. The satellite is 39,542 km away, and the system requires a 12 dB carrier-to-noise ratio. Find (a) the link margin, (b) the link margin if the system bandwidth drops to 100 kHz, and (c) the suitability of a 100-kHz bandwidth system for a personal communications system operating into a suburban home.

Ans:

$C/N = P_{TX} + G - F - L_{total} + G/T_{sys} - 10\log(k_b) - 10\log(B)$, $M = C/N - C_r/N$

(a) $M = 36 - 4 - 195.8 + 25.2 + 228.6 - 75.56 - 12 = 2.44$ dB

(b) 10log(36,000,000/100,000) +2.44 = 28 dB

(c) PCS: $G/T = -25$ dB/K for a 2-dB noise figure for a net margin of 3 dB to include building-penetration losses and multipath effects. The system appears unsuitable for a 100-kHz bandwidth PCS link but may be marginally suitable for rural mobile applications.

5.21 Estimate the per-minute cost of a satellite voice telephone system where the satellite is capable of sustaining 250W RF power and has a $10 billion present-value ten-year cost. Assume that 0.25W of average power is required per voice user for suitable link margin and that sufficient spectrum is available.

Ans:

250W RF can support 1,000 simultaneous users. At least 1,060 physical channels are needed for a 5% grade of service. Assume 10 busy hours per day. So,

Cost/user/min = $10 billion/(60 minutes/hour × 10 hour/day × 31 day/month × 120 month × 1,000 users) = $4.48 per minute

Assume that the system can be reused four times around the world and that it is operating at 50% capacity. In that case,

Cost/user/minute = $4.48/4/0.5 = $2.24 per minute

The is the cost to the satellite owner/service provider. There are additional costs in securing the land-line portions of a satellite call. The price to the subscriber is cost plus the service provider's profits.

5.22 Estimate the monthly cost of satellite paging. Paging consists of four messages per day in the busy hour, and each paging message is equivalent to 2 sec of voice time at 10 dB more link margin than voice calls. The satellite is capable of sustaining 250W RF power, which it shares between voice and paging, and has a $10 billion present-value 10-year cost. Assume that 0.25W average power is required per voice user for suitable voice link margin and that sufficient spectrum is available.

Ans:

Link margin here is described in terms of the voice users. 250W RF can support 1,000 simultaneous voice users. Assume 10 busy hours per day. So,

Cost/user/minute = $10 billion/(60 minutes/hour × 10 hour/day × 31 day/month × 120 month × 1,000 users) = $4.48 per minute for voice

Assume that the system can be reused four times around the world and that it is operating at 50% capacity. So,

Cost/user/minute = \$4.48/4/0.5 = \$2.24 per minute for voice

Paging requires 10 dB more equivalent satellite link resources, either in power or in data rate, so equivalent cost per minute is \$22.40. Average paging use per month is 2 sec × 4 messages/day × 31 days/60 seconds per minute = 4.13 equivalent minutes per month at \$22.40 per minute = \$92.58 per user per month. The cost is to the service provider for the use of satellite infrastructure.

References

[1] Clarke, Arthur C., "Extra-Terrestrial Relays," [reprint from *Wireless World*, 1945], *Microwave System News and Communications Technology*, August 1985, pp. 59–67.

[2] Bate, R. R., D. D. Mueller, and J. E. White, *Fundamentals of Astrodynamics*, New York: Dover Publications, 1971.

[3] Taff, L. J., *Celestial Mechanics: A Computational Guide for the Practitioner*, New York: John Wiley and Sons, 1985.

[4] *InstantTrack Satellite Tracking Software*, AMSAT, Washington, D.C., 1989.

[5] *MAC Doppler PRO*, AMSAT, Washington, D.C., 1989.

[6] *SatPC32*, AMSAT, Washington, D.C., 1994.

[7] *NOVA Satellite Tracking Program,* available at www.nlsa.com.

[8] *OrbiTrack Orbit Tracking Software*, BEK Developers, St. Petersburg, FL, 1992.

[9] Morgan, W. L., and G. D. Gordon, *Communications Satellite Handbook*, New York: John Wiley and Sons, 1989.

[10] Nelson, R. A.. "Satellite Constellation Geometry," *Via Satellite*, Vol. 10, No. 3, March 1995, pp. 110–122.

[11] Vittore, V., "Will LEOs Get Off the Launching Pad?" *Telephone Engineer and Management* (supplement), November 1993, pp. 28–29.

[12] Frieden, R., "Satellite-Based Personal Communication Services," *Telecommunications*, December 1993, pp. 25–28.

[13] Martin, J., *Communications Satellite Systems*, Englewood Cliffs, NJ: Prentice Hall, 1978.

[14] Elbert, B. R., *Introduction to Satellite Communication*, Norwood, MA: Artech House, 1987.

[15] Stutzman, W. L., "Prolog to the Special Section on Propagation Effects on Satellite Communication Links," *Proc. of the IEEE*, Vol. 81, No. 6, June 1993, pp. 850–855.

[16] "Project Cyclops," *NASA Report CR 114445*, NASA/Ames Research Center, Moffett Field, CA, July 1973.

[17] Davidson, A., "Land Mobile Radio Propagation to Satellites," *Proceedings of the 1991 Antenna Applications Symposium*, University of Illinois, Allerton Park, IL, September 25–27, 1991.

[18] Hess, G. C., "Land-Mobile Satellite Excess Path Loss Measurements," *IEEE Transactions on Vehicular Technology*, Vol. VT-29, No. 2, May 1980, pp. 290–297.

[19] Luebbers, R. J., "Finite Conductivity Uniform GTD versus Knife Edge Diffraction in Prediction of Propagation Path Loss," *IEEE Transactions on Antennas and Propagation*, Vol. AP-32, No. 1, January 1984, pp. 70–76.

[20] Collin, R. E., *Antennas and Radiowave Propagation*, New York: McGraw-Hill, 1985.

[21] Jordan, E. C., and K. G. Balmain, *Electromagnetic Waves and Radiating Systems,* 2nd ed., Englewood Cliffs, NJ: Prentice Hall, 1968.

[22] Siwiak, K., "Hams Test Antennas aboard Space Shuttle Columbia," *QST*, October 1993, pp. 53–55.

6

Radiowave Propagation over a Smooth Earth

6.1 Introduction

The two-ray model for radiowave propagation is introduced with the intention of developing a foundation for the analysis of open-field antenna test sites that are suitable for measuring the field-strength sensitivities of personal communications receivers. The two-path model is also used to illustrate the unique propagation behavior of UWB pulses. Modern personal communications devices are increasingly being designed to the limits of system specifications. Consequently, there is a demand for radio test engineers to measure field-strength performance to increasingly higher accuracies. The radiowave propagation models that are developed and studied in detail in this chapter are applicable to the open-field antenna testing ranges that are still critically important in personal communications device development. The test range analysis explores the effects of range geometry and fixed antenna arrangements. We also analyze the calibration of field strengths at open-field test ranges and the effects on accuracy of using different kinds calibrated gain standards. Finally, the effects of the ground on the fields of horizontally and vertically polarized yagis are studied.

6.2 A Two-Ray Propagation Model for Harmonic Waves

Radio propagation for harmonic waves between two points near the ground involves an expanding spherical wave propagating from the source antenna to the target antenna. Propagation of short-duration UWB pulses is considered in a later section. Because of the air-ground boundary, ground currents are induced, which then reradiate and combine as complex vectors with the source spherical wave. This is shown pictorially in Figure 6.1(a). This general case is handled approximately with great simplification using plane wave theory, as shown in Figure 6.1(b). There, a plane wave is shown traveling between the source and target antennas. A second plane wave reaches the target antenna after a single specular reflection from the assumed perfectly smooth ground. The simplified two-ray approach uses an expanding spherical wave, as in the Friis transmission formula encountered in Section 3.3, modified by plane wave reflection from a smooth ground. Additional terms are added, if required, for approximating the surface wave and higher-order phenomena.

6.2.1 Spherical Wave with Modifiers

For time-harmonic waves (sine waves), the approximation of Figure 6.1(b) is expressed mathematically, as in [1, 2], by the ratio P of received to transmitted power between unity-gain antennas:

$$P = \left| \frac{e^{-jkD}}{2kD} \left[F_d + F_r \left[\Gamma + (1 - \Gamma) A \right] e^{-j\phi} + \ldots \right] \right|^2 \tag{6.1}$$

where k is the wave number, and the leading phasor term is the basic free space expanding spherical wave traveling a distance D from the source to the target.

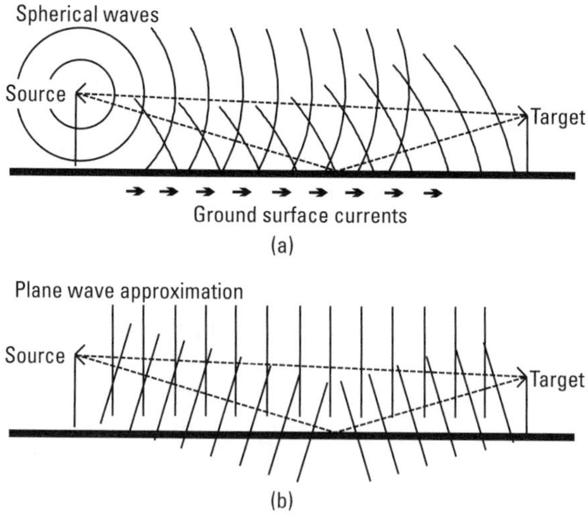

Figure 6.1 Propagation near a planar Earth by (a) an expanding spherical wave, and (b) by the plane waves.

The term F_d is the direct wave modified by the antenna pattern evaluated in the direction of the direct ray. The term F_r includes the antenna pattern evaluated in the reflected ray direction and may include an amplitude correction for the added distance traveled by the reflected ray compared to the direct ray. Γ is the specularly reflected plane wave ground reflection coefficient, while the term including factor A is the dominant surface wave contribution. The phasor represents the additional phase delay, ϕ, for the reflected ray path. The unspecified additional terms in (6.1) include induction fields and secondary effects of the ground and account for the remaining differences between this two-ray approximation and the complete electromagnetic problem pictured in Figure 6.1(a). Usually, F_d and F_r are taken to be unity, and the surface wave A is ignored.

All of the terms in the brackets are plane wave approximations to a complex spherical wave phenomenon. The third and following term can be expressed otherwise [3] by considering the reflection of spherical waves from a boundary layer. The general problem gives rise to the existence of the Zenneck surface wave for vertically polarized antennas, as well as to a lateral wave, which is associated with propagation along the air-ground boundary.

Written as (6.1), the reflected and surface wave terms are modifiers to the basic free space propagation law, which was encountered in Section 3.3 as the Friis transmission formula. The terms in (6.1) can be rewritten using the geometry pictured in Figure 6.2.

The incidence angle on the ground, with reference to Figure 6.2, is given by

$$\theta = a\tan\left(\frac{H_1 + H_2}{d}\right) \tag{6.2}$$

The direct path and reflected path lengths D and R in terms of the antenna heights are

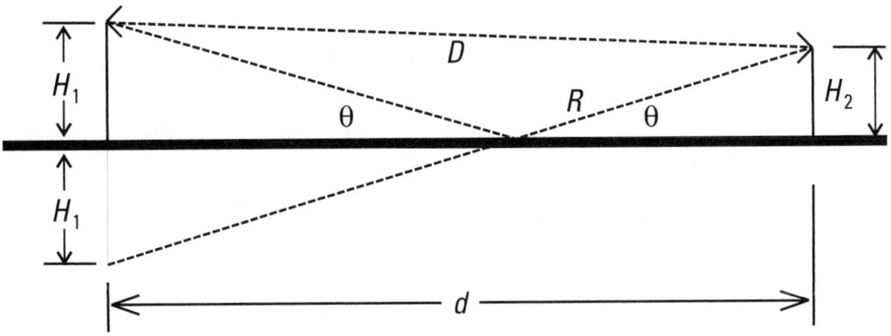

Figure 6.2 Geometry for the two-ray propagation model.

$$D = \sqrt{d^2 + (H_1 - H_2)^2} \tag{6.3}$$

and

$$R = \sqrt{d^2 + (H_1 + H_2)^2} \tag{6.4}$$

from which the path phase difference is computed as

$$\phi = k(R - D) \tag{6.5}$$

A similar expression will be written later in terms of time delay for UWB short pulses.

6.2.2 Plane Wave Reflection Coefficients

The ground reflection coefficients for vertical and horizontal polarizations are approximated by the plane wave reflection coefficients [4, 5] expressed here as

$$\Gamma(\theta) = \frac{\sin(\theta) - X}{\sin(\theta) + X} \tag{6.6}$$

where, for vertical polarization, X is

$$X_v = \frac{\sqrt{\varepsilon_g - \cos^2(\theta)}}{\varepsilon_g} \tag{6.7}$$

and, for horizontal polarization, X is

$$X_h = \sqrt{\varepsilon_g - \cos^2(\theta)} \tag{6.8}$$

The incidence angle θ is measured with respect to the ground as pictured in Figure 6.2.

The forms of (6.7) and (6.8) were chosen to make (6.1) and (6.6) independent of polarization. *The reflection coefficient given by (6.6) approaches –1 as the angle of incidence approaches 0 for any finite value of ground conductivity, even for solid copper.* The Brewster angle for which the vertical polarization reflection coefficient is a minimum occurs when $\sin(\theta) = |X_v|$. This does not occur for horizontal polarization because X_h is always greater than unity.

The dielectric constant of the ground is

$$\varepsilon_g = \varepsilon_1 - j\frac{\sigma_1}{\varepsilon_0 2\pi f} \tag{6.9}$$

in terms of frequency f in hertz, the ground relative dielectric constant ε_1, and the ground conductivity σ_1.

The ground dielectric constant is in the vicinity of $\varepsilon_1 = 7$ but may range from about 3 for loose gravel to nearly 25 for moist earth. The conductivity is typically $\sigma_1 = 0.005$ S/m and may range from 0.0001 to as much as 0.03 S/m for moist earth. The effect of the ground parameters on propagation will be detailed in Section 6.3, where the open-field test site is studied in detail. Expression (6.6) can be used in (6.1) to obtain a two-ray propagation model, where the ground is a homogeneous half-space with parameters ε_1 and σ_1.

6.2.3 Two-Layer Ground Model

Although a homogenous half-plane model of the Earth is usually sufficient, especially for shallow angles of incidence of the ground, a two-layer model can be used when such detail is important. The two-layer ground is modeled by first defining the angle of incidence θ_s on the buried layer in terms of the surface-layer incidence angle θ and the surface-layer parameters:

$$\theta_s = a\cos\left(\frac{\cos(\theta)}{\sqrt{\varepsilon_g}}\right) \tag{6.10}$$

The reflection coefficient, G, at the air-ground interface is

$$G = \frac{\Gamma_1(\theta) + \Gamma_2(\theta_s)e^{-2jk\sqrt{\varepsilon_g}\sin(\theta_s)t}}{1 + \Gamma_1(\theta)\Gamma_2(\theta_s)e^{-2jk\sqrt{\varepsilon_g}\sin(\theta_s)t}} \tag{6.11}$$

where k is the free space wave number. The thickness of the surface ground layer, as shown in Figure 6.3, is t. The reflection coefficient Γ_1 is from (6.6), and θ_s is from (6.10). Γ_1 is calculated using the surface-layer ground parameters ε_1 and σ_1. The lower-layer reflection coefficient Γ_2 is computed using the ground parameter ε_L defined by

$$\varepsilon_L = \frac{\varepsilon_2 - j\dfrac{\sigma_2}{\varepsilon_0 2\pi f}}{\varepsilon_g} \tag{6.12}$$

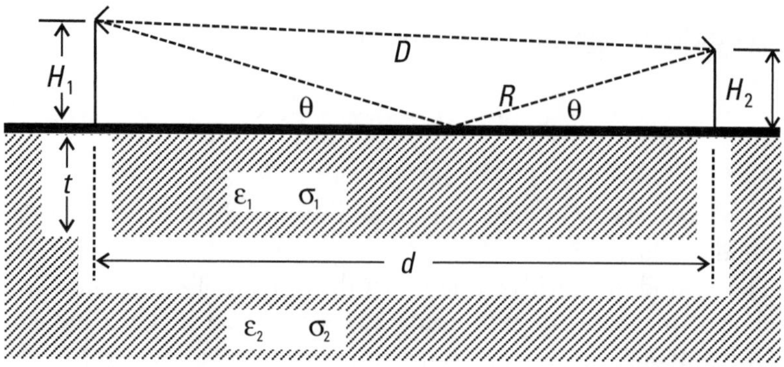

Figure 6.3 Propagation model with a two-layer ground.

where the lower-layer ground parameters are ε_2 and σ_2, and where ε_g is given by (6.9).

The two-layer ground reflection coefficient, *G*, from (6.11) can be used in place of Γ in (6.1) to obtain a two-ray propagation model, where the ground is modeled by two layers. The upper layer has thickness *t* and ground parameters ε_1 and σ_1, while the lower, buried layer has ground parameters ε_2 and σ_2 that extend down indefinitely.

Expression (6.11) is derived using an infinite set of reflections and rereflections between the air-ground boundary and the interlayer boundary. This is, of course, not physically possible for short distances *d*, and (6.11) becomes approximate for the spherical wave case, with the restriction that the upper layer must be thin relative to the range distance. L. M. Brekhovskikh considered the more complete, and much more complex, treatment of the spherical wave reflected from a boundary [3].

6.2.4 Surface Wave Factor

The surface wave factor *A* in (6.1) is a function of frequency, ground parameters, polarization, and incidence angle. It is always smaller than unity and decreases with increasing distance and frequency, as seen by the following expression:

$$A = \frac{-1}{1 + jkd\,(X + \sin(\theta))^2} \tag{6.13}$$

This approximation [1, 2] to *A* is accurate as long as $|A| < 0.1$. Figure 6.4 shows the magnitude of the effect of the surface wave contribution for both vertically and horizontally polarized path loss by comparing the ratio expressed in

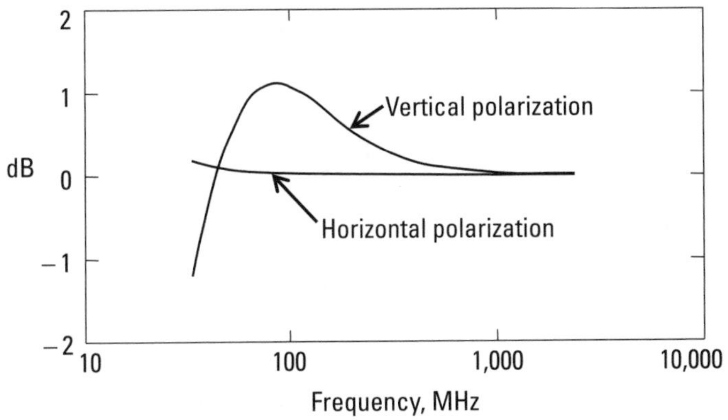

Figure 6.4 Error in signal strength due to omitting the surface wave contribution.

decibels of path attenuation calculated from (6.1) without the surface wave term A present relative to the path attenuation including the surface wave. Two cases are shown as a function of frequency, both for 1,000m distance with antennas at each end of the link at a height of 3m above ground. In Figure 6.4, we see that the surface wave term contribution to the path loss generally decreases with frequency above 100 MHz. We are justified in ignoring the surface wave factor for horizontal polarization at VHF and higher frequencies but must consider the effect with vertical polarization when accuracies on the order of 1 dB are important.

[6-2a.mcd] Using (6.1) for path attenuation with (6.13) to model the surface wave, and using (6.2) to (6.9) to model a homogenous ground, calculate the magnitude of the surface wave contribution for a 30m-long antenna test range at frequencies from 150 to 1,500 MHz.

6.2.5 Grazing Angle of Incidence

When the reflected ray grazes the Earth, that is, the incidence angle θ is small, the reflection coefficient may be replaced by –1, and terms of higher order than the surface wave can be ignored. The exponential term in ϕ of (6.1) can be written in terms of the Euler identity (i.e., $\exp(-j\phi) = \cos\phi - j\sin\phi$), and the transmission power ratio between isotropic antennas separated by dm, including the free space and surface wave terms, may be written as

$$P_{2-ray} = \left| \frac{1}{2kd}\left[1 - \cos(\phi) + j\sin(\phi) + 2A\right] \right|^2 \tag{6.14}$$

where k is the wave number. The phase difference ϕ is approximated as

$$\phi = k(R - D) = k\frac{2H_1H_2}{d}$$

For the conditions under which $\Gamma = -1$, that is, where $\theta \approx 0$ and the distance d is large, the surface wave term from (6.13) becomes $A = j/(kdX^2)$, and the two-ray path propagation function can then be written approximately as

$$P_{2-ray} = \left| \frac{\sin\left[k\frac{2H_1H_2}{d}\right]}{2kd} + \frac{1}{(kXD)^2} \right|^2 \quad (6.15)$$

Ignoring for the moment the surface wave contribution, this simplified form of the two-ray propagation formula is interesting because it readily reveals the basic form of the field intensity as a function of height. Evidently, from the sinusoid in (6.15), the field intensity is a standing wave in the vertical dimension. The first maximum occurs when the argument of the sine in (6.15) equals $\pi/2$, or when $H_1H_2 = d\lambda/8$. This expression is often further approximated at distances large enough that the sine in (6.15) can be replaced by its argument, giving

$$P_{2-ray} = \left| \frac{H_1H_2}{d^2} + \frac{1/(kX)^2}{d^2} \right|^2 \quad (6.16)$$

The surface wave contribution, with X given by (6.7) or (6.8), appears like a "minimum height" [1] and is not important for heights greater than a few wavelengths. The resulting field intensity according to (6.16) depends on the square of the antenna heights and on the inverse fourth power of distance d for this approximation to the two-ray path propagation formula. This inverse fourth-power law, usually stated without the surface wave component, is often used as a basis for radiowave propagation models.

[6-2b.mcd] Use (6.1) then (6.16) to calculate the field strength as a function of height for a distance d = 100m, H_1 = 30m, at 2.5 GHz. Repeat the calculation for a frequency of 150 MHz.

6.3 An Open-Field Test Range Model

The two-ray path propagation model developed in the previous sections can be applied to model and analyze open-field antenna test ranges and receiver field-strength sensitivity test ranges. Starting with the form of (6.1) for the path attenuation and using the geometry of Figure 6.2, we can express the power attenuation on an open-field antenna test range by

$$P_R = \left| \frac{1}{2kd} \left[\left[\frac{d}{D} \right]^{n+1} + \left[\frac{d}{R} \right]^{n+1} \left(\Gamma e^{-j\phi} + [1-\Gamma] A e^{-j\phi} \right) \right] \right|^2 \qquad (6.17)$$

The leading term in (6.17) is recognized as the free space path loss between two points separated by distance d along the ground. The first term in the brackets is the direct ray modified to account for the proper distance D relative to d. The second term includes a similar amplitude correction for distance R, and the last term includes approximation (6.13) for the surface wave component. The exponents n approximate the antenna pattern in the ray direction with the nth power of the cosine of the angle with respect to the horizontal. The ground reflection coefficient Γ can be either the homogeneous half-space ground model given by (6.6) or the two-layer ground model expressed by (6.11). The ray path phase differential ϕ is from (6.5).

Often, directional antennas are used at one end of an antenna test range, here presumed to be at height H_1, as part of the fixed and permanent test range design. The radio device under test is placed at a separation d along the ground and at height H_2. The directional antenna is assumed to be parallel to the ground. We model the elevation plane beam characteristics of this fixed directional antenna and the antenna under test by the nth power of the cosine of the angle between the ray direction (D or R) and the ground. For example, the pattern for a dipole can be approximated by matching the nth power of $\cos(\theta)$ at θ equal to the dipole half-power half-beamwidth of 39°. Thus, for a single dipole, $n = 1.4$. This is a simple and expeditious manner of accounting for vertical plane beamwidth.

It is easy to show, by inspection of Figure 6.2, that the cosine of the required angle is equal to (d/D) for the direct ray and to (d/R) for the reflected ray. The unity term in the exponent $(n + 1)$ of (6.17) accounts for the spherical wave expansion amplitude term relative to the distance d, while n is chosen to match the combined pattern characteristics of the directional fixed antenna and the antenna under test. Table 6.1 lists some useful values of n, which may be encountered during antenna testing. With a dipole at each end of the test range, the exponents add up to 2.8. Similarly, a corner reflector can be modeled with $n = 2.2$, which in combination with the dipole under test results in $n = 3.6$, shown in Table 6.1.

The case of "omnidirectional" antennas refers to omnidirectionality in the vertical plane, as may be encountered between two horizontally polarized dipoles. A common testing configuration involves a corner reflector antenna as

Table 6.1
Parameter *n* for Various Fixed and Test Antenna Combinations

Fixed Antenna (3-dB beamwidth)	**Antenna Under Test (3-dB beamwidth)**	***n***
Omnidirectional, (n/a)	Omnidirectional, (n/a)	0
Vertical dipole, (78°)	Vertical dipole, (78°)	2.8
Corner reflector, (62°)	Vertical dipole, (78°)	3.6
Log periodic, (53°)	Vertical dipole, (78°)	4.5
Corner reflector, (62°)	17-dBi horn, (29°)	13.2

the fixed antenna and a dipole as a range calibration standard at the test location. Sometimes a standard-gain horn antenna is used during range field-strength calibration. As will be shown in Section 6.3.5, to minimize calibration errors, the beamwidth of the calibration antenna at the test position should match the beamwidth of the antenna under test, particularly if the distance between the specular point and the test location is short.

6.3.1 A Two-Ray Model of an Open-Field Test Site

By way of a practical example, a test range suitable for use as a receiver field-strength sensitivity test range will be described and modeled analytically here. This range is studied experimentally in Chapters 9 and 10, when receiver field-strength sensitivity measurements are introduced. For this example, the range under analysis has a fixed corner reflector antenna at height H_1 = 3.2m. Referring to Figure 6.2, the range is d = 45.5m long between the fixed antenna and the test location of the test point. A drawing of the type of range under consideration here and used as a receiver field-strength sensitivity test site is shown in Figure 9.3. We are interested in studying (1) the field strength at the test location as a function of ground parameters, and (2) the field strength versus height at the test location. We will also examine the field-strength calibration problem.

An antenna test site may be used with the fixed antenna either transmitting or receiving. By the theorem of reciprocity [4, 5], the characteristics of the range are identical independently of which antenna transmits and which antenna receives the signals. Here, we use the fixed antenna as a transmitter. For a numerical example, we assume a generator connected to a length of coaxial cable with a fixed attenuator in line and connected to a transmitting corner reflector antenna. The effective radiated transmitted power relative to an isotropic level can be modeled for this configuration as

$$P_{eirp} = P_{tx} + G - L - 0.06\sqrt{f} \qquad (6.18)$$

P_{tx} represents the transmitter or signal generator power, here in decibels relative to 1 mW (dBm), and L is a fixed attenuator in decibels. G represents the gain relative to an isotropic reference of the corner reflector antenna. The last term is the coaxial cable loss, of the form of (3.16), as a function of frequency f in megahertz, for the particular example presented here. The power received by a unity-gain antenna in the test location is found from expressions (6.17) and (6.18): P_{eirp} + 10log(P_R) dBm. Since we are interested in electric field strength, the received power must now be related to electric field strength assuming an effective aperture, A_e, of a unity-gain (isotropic) antenna as described in Section 3.3.1. For an effective received power PW, the field strength in volts per meter at the unity-gain aperture is

$$E = \frac{\sqrt{P\eta_0 4\pi}}{\lambda} \qquad (6.19)$$

When the power is in decibels relative to a milliwatt, and the electric field is $E_{dB\mu V/m}$ in decibels relative to one microvolt per meter, field strength and receiver power are related by

$$E_{dB\mu V/m} = P_{eirp} + 10\log(P_R) + 77.2159 + 20\log(f) \qquad (6.20)$$

Received power P_R is from (6.17) and frequency f is in megahertz. Expression (6.20) in combination with (6.18) allows us to calculate the electric field strength directly in terms of our hypothetical transmitter power. We are now positioned to study the hypothetical open-field range as a function of ground parameters, field profile, and calibration methods. This hypothetical study is based on a practical case and will be related to experimental measurements in Chapter 9.

6.3.2 Field Strength Versus Ground Parameters

We can apply the path attenuation model (6.17) with the ground modeled by (6.6) and the antenna beamwidth parameter set to n = 3.6 to represent a corner reflector at the fixed end with a dipole in the test position. Figure 6.5 shows the path attenuation at 170 MHz as the ground dielectric constant is varied from 3 to 25. The convention used here is that increasingly larger negative decibel values represent increasingly greater path attenuation. Varying the dielectric constant from 3 to 25 results in only ±2 dB variation in path attenuation.

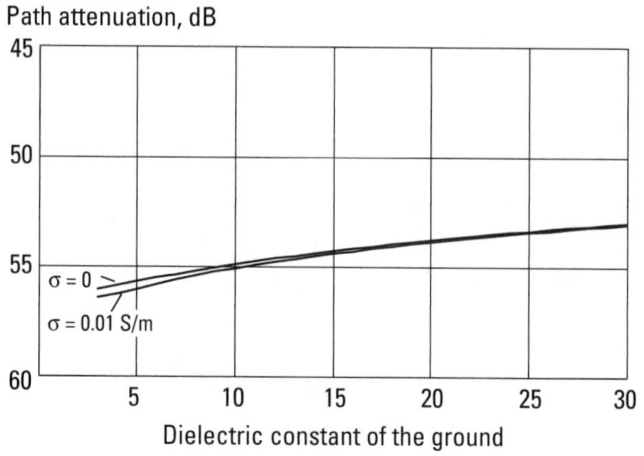

Figure 6.5 Test-site attenuation versus ground dielectric constant.

Figure 6.6 repeats this calculation for $\varepsilon = 3$ and $\varepsilon = 25$, now with ground conductivity varying form 0.001 to 100 S/m. Again, for the range of conductivities normally encountered in practice ($\sigma < 0.01$ S/m) the path attenuation variation remains small. Observe, however, in the range of conductivities $\sigma > 0.2$ that the path attenuation is a strong function of conductivity. On a test range complemented with a ground screen, the ground screen quality is an important contributor to accuracy and repeatability on such a range, perhaps more so than if an Earth ground were used.

6.3.3 Field-Strength Profile on a 45m Range

The vertical variation of the field at the test location can be studied by using the path attenuation expression (6.17), with (6.6) used to represent either a

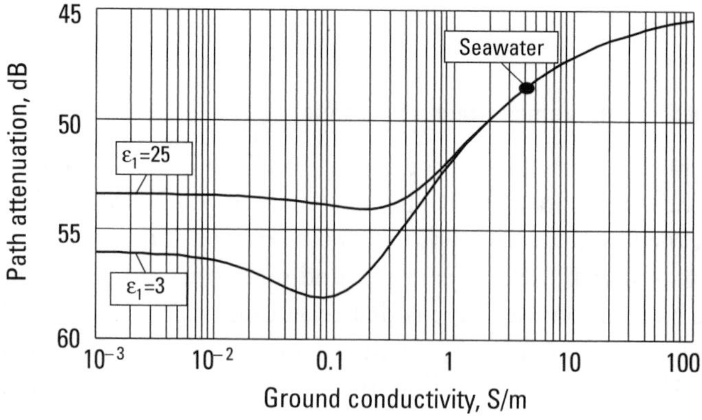

Figure 6.6 Test-site attenuation versus ground conductivity.

homogeneous Earth ground or a typical "ground plane" range often used for spurious emissions measurements. Figure 6.7 shows the field profile for our hypothetical 45.5m-long Earth ground range with the transmitting corner reflector fixed at a height of 3.2m. The fields at frequencies between 150 and 930 MHz are smoothly varying and well behaved over heights up to about 2m at the test position. This is an example of a well-designed receiver-sensitivity test range, which will be studied experimentally in Chapter 9.

Another test range that might be used in spurious emissions testing comprises a conductive ground screen ($\varepsilon = 1, \sigma = 20$ S/m) and a measuring distance of 10m. The vertical profile of the fields on this range with a fixed antenna height of 3.2m is shown in Figure 6.8. While suitable for finding peak emissions levels in electromagnetic interference (EMI) measurements, this range is completely inappropriate for measurements of body-worn radio devices because of the many nulls in the height range below 2m. The performance of this range could be improved for radio performance measurements by omitting the ground screen and lowering the fixed antenna height. Omitting the ground screen raises the nulls by a half-period, lowering the fixed antenna height stretched the vertical null-to-null spacing as might be expected from an inspection of (6.15).

6.3.4 Calibrating a Test Site

The range calibration factor is defined as the field strength measured at the test location resulting from a transmitter or signal generator power of 1 mW. The factor is site specific since the relationship between radiated power and signal generator power depends on the specific equipment utilized. Calculation of range performance requires a transmitting model—(6.18) in our specific example—and a generalized propagation model—(6.17). The field strength at the

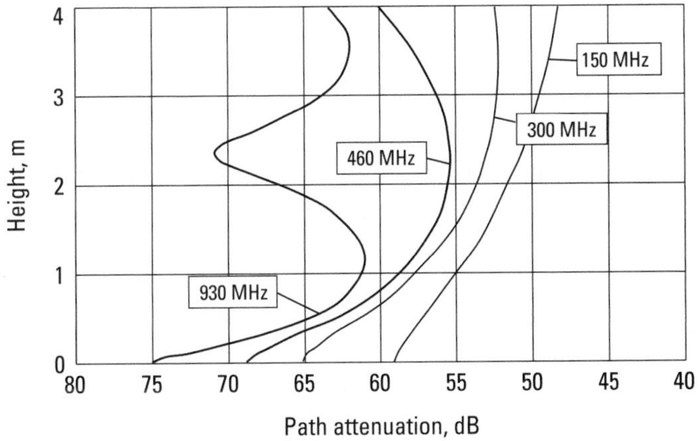

Figure 6.7 Calculated field profile at the test position on a 45.5m test range.

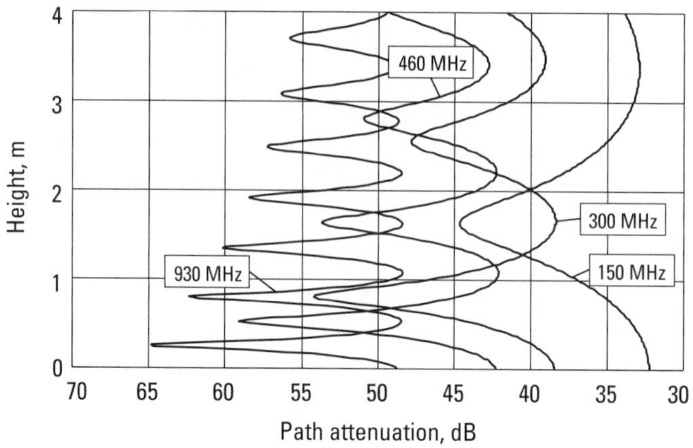

Figure 6.8 Field profile on a 10m test site having a conducting ground screen.

testing location is found once the ground parameters are specified. Expression (6.20), along with the appropriate ground model, can be used to predict the absolute field strength on an antenna measuring site. The ground parameters must be known in order to get very accurate results, but a reasonable guess (e.g., $5 < \varepsilon < 12$) will yield an accuracy within a few decibels, as suggested in Figures 6.5 and 6.6.

Measured calibration factors are available [6] for the range studied here, and if needed, ground parameters may be determined experimentally [7], using a standing wave procedure. A best-fit set of ground parameters for the range under study was derived by comparing calculations with measurements (see Tables 6.2 and 9.1).

Table 6.2 shows calculated calibration factors using (6.20) with (6.18) and (6.17), with $G = 10$ dB and $L = 6$ dB. Three different ground models are used. The two-layer model ground parameters and layer thickness were found by matching calculated and experimental results and minimizing the sum of the squares of the difference using the iterative Levenberg-Marquardt method [5, 6, 8, 9]. The third column in Table 6.2 represents the best-fit to experiment using a homogeneous ground. The fourth column reports the calculated results for a ground dielectric constant at the upper limit of our "reasonable guess" range. Measure values are shown in the last column, including standard deviations of multiple measurements.

The calculated and measured calibration field strengths differ by no more than 2.6 dB for the "worst" ground parameter guess of $\varepsilon = 12$. This is consistent with the behavior seen in Figures 6.5 and 6.6. In any case, calibration of the range and experimental determination of the calibration factor should be included in every measurement procedure. The experimental study of antenna test ranges in Chapter 9 shows that typical measurement accuracies of within 1

Table 6.2
Calculated and Measured Calibration Factors, Decibels Relative to 1 μV/m. *From* [10].

Frequency (MHz), h	Calculated: $\varepsilon_1 = 5.2$, $t_1 = 0.66$m, $\varepsilon_2 = 11.9$	Calculated: $\varepsilon_1 = 5$	Calculated: $\varepsilon_1 = 12$	Measured
150, $h = 1.0$m	66.4	67.8	69.1	66.5 ($\sigma_{cal} = 0.40$)
$h = 1.4$m	68.3	69.5	70.3	68.2
170, $h = 1.0$m	68.1	68.3	69.4	68.2 ($\sigma_{cal} = 0.42$)
$h = 1.4$m	70.1	70.1	70.7	69.9
280, $h = 1.0$m	70.9	70.7	71.0	71.0
$h = 1.4$m	73.1	72.8	72.7	73.5
470, $h = 1.0$m	73.1	73.7	73.4	73.4 ($\sigma_{cal} = 0.22$)
$h = 1.4$m	75.3	75.6	75.0	75.1
929, $h = 1.0$m	76.8	76.5	75.9	76.9 ($\sigma_{cal} = 0.35$)

dB are possible with careful experimental techniques, which include careful site calibration. Figure 6.9 shows this calculated calibration factor using (6.20) with (6.17) and (6.18) with the ground parameters of column 2 in Table 6.2 for the test site described in Section 6.3.3. The calibration factors for horizontal and vertical polarizations are different, and the ripple versus frequency is due to the layered ground.

6.3.5 Effect of the Calibration Gain Standard

Actual field calibration involves a gain standard that has a finite length and an aperture illumination. The aperture illumination is uniform with height for a vertically polarized horn antenna and a cosine for a dipole. If the field strength versus height is $E(h)$, and the calibration gain standard illumination is $g(h)$, the calibration figure is obtained by integrating over the height $2b$ of the calibration standard weighted by the aperture illumination:

$$E_{cal}(h) = \frac{\int_{-b}^{b} E(h+x)g(x)dx}{\int_{-b}^{b} g(x)dx} \tag{6.21}$$

The calibration factor $E_{cal}(h)$ obtained in this manner, which models practice, is slightly different from the point value of the field $E(h)$.

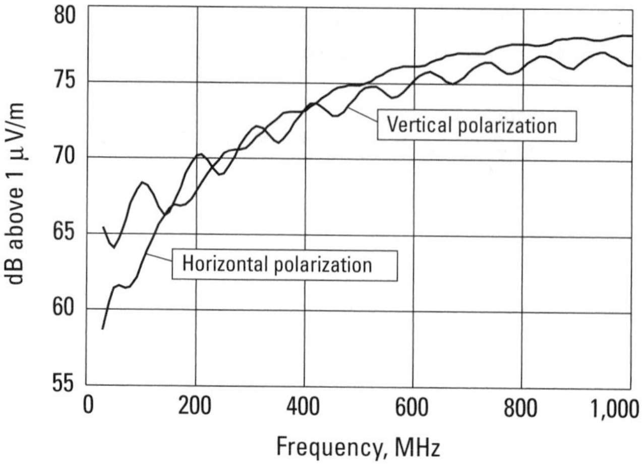

Figure 6.9 Calibration factors for horizontal and vertical polarizations.

From (6.21) applied to the open-field test range developed earlier, the calibration factor determined by using a horn with an 0.61m vertical aperture (n = 13.2), relative to that determined using a dipole (n = 3.6), is 0.44 dB. The horn calibration would report a stronger field by 0.44 dB than would be determined with a dipole calibration. This has the effect of making body-worn radio receivers measured on a horn-calibrated range appear nearly a half-decibel more sensitive than if the calibration were made with a standard dipole. At well-designed test sites, such as described above, measurement accuracies are in the range of a decibel, so a half-decibel is important. At an inappropriately designed range, the difference could be greater. The "correct" calibration has a gain standard with nearly the same vertical beamwidth as the radio antenna under test. For most body-worn radios, the appropriated gain standard is a vertical dipole.

[6-3.mcd] Expression (6.17) provides a method of including antenna elevation plane beamwidth in test range analysis. Use (6.17) to estimate the field profile and (6.21) to model the effects of a dipole calibration standard, and compare the results using a horn with a 0.61m vertical aperture. Use range length d = 45.5m, H_1 = 3.2m, and the nominal measurement height H_2 = 1m. The ground parameters are ε = 7 and σ = 0.005 S/m.

Calibrated gain is often stated at specific distances, as for example, in EMI/EMC (Electromagnetic Interference/Electromagnetic Compatibility) measurements, where the standard measurement distances include 1m and 3m. These distances are in the quasi–near field of the measurement antennas. Figure 6.10 shows the gain calibration for a specific serial number of an AH

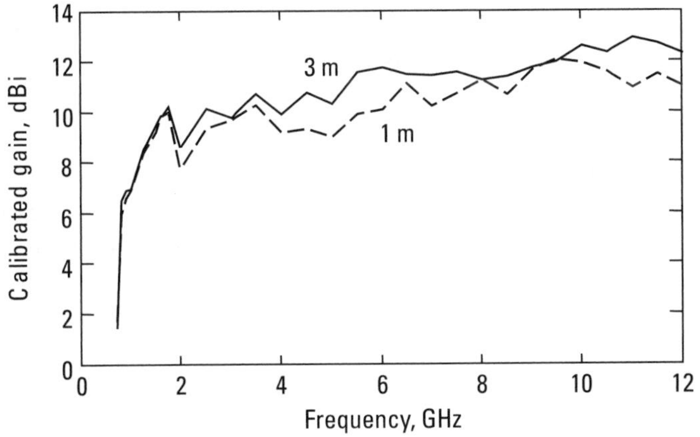

Figure 6.10 Calibrated gain at 1m and 3m measurement distances.

Systems model SAS-571 dual-ridge horn antenna. The calibrated gain can vary up to 2 dB between measurements at distances of 1m and 3m.

6.4 UWB Pulse Propagation with a Ground Reflection

We first establish a transmission formula for pulses in free space, then apply it to the fully deterministic two-ray path model. In summary, propagation starts off initially following inverse square law, but without the interference pattern exhibited by sine waves. Then, in the transition region, the direct and reflected signals partially overlap, resulting in an interference pattern that depends on the specific shape and characteristics of the UWB pulse signal. Finally, beyond the transition distance, the direct path pulse and the polarity-inverted reflected pulse increasingly overlap as distance is increased resulting in inverse fourth-power propagation, just like for sine waves. This model is applicable to propagation between antennas that are generally in the clear and can be accurately described by an unobstructed direct path plus a ground reflected path.

6.4.1 UWB Pulse in Free Space

In free space, the UWB energy expands spherically in proportion to the square of the distance, but, as is shown in Chapter 11, the wave shape in space may depend on the propagation direction relative to the antenna axis. By considering the energy in the pulse, we will avoid difficulties arising from the pulse wave shapes. Energy is captured by a unity-gain antenna having capture area

$$A_e = \lambda^2 / 4\pi \tag{6.22}$$

where the wavelength $\lambda = c/f_c$ written in terms of the speed of propagation c and a frequency f_c. The effective areas of the transmit and receive antennas are identical, that is, $A_e = A_t = A_r$.

The effective transmission frequency f_c needs to be carefully defined since the UWB pulse bandwidth can exceed 100%. For constant-gain antennas, the capture area is found from (6.21) with wavelength replaced by the equivalent frequency term. Integration over the pulse frequency band, weighted by $\underline{U}(f)$, the normalized pulse power spectral density, gives

$$A_e = \frac{1}{f_U - f_L} \int_{f_L}^{f_U} \frac{c^2}{4\pi f^2} U(f) df \tag{6.23}$$

We take $U(f)$ to be symmetrical in frequency, and with no loss in generality, let $U = 1$. So,

$$A_e = \frac{c^2}{4\pi f_L f_U} \tag{6.24}$$

and we define

$$f_c = \sqrt{f_L f_U} \tag{6.25}$$

From (6.25), we see that the geometrical rather than the arithmetical mean defines the "center frequency" for UWB pulse propagation between antennas with constant (unity) gain as a function of frequency.

The free space transmission of UWB pulses between infinitesimal antennas can now be found by first choosing a current pulse that is supplied to the antenna feed point:

$$I_T(t) = \Delta h \exp\left[-(\pi t B)^2 \log(e) / 2\right] \cos\left(2\pi f_C t\right) \tag{6.26}$$

where B is the "10-dB bandwidth" as defined by the FCC [11]. Other short pulse current shapes can be used with no loss of generality. The antenna here is a hypothetical infinitesimal dipole of length Δh having radiation resistance (see [5]):

$$R_{rad} = \frac{\eta_0}{6\pi} \left(\frac{\Delta h 2\pi f_c}{c} \right)^2 \tag{6.27}$$

The transmitted pulse energy is found by integrating the square of the pulse current over time:

$$W_T = R_{rad} \int_{-\infty}^{\infty} [I_T(t)]^2 \, dt \tag{6.28}$$

The magnetic far fields due to this infinitesimally short radiating current I_T oriented along the z-axis are found from the Boit-Savart law generalized to time-varying currents:

$$H_\phi(d,t) = \frac{\Delta h \sin(\theta)}{4\pi dc} \frac{\partial}{\partial t} I_z(t) \tag{6.29}$$

The corresponding electric far field is

$$E_\theta = -\eta_0 H_\phi \tag{6.30}$$

The receiving antenna is also an infinitesimal "field probe" dipole of length Δh oriented along the z-axis. We can deduce the induced voltage from the Lorentz force law, which relates the force **F** exerted on a charge q by incident electric (**E**) and magnetic (**H**) fields on an antenna of length Δh.

$$\mathbf{F} = \mathbf{E}q + q\mu_0 \mathbf{v} \times \mathbf{H} \tag{6.31}$$

The antenna velocity **v** here is zero, so

$$V_R(t) = -\int_{-\Delta h/2}^{\Delta h/2} \frac{1}{q} \mathbf{F} \cdot \mathbf{dl} = -\int_{-\Delta h/2}^{\Delta h/2} \mathbf{E}^{\mathbf{inc}} \cdot \mathbf{dl} = -E^{inc}(t)\Delta h \tag{6.32}$$

Using (6.29), (6.30), and (6.32), we find that the open-circuit voltage at the terminals of the ideal infinitesimal receiving dipole is

$$V_R(t) = \frac{\eta_0 (\Delta h)^2}{4\pi dc} \frac{\partial}{\partial t} I_T(t) \tag{6.33}$$

The energy associated with the time derivative of the current through a 1Ω load is

$$W_{ddt} = \int_{-\infty}^{\infty} \left(\frac{\partial}{\partial t} I_T(t) \right)^2 dt \tag{6.34}$$

which, for the current given by (6.26), is

$$W_{ddt} = \int_{-\infty}^{\infty} \left[-\exp\left[\frac{-(tB\pi)^2}{2\ln(10)}\right]\left[\frac{t(B\pi)^2}{\ln(10)}\cos(2\pi f_c t) + 2\pi f_c \sin(2\pi f_c t)\right]\right]^2 dt \qquad (6.35)$$

This expression does not have a closed-form solution, but it can be shown by numerical methods [12] that for bandwidths under about 150%,

$$W_{ddt} = W_T (2\pi f_c)^2 / R_{rad} \qquad (6.36)$$

gives sufficient accuracy. Note that $(2\pi f)^2 = -(j\omega)^2$, and $j\omega$ is the time-derivative operator for time-harmonic (sine wave) solutions to Maxwell's equations. The energy delivered to a load resistance R_{rad} by the receiving antenna is

$$W_R = R_{rad} \int_{-\infty}^{\infty} \left[\frac{1}{2} V_R(t)\right]^2 dt \qquad (6.37)$$

Using (6.27) to (6.29), (6.33), and (6.36), the ratio of received to transmitted energy is

$$\frac{W_R}{W_T} = \frac{c^2 G_T G_R}{(4\pi f_c d)^2} \qquad (6.38)$$

where a factor $(1.5)^2$ was replaced by $G_T G_R$, the product of the gains of the pair of infinitesimal dipoles. We now make the substitutions $\lambda = c/f_c$ and $A_r A_t = G_T G_R/(4\pi)^2$ to obtain the Friis transmission formula that we encountered in Chapter 3 (see [13]) to obtain the pulse energy transfer between two "constant-gain" antennas:

$$\frac{W_R}{W_T} = \frac{A_r A_t}{\lambda^2 d^2} \qquad (6.39)$$

When antenna gains G_T and G_R are equal to 1, the result is also recognized as the free space propagation formula.

6.4.2 Ground Reflection with a UWB Pulse

Propagation over a smooth surface involves a reflection from the ground, as shown in Figure 6.2. The direct path and reflected path lengths D and R are from (6.3) and (6.4). The differential time delay between the reflected path and the direct path over a plane Earth is

$$\Delta t = (R - D) / c \tag{6.40}$$

where c is the speed of propagation. The time delay in (6.40) is comparable to the equivalent sine-wave phase delay in (6.5). Figure 6.11 shows the composite direct and reflected signals for the time-harmonic (sine wave) case and for the UWB pulse. The direct and reflected path sine waves both persist forever, forming a vertical standing wave pattern whose magnitude is depicted on the right-hand side of Figure 6.2(c). UWB pulse propagation is a transient phenomenon. The direct path pulse arrives first, followed Δt later by the reflected path pulse.

The ground reflection coefficient for the cases of interest (shallow incidence angles) is very nearly –1, so the reflected signal undergoes a simple polarity inversion. Reflected signals that arrive by paths having differential delays greater than the pulse signal duration do not overlap in space-time and can add to the total received energy if matched filter techniques are employed. This is depicted as A in Figure 6.12, with the corresponding path attenuation shown in Figure 6.13. The time-harmonic signal at this range of distances exhibits characteristics of the fully interfering pattern between direct and reflected paths. Point B is the worst-case destructive interference, while C shows the maximum constructive enhancement overlap between the direct and reflected pulses. Finally,

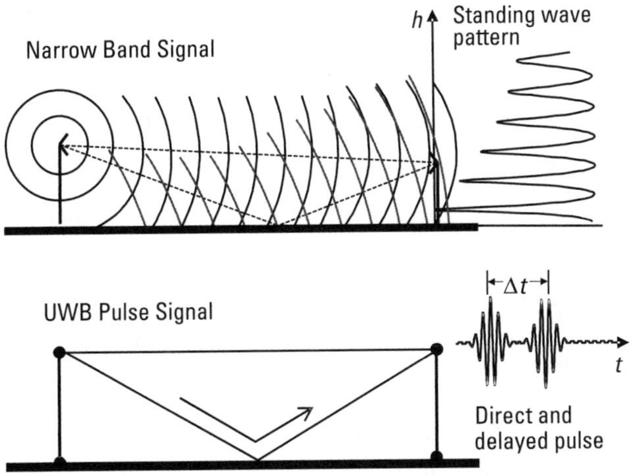

Figure 6.11 Direct and reflected signals in two-path propagation.

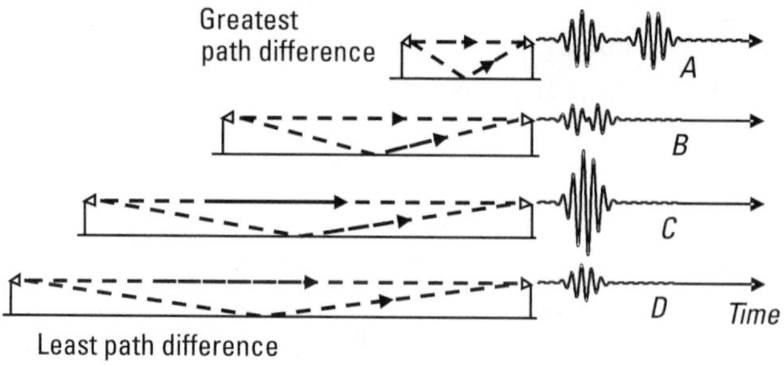

Figure 6.12 UWB propagation near ground.

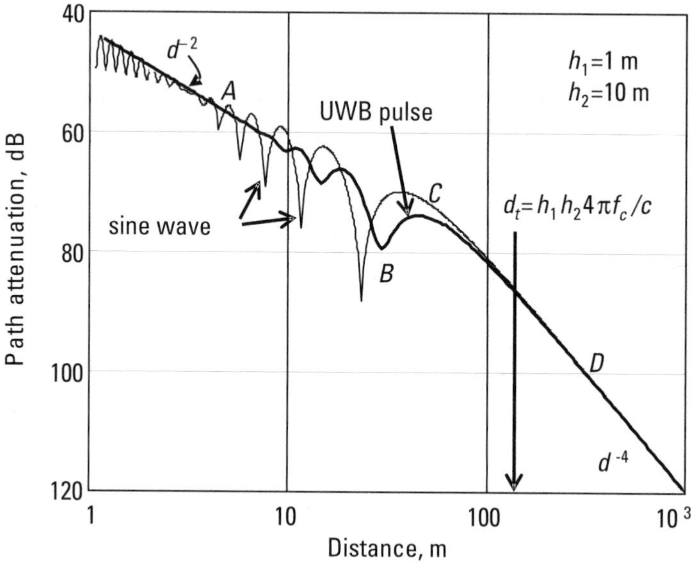

Figure 6.13 UWB and harmonic wave propagation near ground.

at *D*, the pulses overlap in increasingly destructive interference exactly like the sine wave. This behavior is captured by

$$\frac{W_R}{W_T} = \frac{1}{W_T}\frac{1}{4R_{rad}}\int_{-\infty}^{\infty}\left[V_R(t) - V_R(t - \Delta t)\right]^2 dt \tag{6.41}$$

where W_T is the transmitted energy from (6.28).

It is easy to show that the two-path propagation law in the overlap region beyond the break-point distance d_t is inverse fourth power and given by

$$W_R / W_T = \left(H_1 H_2 / d^2\right)^2 \quad (6.42)$$

which is the same as for time-harmonic signals. It is noteworthy that there is no frequency dependency in (6.42) for constant-gain antennas. The break-point distance d_t between the inverse square law region governed by (6.38) and the inverse fourth-power law governed by (6.42) is found by equating (6.38) and (6.42) and solving for distance:

$$d_t = 4\pi f_m H_1 H_2 / c \quad (6.43)$$

The break-point distance d_t depends on the antenna heights as well as the effective UWB pulse center frequency.

6.4.3 UWB Pulses Sent at High Repetition Rate

The two-path propagation analysis of UWB short pulses was carried out for an isolated pulse along with its reflection from a smooth ground. We note in Figure 6.13 that the signal-level peaks and valleys were not as extensive for UWB pulses as for the sine wave signal. In many cases of interest, UWB pulses are transmitted at a high pulse-repetition frequency, often with no gaps between successive pulses. The pulses are polarity encoded in direct sequence spread spectrum fashion, where the polarity encoding appears random. Figures 6.12 and 6.13 show the behavior of isolated pulses in a two-path propagation scenario, whereas Figure 6.14 shows the probability density distribution (

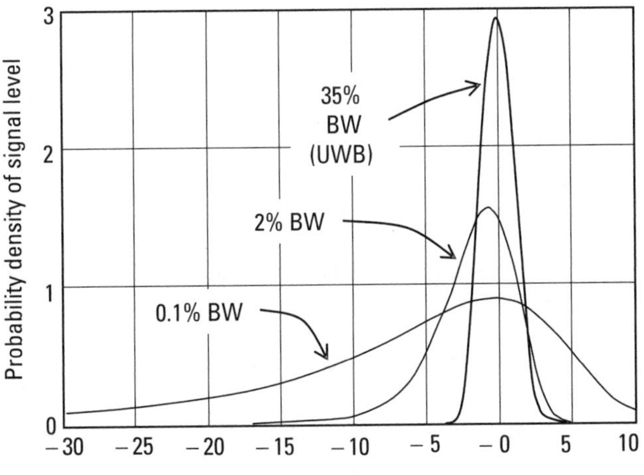

Figure 6.14 Probability densities of narrowband and UWB signals.

normalized to the peak value) of a narrowband signal subjected to Rayleigh fading compared with the distribution for a pulsed UWB signal. The details of the statistical distribution of signals are presented in Chapter 8. The pictured narrowband signal has a 0.1% relative bandwidth, and a moderately wideband signal has 2% bandwidth, while the UWB signal has a 35% bandwidth. The tight statistical distribution of the UWB signal levels is why UWB is sometimes referred to as "fade-free radio." Multipath and fading are considered in detail in Chapter 8.

6.5 Summary

A simple propagation model involving a planar Earth was shown, including the effects of a layered Earth and of antenna beamwidths. An open-field antenna measurement test site can be analyzed very accurately with a two-layer ground model and reasonably well with a relatively simpler homogeneous Earth model. In the analysis of potential errors at an open-field antenna measurement test site, we determined that field-strength calibration errors can result from the use of gain antennas on short antenna ranges when the beamwidth of the calibration antenna does not match the beamwidth of the antenna under test. We also saw that gain standard calibrations are distance dependent, especially when measurements are carried out in the quasi–near field of the test antenna. Finally, we investigated the special properties of short-pulse UWB propagation near the ground.

Problems

6.1 Find F_d and F_r—see (6.1)—for a vertical dipole on a range with transmitter antenna height of 3m, receiver antenna height of 1m, and antenna separations of 5m, 10m, 30m, and 45m.

6.2 Show that the reflection coefficient for a copper ground approaches –1 as the incidence angle approaches 0.

6.3 Calculate the approximate surface wave contribution to path loss at 1, 10, 100, and 1,000 MHz for both horizontal and vertical polarizations.

6.4 Using the two-ray grazing angle of incidence path loss approximation, find the apparent magnitude of "minimum antenna height" for vertically polarized antennas over seawater ($\varepsilon = 80$, $\sigma = 4$ S/m) and over ground ($\varepsilon = 5$, $\sigma = 0.001$ S/m), when $H_1 = H_2 = 5$m, and the frequency is 100, 300, and 1,000 MHz.

Ans:

$H_{min} = 1/(kX) = \lambda\varepsilon / \sqrt{\varepsilon - 1}$ sea water: $H_{min} = 80.6, 15.9, 3.1\text{m}$

ground: $H_{min} = 6.9, 2.3, 0.7\text{m}$

6.5 Using (6.17), estimate the variation for a calibration done using a corner reflector transmitter and a horn antenna compared with a measurement using a dipoles at each end. The antenna heights are 5m and 1m, respectively, and the range distance is 15m.

6.6 Find the path loss at 100, 300, and 1,000 MHz between isotropically radiating antennas on a 45m Earth ground (ε = 5, σ = 0.001 S/m) test site with antenna heights of 3m and 1m. Compare the results to a two-ray determination of path loss and explain the differences.

6.7 Estimate the calibration factor at 170, 460, and 930 MHz on a 45m range with an 8-dBi transmitting yagi at 3m height and the test point at 1m height, using the two-ray approximation. Assume 8 dB of line losses between the generator and the yagi. Solve for both vertical and horizontal polarizations.

6.8 Explain why the two-ray approximation cannot be used to estimate the calibration factor on a 10m-long range with a 3m transmitting antenna height and a 1m field point height.

6.9 An antenna range transmitter generates 1 mW, which is supplied through a cable/attenuator having 7 dB loss to an 8-dBi gain antenna. A dipole is used at the field point and connects to a calibrated receiver through a cable having 10 dB of loss. Find the site attenuation at 930 MHz if the calibration factor for this range is 77 dBμV/m.

6.10 An antenna range transmitter generates 1 mW, which is supplied through a cable and attenuator having 7 dB loss to an 8-dBi gain antenna. A dipole is used at the field point and connects to a calibrated receiver through a cable having 10 dB of loss. If the site attenuation at 930 MHz is 62 dB, find (a) the calibration factor for this range, and (b) the signal level at the calibration receiver.

6.11 A 45m-long antenna test range with a permanently mounted transmitter antenna at a height of 3m is calibrated at 930 MHz following a severe rain storm. There remains some standing water on the range surface. The calibration factor measures 17 dB higher than the historical figure for that range. Explain the possible causes for the discrepancy.

6.12 A circularly polarized, crossed-dipole transmitting antenna is situated 3m above ground on a 30m Earth ($\varepsilon_g = 5$, $\sigma = 0.001$ S/m) ground test site. Find the polarization axial ratio as a function of height from 0.3m to 1.8m at 170, 470, 930, and 1,620 MHz.

6.13 A test site 3m in length with an aluminum floor is proposed for testing the performance of a circularly polarized PCS device. Find the polarization axial ratio at 1,600 and 2,500 MHz if the transmitting and receiving antennas are each 1m above the ground plane. How can the measurement facility be improved for testing of circularly polarized antennas?

6.14 A 10m-long test site with an metal ground plane floor is proposed for testing the performance of a circularly polarized PCS devices at 1,620 and 2,510 MHz. A pair of phased-crossed dipoles is used for the transmitting antenna. A test engineer suggests that the dipoles should be oriented at 45° with respect to the ground so that they will be affected the same way by the ground reflection. Will this result in an acceptable polarization axial ratio? Why?

6.15 What is the wave impedance near ground level for a vertically polarized wave traveling over a horizontal conductive metal surface?

6.16 A dipole is parallel to, and five wavelengths above, the ground. Find an expression for the wave impedance directly below and away from the dipole for the vertically traveling wave. Derive a method for estimating the ground parameters in terms of the fields under the dipole.

6.17 Figure 6.P1 shows a receiver test site configured for calibrating field strength using a dipole at C. Cables A and B have 7 and 8 dB attenuation, respectively, and the transmit antenna has 8.5-dBi gain. Find (a) the site attenuation if the ground parameters are $\varepsilon = 5.5$, $\sigma = 0.001$ S/m and (b) the calibration factor and the signal strength reading at receiver RX at 170, 300, and 940 MHz.

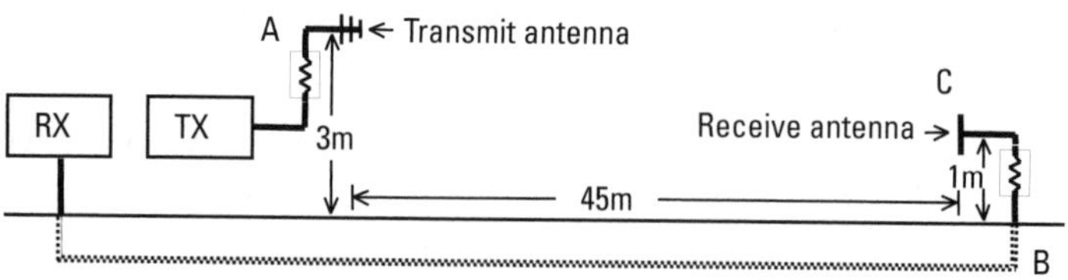

Figure 6.P1 A test-site configuration.

6.18 In Figure 6.P1, cable A attenuates 7.3 dB, the transmit antenna has 8.4-dBi gain, and cable B attenuates 9.5 dB. With 0 dBm generated at transmitter TX, the receiver RX indicates –71.3 dBm. Find (a) the calibration factor and (b) the site attenuation at 902 MHz.

6.19 When UWB pulse energy is collected as shown in (6.41), in the region (A in Figure 6.14) where Δt is greater than a pulse width, the direct and reflected pulse energy is added. Is this the maximum signal that can be received in region A?

Ans:

An additional 3-dB signal improvement may be achieved by a rake-type receiving system. This begins to occur naturally in region C where the propagation attenuation is almost 3 dB less than the extension of the square law behavior in A.

References

[1] Bullington, K., "Radio Propagation Fundamentals," *Bell System Technical Journal*, Vol. 36, May 1957, pp. 593–626.

[2] Bullington, K., "Radio Propagation for Vehicular Communications," *IEEE Transactions on Vehicular Technology*, Vol. VT-26, No. 4, November 1977, pp. 295–308.

[3] Brekhovskikh, L. M., *Waves in Layered Media*, trans. R. T. Beyer, 2nd ed., London: Academic Press, 1980.

[4] Collin, R. E., *Antennas and Radiowave Propagation*, New York: McGraw-Hill, 1985.

[5] Jordan, E. C., and K. G. Balmain, *Electromagnetic Waves and Radiating Systems*, 2nd ed., Englewood Cliffs, NJ: Prentice Hall, 1968.

[6] Siwiak, K., L. Ponce de Leon, and W. M. Elliott III, "Pager Sensitivities, Open Field Antenna Ranges, and Simulated Body Test Devices: Analysis and Measurements," *First Annual RF/Microwave Non-Linear Simulation Technical Exchange*, Motorola, Inc., Plantation, FL, March 26, 1993. (available from the author).

[7] Babij, T. M. "Measurement of the Reflection Coefficient of Earth at Frequencies above 100 MHz" (in Polish; abstract in English and Russian), *Prace Instytutu Metrologii Elektrycznej Politechniki Wroclawskiej [Proceedings of the Electrical Measurements Institute of the Wroclaw Technical University]* (Warsaw, Poland), No. 1, 1970, pp. 43–54.

[8] *Mathcad User's Guide* (Versions 2.0–7.0 Plus), Cambridge, MA: MathSoft, Inc., 1988–1997.

[9] More, J. J., B. S. Garbow, and K. E. Hillstrom, *User's Guide to Minpack I*, Argonne National Laboratory, Pub. ANL-80-74, 1980.

[10] Siwiak, K., and W. M. Elliott III, "Use of Simulated Human Bodies in Pager Receiver Sensitivity Measurements," *SouthCon/92 Conference Record*, Orlando, FL, March 11, 1992, pp. 189–192.

[11] U.S. 47 C.F.R. Part 15(f), Ultra-Wideband Operations, FCC Report and Order, April 22, 2002.

[12] Siwiak, K., and D. McKeown, *Ultra-wideband Radio Technology*, London: John Wiley and Sons, 2004.

[13] Friis, H. T. "A Note on a Simple Transmission Formula," *Proc. of the IRE*, Vol. 34, 1946, p. 254.

7

Radiowave Propagation: Urban and Suburban Paths

7.1 Introduction

The previous chapters have investigated the manner in which electromagnetic waves are launched from fixed-site antennas, and they have surveyed the characteristics of the radio frequency spectrum. The frequencies from 30 MHz to about 10 GHz were identified as prime radio frequency spectrum for personal communications services. The last chapter presented radiowave propagation in terms of a two-ray plane wave model that very accurately predicts the field strengths of well-designed open air test sites. One limiting case of the two-ray analysis, for large distance-to-antenna-height ratios, predicts an inverse fourth power with distance propagation law that, when surface waves are neglected, is independent of frequency. This chapter investigates radio wave propagation primarily in the range of 30 MHz to 10 GHz, with a focus on the urban and suburban environments, and includes in-building propagation.

Starting with purely analytical approaches, a theoretical urban model is derived based on an assumed regularity in the urban environment. Next, empirical models are presented that rely essentially on curve-fitting measured radiowave propagation behavior with measurable parameters describing the urban and suburban environment. The propagation behavior of waves within, near, and into buildings is studied with theoretical models as well as measurements. We start with a theoretical model that relates the physical basis for propagation power laws greater than free space law based on the observed behavior of multipath. We see another approach that accounts for in-building diffraction effects, and we introduce a fully deterministic in-room ray-tracing model for the personal-area space. Then, cross-coupling of polarization is investigated with urban models.

7.2 Theoretical Models for Urban Propagation

The purely theoretical treatment of suburban or urban propagation is basically an intractable electromagnetics problem. The sheer size and general nonavailability of a detailed geometric description of the coverage area, together with the numerous boundary conditions (some of which are transitory), force us to use approximate methods. We saw in Chapter 6 that propagation in the presence of even a simple planar ground benefited from plane wave approximations to keep the solution numerically reasonable. The urban and suburban models presented here also rely on physical generalizations that allow the solution of a far simpler electromagnetics problem to be effectively applied to the more complicated urban geometry.

The urban and suburban problems are complicated because the fields in the immediate vicinity of the portable or mobile radio are a superposition of localized multipath scattering. The signal strength varies from peak levels of a few decibels above the mean or median level to tens of decibels below the peaks in deep fades. Consequently, we rely on a statistical description of the signal levels in the vicinity of the portable or mobile radio that states the local average and a description of the variation. The signals in a local vicinity are described in terms of (1) a mean or median signal level, (2) a statistical distribution of levels, and (3) a measure of temporal and frequency spreading. In this section, we will consider models which provide the mean or median signal level. The statistical description of signals and the effects of temporal and frequency spreading will be covered in Chapter 8.

The theoretical models considered here are by no means exhaustive. There are many excellent models, including the Longley-Rice model [1, 2], the TIREM model [3], the Bullington model [4, 5], and Lee's model [6–8], as well as several proprietary models, that are well described elsewhere. These will not be considered further here. We will begin with an urban model that considers propagation from a fixed-site antenna in a city of buildings of nearly uniform height organized into rows of streets to a final diffraction of the rooftop field down to a street-level mobile or portable radio.

7.2.1 Diffracting Screens Model

It can be argued that, except for a small high-rise downtown region, urban and suburban areas are relatively homogeneous. The theoretical model described here is based on a geometrical generalization for which solutions to the electromagnetics problem can be readily obtained. J. Walfisch and H. L. Bertoni [9, 10] modeled the rows of city buildings as a series of absorbing diffracting screens of uniform height. The forward diffraction, along the screens, with a final diffraction down to street level gave an overall propagation model for the case of an elevated fixed antenna above the building roofline to a location at street level as pictured in Figure 7.1. Since absorbing screens are used, this model is essentially polarization independent.

L. R. Maciel, Bertoni, and H. H. Xia [11] extended the Walfisch-Bertoni model to allow the fixed-site antenna to be below, as well as above, the rooftop level, as pictured in Figure 7.2, but the restriction of a nearly homogeneous neighborhood of uniform-height buildings still applies. The approach is similar to that of Walfisch and Bertoni and applies a diffraction function to the fixed-site antenna for the case of the fixed-site antenna below the average roof-top level. Table 7.1 lists the parameters for the model.

The resulting expression for path propagation L_{ds}, based on the models of Maciel, Bertoni, Xia, and Walfisch [9–11], is written in terms of the free space loss F, excess losses Le_2 due to diffraction along the rooftops, and a final diffraction Le_1 down below roof-top level. A correction term is included for Earth curvature, provided that the distance does not approach the radio horizon. The average signal is

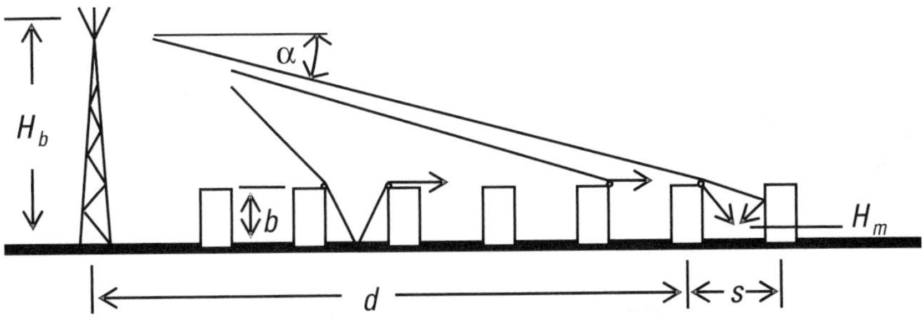

Figure 7.1 Wave propagation in a homogeneous urban region.

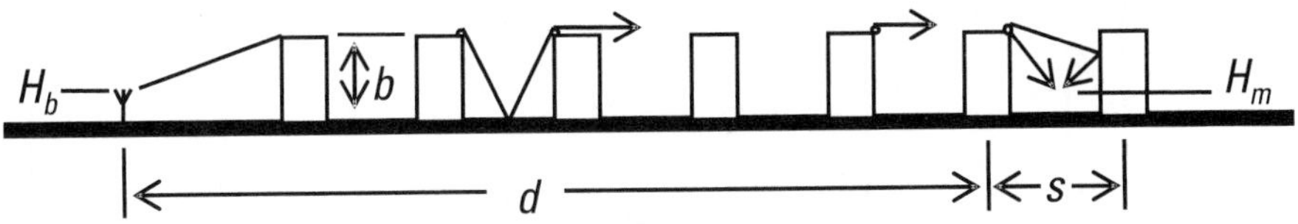

Figure 7.2 Suburban propagation between two sites below roof level.

Table 7.1
Parameters for the Diffracting Screens Model

Parameter	Definition
L_{ds}	Diffracting screens propagation, average signal (dB)
H_b	Fixed-site antenna height (m)
H_m	Mobile antenna height (m)
b	Building height (m)
s	Separation between rows of buildings (m)
w	Distance from mobile to building on street (m)
d	Range (km) (not beyond radio horizon)
f	Frequency (MHz)

$$L_{ds} = -F - Le_1 - Le_2 - 18\log\left[\frac{17H_b + d^2}{17H_b}\right] \tag{7.1}$$

The first term F is free space propagation loss

$$F = 32.4479 + 20\log(fd) \tag{7.2}$$

which is specialized for frequency in megahertz and distance in kilometers.

The geometry for the rooftop diffraction term Le_1 is shown in Figure 7.3, and the parameters are identified in Table 7.1. This diffraction coefficient is based on an absorbing screen model of the rooftop edge visible to the mobile. The loss Le_1 is

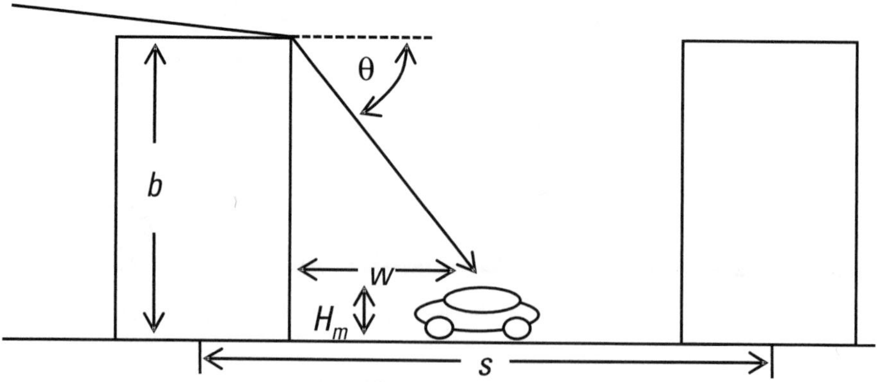

Figure 7.3 Local vicinity of a mobile radio in a suburban area.

$$Le_1 = -10\log\left[\frac{G_m(\theta)}{\pi k\sqrt{(b-H_m)^2+w^2}}\left[\frac{1}{\theta}-\frac{1}{2\pi+\theta}\right]^2\right] \tag{7.3}$$

where $G_m(\theta)$ is the mobile antenna gain in the roof edge direction, k is the wave number, and θ is the angle from the roof edge to the mobile found from

$$\theta = \text{atan}\left[\frac{b-H_m}{w}\right] \tag{7.4}$$

Factor Le_2, including G_b, the fixed-site antenna gain (usually taken to be unity) in the roof direction, is

$$Le_2 = -10\log\left[G_b Q^2\right] \tag{7.5}$$

and Q is either Q_E or Q_L, depending on whether the fixed-site antenna is above or below the rooftop level. Practically, Q_E is chosen when the fixed-site antenna height H_b is more than $\sqrt{\lambda s}$ above rooftop level b, and Q_L is chosen when H_b is below rooftop level by more than $0.5\sqrt{\lambda s}$. Intervening values are linearly interpolated from Q_E and Q_L, evaluated at $b+\sqrt{\lambda s}$ and $b-0.5\sqrt{\lambda s}$, respectively. Q_L is given by

$$Q_L = \frac{\dfrac{s}{d1000-s}}{\sqrt{2\pi k\sqrt{(b-H_b)^2+s^2}}}\left[\frac{1}{\text{atan}\left[\dfrac{b-H_b}{s}\right]}-\frac{1}{2\pi+\text{atan}\left[\dfrac{b-H_b}{s}\right]}\right] \tag{7.6}$$

and is used in place of the similar expression derived by Walfisch-Bertoni for the case where the fixed-site antenna is above rooftop level [10]. Above the rooftop level, the expression Q_E is

$$Q_E = 2.35\left[\text{atan}\left[\frac{H_b}{d1000}\right]\sqrt{\frac{s}{\lambda}}\right]^{0.9} \tag{7.7}$$

For small angles, $\alpha = H_b/(1{,}000d)$, giving rise to a 0.9 power with distance behavior, which, on top of the inverse d dependence of a free space propagation of a spherical wave, gives a total field dependence of $[1/d]^{1.9}$, corresponding to an inverse 3.8 power law with distance behavior. The path attenuation is shown in Figure 7.4 for the parameters: fixed-site antenna height of 100m, mobile

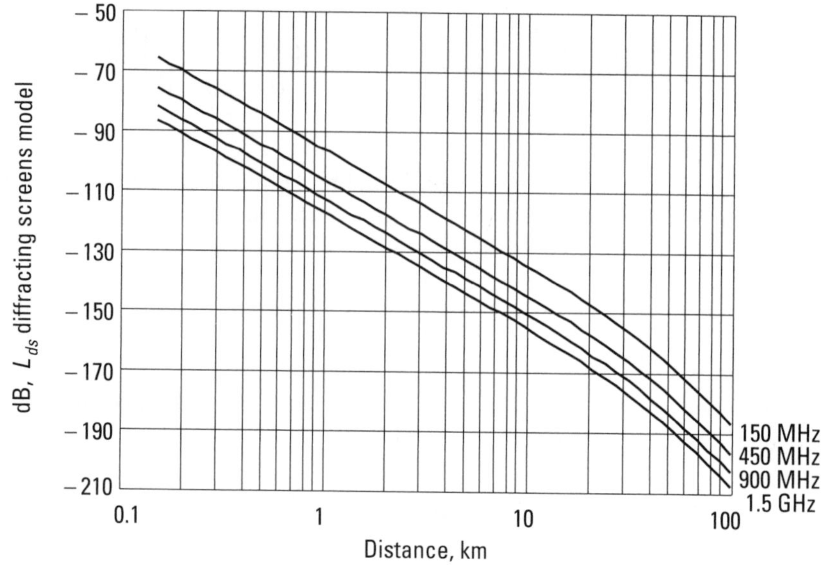

Figure 7.4 Path attenuation using the diffracting screen model with b = 15m, s = 40m, and w = 20m.

height of 1.5m, and model parameters, b = 15m, s = 40m, w = 20m, and a unity-gain mobile. The results are not valid close to, or beyond, the radio horizon, which, from a height of H and for a normal Earth refraction factor of K = 4/3 (see Chapter 4), is,

$$d_{hor} = 4.124\sqrt{H} \tag{7.8}$$

Here, $H = H_b - b$ in meters, and d_{hor} = 38 km. At distances beyond 20 km, the Earth's curvature is seen in Figure 7.4 to begin to have an effect. Beyond the radio horizon, losses are greater than predicted by the model.

The diffracting screens model accounts for fixed-site antennas lower than rooftop level, as well as those elevated above the roofline. Figure 7.5 shows the calculated path attenuation as a function of fixed antenna height for a distance of d = 10 km. The signal level is seen to increase dramatically as the fixed-site antenna height is increased above the rooftop height of 15m. Personal communications systems that are likely to use antennas at heights below rooftop level are also likely to be low-power microcellular systems that tend to be interference rather than propagation limited.

The problem of propagation over irregular buildings has been attacked by L. E. Vogler [12] and K. Furutsu [13]. The solutions are expressed in terms of multiple integrals whose dimensions are equal to the number of edges. The

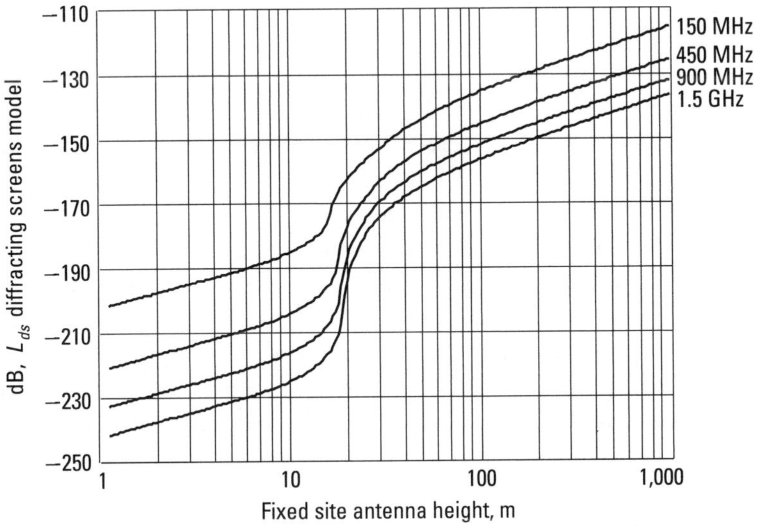

Figure 7.5 Path attenuation with varying fixed-site antenna height.

number of computer resources required to solve even moderate-sized problems involving just a few buildings is prohibitive, so hybrid techniques combining this with the Walfisch [10] solution model have been proposed. More efficient numerical treatments of propagation over irregular buildings have been presented by S. R. Saunders and F. R. Bonar [14].

[7-2a.mcd] Find the path attenuation using the diffracting screens model with b = 15m, s = 40m. Vary the distance d from 5 to 20 km and study the effect of distance on the height gain of the transmitting antenna. Compare this with the height gain from a two-ray model.

7.2.2 COST 231 Model

The European Research Committee COST 231 (evolution of land-mobile radio) created an urban model based on the work of Walfisch-Bertoni [10] and of F. Ikegami et al. [15], along with empirical factors [16]. The basic COST 231 model uses Walfisch-Bertoni results to account for the urban environment along with Ikegami's correction functions for dealing with street orientation. Empirical corrections were also added to deal with fixed-site antennas below rooftop level. The model was applied to the 800–1,800-MHz bands and tested in the German cities of Mannheim and Darmstadt. It was found to require considerable improvements when the fixed-site antenna was at or below rooftop level. The influence of street orientation was found to be minimal.

7.2.3 Diffraction over Knife-Edge Obstacles

We turn our attention now to radiowave propagation along paths that include an obstacle such as a fence or a hill high enough to obstruct the reflected ray and possibly to obstruct the direct ray connecting the transmitting and receiving points. The propagation over such a path has been approximated by replacing the obstruction with a perfectly absorbing knife-edge. The fields from the source are stated in the plane of the obstruction and, using the principle of Huygen's secondary sources, are reradiated to the receiving point. The obstruction is treated like a perfectly absorbing "knife-edge," and only the nonabsorbed secondary sources are involved in the reradiation process. The problem is appealing from an analytical point of view because a simple solution is available, but it suffers in practical application because the modeled geometry rarely occurs practically. It is presented here to illustrate the principle of its solution.

The knife problem has been analytically extended to multiple knife-edges [13, 17, 18], but the solutions are very computationally intensive, suffer from lack of general experimental verification, and will not be considered further here. With reference to the geometry of Figure 7.6, the problem is solved by expressing the field from an antenna at height H_1 in the plane f of the obstruction by using the method of secondary sources [19, 20] to modify the field propagated to H_2.

The integral expression for the secondary sources is given [6, 19, 20] as

$$F(u) = \frac{1}{2} - \frac{1+j}{2}\int_0^{-u} \exp\left[-j\frac{v^2}{2}\pi\right]dv \tag{7.9}$$

The appropriate geometry for the problem from Figure 7.6 gives the expression for u,

$$u = \frac{(H_1 - f)D_2 + (H_2 - f)D_1}{D_1 + D_2}\sqrt{\frac{2(D_1 + D_2)}{\lambda D_1 D_2}} \tag{7.10}$$

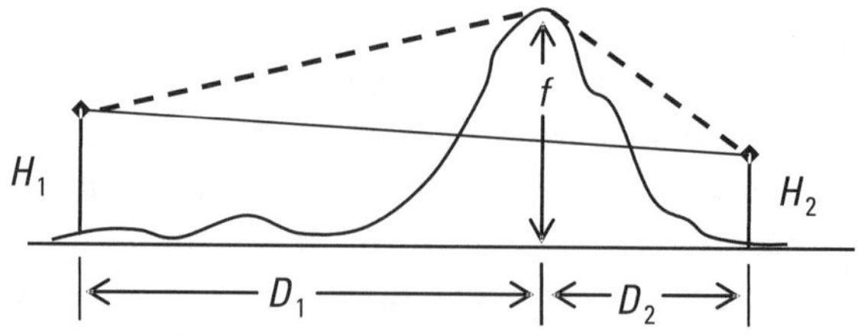

Figure 7.6 An obstruction in the propagation path.

where λ is the wavelength. The parameter u is a normalized knife-edge clearance factor. The term preceding the square root in (7.10) is the clearance height over the obstruction. The general behavior of the knife-edge diffraction integral $F(u)$ with clearance parameter u is shown in Figure 7.7. When u is zero, the ray path connecting the transmitting and receiving antennas just grazes the knife-edge. Half of the incident radiation is absorbed (or half of the secondary sources are absorbed) for a net propagation loss of 6 dB in addition to the free space loss. Expression (7.9) is applied as a multiplier to the basic free space propagation law in cases where the specular reflection is blocked and the direct ray is either clear or blocked by the knife-edge obstruction. Numerical evaluations of (7.9) are readily [21–23] available.

[7-2b.mcd] Find the path attenuation using the free space term modified by knife-edge diffraction for a path $D_1 + D_2 = 1$ km, where $H_1 = H_2 = 100$m. The knife-edge height varies from 0m to 300m. Use frequencies 150, 460, and 930 MHz.

The knife-edge diffraction problem has been extended to the cases where there are specular reflections on one or on each side of the dominant obstacle [18]. Improvements have proposed applying wedge diffraction in the form of the geometric theory of diffraction modified to include finite conductivity and local surface roughness effects [24]. Improvements over the simple knife-edge diffraction were reported, particularly for the case of large wedge angles and where geometries near grazing angle to the wedge are present.

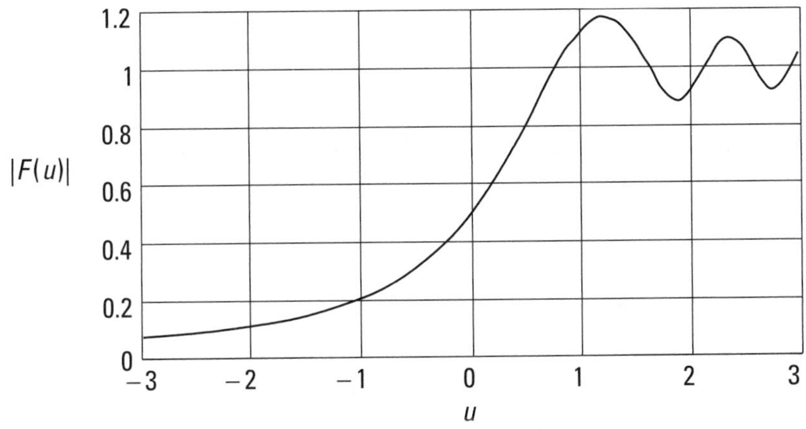

Figure 7.7 Diffraction factor for field strength beyond a knife-edge obstruction.

7.3 Empirical Models for Urban Propagation

Propagation in urban and suburban areas is different from the two-ray path in that a single specular ground reflection rarely exists. Often, even the direct ray path is obscured as mobiles are often below building roof level on city streets. Several empirical models have been developed based on measured data and use curve-fit equations to model propagation in areas of "definable" urbanization. Often empirical models are city specific and tied to urban land-use maps. The London model of M. F. Ibrahim and J. D. Parsons [25, 26] is an example of such a model. Another, more generalized, hence more commonly used, empirical model is that of Y. Okumura and colleagues [27]. Their model is based on extensive measurements in the Tokyo, Japan, environs. Because the Okumura data is well documented and widely known, the model has been extensively adopted around the world using "correction factors" to force-fit applicability to regions other than Tokyo.

7.3.1 Okumura Signal Prediction Method

Okumura and colleagues [27] measured signal strengths in the vicinity of Tokyo, over a wide range frequencies, at several fixed-site and mobile antenna heights, and over various irregular terrains and environmental clutter conditions. They then generated a set of curves relating field strength to distance for a range of fixed-site heights at several frequencies. Curves were then generated that extracted various behaviors in several environments, including the distance dependence of field strength in open and urban areas, the frequency dependence of median field strength in urban areas, and urban versus suburban differences. This led to curves giving a "suburban correction factor," variation of signal strength with fixed-station antenna height, and the dependence of mobile antenna height on signal strength. Additionally, corrections were extracted for various types of terrain and foliage. The tests were carried out at 200, 435, 922, 1,320, 1,430, and 1,920 MHz. Behavior was extrapolated and interpolated to frequencies between 100 and 3,000 MHz. The completeness of the study has made the model a standard in the field [28], but since the data is only available as curves, it is inconvenient to use, and formulas have been devised to fit the Okumura curves.

7.3.2 Hata and Modified Hata Formulas

M. Hata [29] prepared a simple formula representation of Okumura's measurements in the form Loss = $A + B \log(d)$, where A and B are functions of frequency, antenna height, and terrain type, and d is the distance. Hata's formula was limited to a frequency range of 100 to 1,500 MHz, distances of between 1 and 20 km, base antenna heights between 30m and 200m, and vehicle antenna heights from 1m to 10m. The basic formula for the median path loss was adopted by the CCIR (Consultive Committee for International Radio) [28] in the form

$$L_{ccir} = \begin{bmatrix} 69.55 + 26.16\log(f) - 13.82\log(H_b)\ldots \\ +[44.9 - 6.55\log(H_b)]\log(d) + a_x(H_m) \end{bmatrix} \quad (7.11)$$

where f is frequency in megahertz, d is distance in kilometers, and H_b is the base station height in meters. A mobile height correction function, $a_x(H_m)$, is applied for mobile antenna heights. In a "medium" city, Hata's mobile height correction takes the form

$$a_m(H_m) = [0.7 - 1.1\log(f)]H_m + 1.56\log(f) - 0.8 \quad (7.12)$$

while in a large city, at 200 MHz and below, it takes the form

$$a_2(H_m) = 1.1 - 8.29\log^2[1.54H_m] \quad (7.13)$$

At 400 MHz and above, Hata specifies,

$$a_4(H_m) = 4.97 - 3.2\log^2[11.75H_m] \quad (7.14)$$

In suburban areas, Hata gives the path loss $L_{ccir} - L_{ps}$, where

$$L_{ps} = -2\log^2\left[\frac{f}{28}\right] - 5.4 \quad (7.15)$$

and in open areas, as $L_{ccir} - L_{po}$, where

$$L_{po} = -4.78\log^2(f) + 18.33\log(f) - 40.94 \quad (7.16)$$

Modifications can be made to the Hata formulas to improve accuracy relative to the Okumura curves. Using the assignments in Table 7.2, the accuracy of the Hata formulas can be enhanced over the entire range of validity of the Okumura curves.

Transition functions can now be defined by

$$F_1 = \frac{300^4}{f^4 + 300^4} \quad (7.17)$$

Table 7.2
Parameters for the Modified Hata Model

Parameter	Definition	Range of Validity
L_{mh}	Modified Hata propagation median (dB)	—
H_b	Base antenna height (m)	30–300
H_m	Mobile antenna height (m)	1–10
U	0 = small/medium, 1 = large city	0–1
U_r	0 = open area, 0.5 = suburban, 1 = urban area	0–1
B_1	Percentage of buildings on the land (B_1 = 15.849 nominally)	3–50
d	Range (km) (not beyond radio horizon)	1–100
f	Frequency (MHz)	100–3,000

$$F_2 = \frac{f^4}{300^4 + f^4} \tag{7.18}$$

Correction for the Earth's curvature, but propagation not beyond the radio horizon, is included as

$$S_{ks} = \left[27 + \frac{f}{230}\right]\log\left[\frac{17(H_b + 20)}{17(H_b + 20) + d^2}\right] + 1.3 - \frac{|f - 55|}{750} \tag{7.19}$$

The term S_{ks} is a departure from Hata's formula but improves accuracy with respect to the Okumura curves for the larger distances. The suburban/urban correction can be linearly transitioned using the urbanization parameter U_r:

$$S_o = (1 - U_r)\left[(1 - 2U_r)L_{po} + 4U_r L_{ps}\right] \tag{7.20}$$

Combining the height correction functions (7.12) to (7.14) with frequency transition functions (7.17) and (7.18) and a small/large city parameter U, an overall height correction a_x can be written as

$$a_x = (1 - U)a_m(H_m) + U\left[a_2(H_m)F_1 + a_4(H_m)F_2\right] \tag{7.21}$$

One additional term accounting for the percentage of buildings on the land in the immediate grid under consideration is

$$B_o = 25\log(B_1) - 30 \tag{7.22}$$

Adding (7.11) and (7.19) through (7.22), the modified Hata formula can be written as

$$L_{mh} = -(L_{ccir} + S_o - S_{ks} + B_o) \tag{7.23}$$

Expression (7.23) can now be tested against points from the curves of Okumura as shown in Figure 7.8. The modified Hata model is seen to be within about 3 dB of the Okumura data points over the frequency range 100 to 3,000 MHz, for distances from 1 to 100 km and fixed-site antenna heights between 30m and 1,000m.

[7-3.mcd] Use expressions (7.11) and (7.14) to study the mobile height gain in a large city for the case where f = 930 MHz, H_b = 100m, d = 5 km, and H_m varies form 1m to 4m. Compare this result to the two-ray propagation model.

A "universal" propagation chart, nearly independent of frequency, can be generated from the modified Hata model in terms of field strength rather than path attenuation. The propagation path loss is stated as the ratio of power

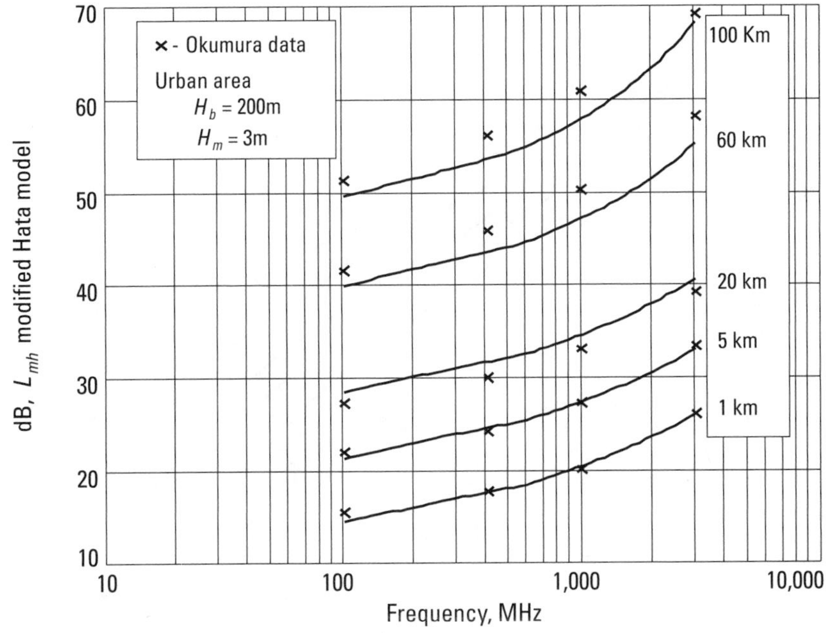

Figure 7.8 Comparison of the modified Hata model with Okumura data.

received by a constant (unity) gain antenna to the EIRP transmitted. In those terms, the received power diminishes as the square of frequency because the receiver antenna aperture is frequency dependent, as seen in Chapter 9. The receiver aperture is omitted when field strength is specified, making the modified Hata formulas roughly frequency independent. Figure 7.9 presents the universal propagation chart by reporting the median field strength for 1 kW transmitted by a dipole antenna in a suburban environment with model parameters $U = 0$, $U_r = 0.5$, and $B_l = 15.849$. Although produced for 930 MHz, the field strength reported in Figure 7.8 is conservative by only 0.8 dB at 460 MHz and 1.6 dB at 160 MHz. Compared with Figure 7.9, the field strength is generally 8 to 9 dB greater in the U.S. suburban environment, where the parameter $U_r = 0.24$.

7.3.3 Ibrahim and Parsons Method: London Model

Ibrahim and Parsons [25, 26] took the approach that propagation in the urban environment depends on such things as the density of buildings, the heights of buildings, and land use in general. Furthermore, although mobile radios are most widely used in urban areas, urban models suffer from an inherent vagueness associated with the qualitative description of the urban environment. The

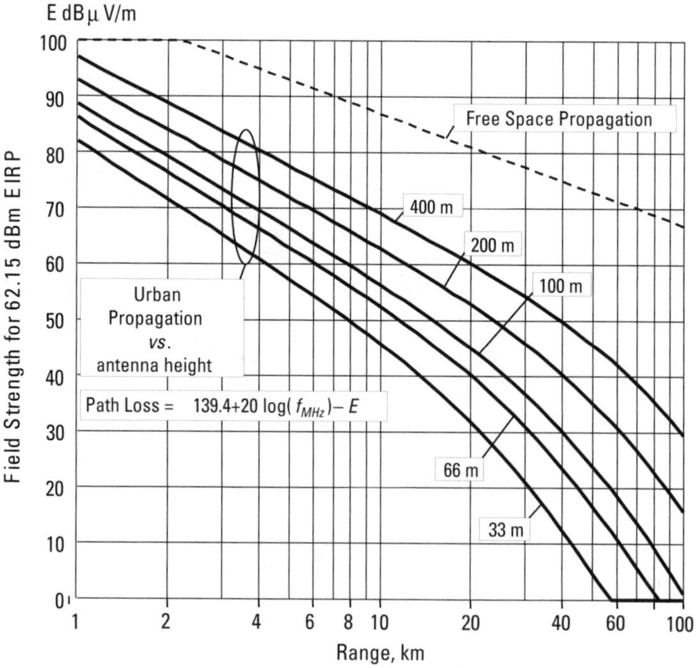

Figure 7.9 Universal urban-area propagation curves.

empirical behavior was extracted from measured data of propagation with regard to such factors as land usage, degree of urbanization, and varying terrain height for the mobile. The data was collected in 500m squares. These are the same parameters available on land-use maps of London, England, and the method can be applied to other cities for which similar land-use maps exist. The basic parameters for the London model are summarized in Table 7.3.

The "best-fit" model based on measurements in London is

$$L_{ip} = -\begin{bmatrix} -20\log(0.7H_b) - 8\log(H_m) + \dfrac{f}{40}\ldots \\ +26\log\left[\dfrac{f}{40}\right] - 86\log\left[\dfrac{f+100}{156}\right]\ldots \\ +\left[40 + 14.15\log\left[\dfrac{f+100}{156}\right]\right]\log(d1{,}000)\ldots \\ +0.265L - 0.37H + 0.087U - 5.5 \end{bmatrix} \tag{7.24}$$

Compared with measurement, the RMS errors produced by this model are 2.1 dB at 168 MHz, 3.2 dB at 455 MHz, and 4.2 dB at 900 MHz. Figure 7.10 shows the case where H_b = 100m, H_m = 1.5m, L = 50%, U = 16%, and H = 0. The power law behavior is greater than inverse fourth at 150 MHz and increases to more than inverse fifth at 900 MHz.

Table 7.3
Parameters for the Ibrahim and Parsons London Propagation Model

Parameter	Definition	Range of Validity
L_{ip}	Ibrahim and Parsons propagation, median (dB)	—
H_b	Base antenna height (m)	30–300
H_m	Mobile antenna height (m)	<3
L	Land-use factor, percentage of grid covered by buildings	3–50
H	Height difference between grid containing the fixed site and grid containing the mobile (m)	—
U	Urbanization factor, percentage of buildings in grid taller than 3 levels; outside city center U = 63.2	0–100
d	Range (km) (not beyond radio horizon)	<10
f	Frequency (MHz)	150–1,000

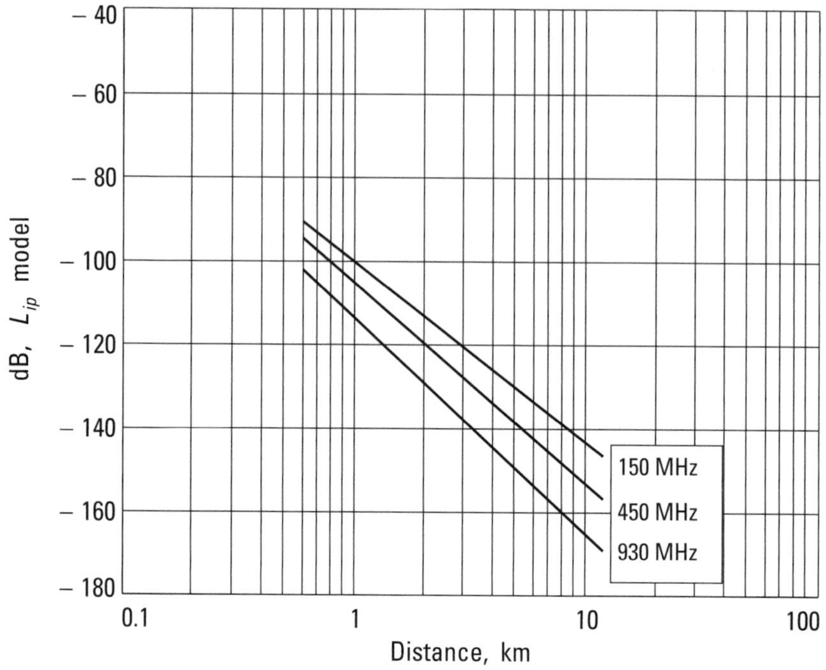

Figure 7.10 Ibrahim and Parsons London propagation model.

7.4 Propagation beyond the Horizon

Propagation beyond the horizon is not a major concern in personal communications problems. The most likely mechanisms for over-the-horizon propagation are troposcatter and meteor scattering, as was mentioned in Chapter 4. The general behavior falls between the theoretical predictions for diffraction over a smooth sphere and diffraction over a knife-edge. The strongest signals over very long paths generally occur for high-gain antennas facing each other along great circle paths. Tropospheric scattering is a factor in beyond-the-horizon transmissions at frequencies from 30 MHz to 10 GHz. Experiments have shown that signals decrease between the seventh and eighth power of distance and have seasonal variations of ±10 dB, which are proportional to seasonal variations in the effective Earth's radius factor K [19]. Table 7.4 presents the parameters for an empirical model of propagation beyond the horizon.

The empirical model is based on a two-ray propagation model to which excess losses have been added:

Table 7.4
Parameters for the Beyond-the-Horizon Model

Parameter	Definition
L_{oh}	Propagation beyond the horizon (dB)
H_1	First antenna height (m)
H_2	Second antenna height (m)
d_{hor}	Sum of distances from each antenna to horizon (km)
d	Distance between antennas (km)
f	Frequency (MHz)

$$L_{oh} = 20\log\left[\frac{H_1 H_2}{(d1{,}000)^2}\right] + 10\log\left[\frac{1+\left[\frac{d}{6d_{hor}}\right]^7}{1+\left[\frac{d}{d_{hor}}\right]^3}\right]\cdots$$

$$+\left[\frac{-d}{13+77\left[\frac{d_{hor}}{d}\right]}\right] + \left[22+\frac{f}{2{,}000}\right]\log\left[\frac{100}{f}\right] \tag{7.25}$$

where the term d_{hor} is the sum of the distances to the radio horizon for antenna heights H_1 and H_2, with atmospheric refraction factor K:

$$d_{hor} = 3.571\sqrt{K}\left[\sqrt{H_1} + \sqrt{H_2}\right] \tag{7.26}$$

Figure 7.10 shows calculation of over-the-horizon propagation for the case H_1 = 10m, H_2 = 150m, and a normal K = 4/3 atmospheric refraction factor. The atmospheric refraction factor was defined in 4.14. The results shown in Figure 7.11 are in agreement with measurements [19].

[7-4.mcd] Find the path attenuation using the tropospheric model of (7.25). Vary the distance from 10 to 700 miles and compute the path attenuation at 40, 100, 1,000, and 5,000 MHz.

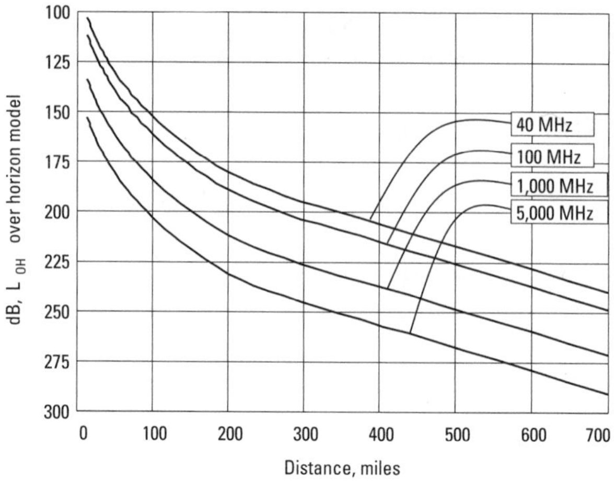

Figure 7.11 Propagation beyond the horizon.

7.5 Propagation within, near, and into Buildings

Personal communications devices are usually where people are, and people spend significant amounts of time in buildings. There are three distinct radiowave propagation situations with respect to buildings. One deals with the propagation of waves inside buildings. Another deals with short-range propagation in a local neighborhood of buildings. This second condition is encountered in microcellular communications systems and can have both ends of the communications link below the local roof level. The third situation deals with finding the additional attenuation that is encountered for the urban or suburban propagation problem to provide coverage into buildings.

7.5.1 Theoretical In-Building Multipath-Based Model

A theoretical basis for propagation in a scattering environment, for both narrowband and UWB signals, comes from an investigation of in-building UWB pulse propagation measurements and a realization that, on average, as a wave expands in a cluttered environment, it attenuates due to several mechanisms. The wave is spherical, so there is an underlying $1/d^2$ component. The wave additionally sheds energy into time in the form of multipath reflections, as was shown in [30] by K. Siwiak and H. Bertoni–based of measurements of S. Yano. The details of multipath are explained in Chapter 8. The wave also attenuates due to absorption by lossy materials—see the α_g attenuation factor in (3.1)—and to polarization cross-coupling in the propagation path, but these effects will not be considered here. The shedding of energy into multipath as measured by delay spread was shown in [31] to increase with the propagation distance. The effect was captured by a delay-spread parameter τ, which increased by some power law u with distance. In buildings, u = 1, while outside in an urban environment, u was approximately 0.5. When combined with the free space term, this yields an overall power law of n, where $u = n - 2$. The resulting theoretically based Siwiak-Bertoni-Yano (SBY) propagation model for median losses can thus be written as free space propagation modified by a multipath scattering term:

$$L_{SBY} = 10\log\left[\left(\frac{c}{4\pi d f_c}\right)^2\left(1 - e^{-(d_t/d)^{n-2}}\right)\right] \tag{7.27}$$

Although derived from observations of UWB pulse transmissions, (7.27) can easily be shown to apply equally well to narrowband waveforms. The model, valid for narrowband as well as UWB signals, predicts free space propagation that transitions to inverse n-power law beyond a break-point distance d_t. For $d >> d_t$, the modifying term reduces to $1/d^u$, which, combined with the free space term, yields an overall power law of n. In-building measurements suggest that $u = 1$, so $n = 3$, while outside $u = 0.5$, resulting in $n = 2.5$. The model also predicts that, on average, the maximum available rake gain in multipath is distance dependent and equal to

$$G_{rake} = -10\log\left(1 - e^{-(d_t/d)^{n-2}}\right) \tag{7.28}$$

because that is the total energy component shed to multipath.

7.5.2 Theoretical In-Building Ray-Tracing Model

Another model of propagation within buildings is governed by two principal mechanisms pictured in Figure 7.12. These are attenuation due to walls and diffraction from obstacles near the floor and in the plenum. Additionally, there is often diffraction around corners. A model of signal propagation was derived by W. Honcharenko and H. Bertoni [32, 33] by realizing that the signal diffracts from a cluttered region in the plenum of a building and a cluttered region near the floor. There is a "clear space" of height W_c meters between the two cluttered regions. The diffraction factor for wave propagation in a building with a clear space W_c results in excess loss compared with a free space path. Rays are drawn between the transmitter and the receiver, with transmission and reflection coefficients attached to rays that, respectively, pass through and are reflected by walls.

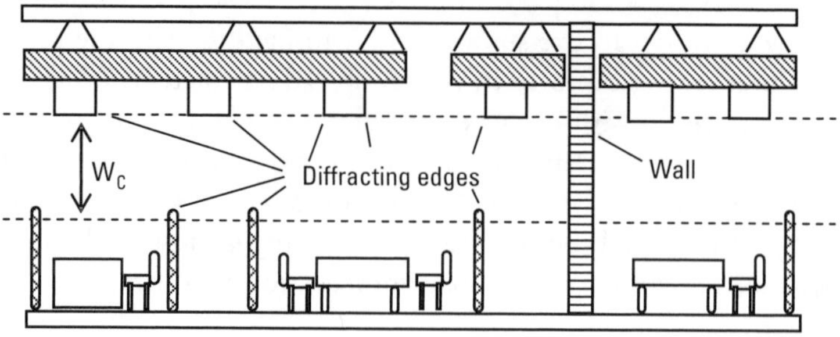

Figure 7.12 The propagation environment in a building.

Path loss within buildings [32] takes the form of the sum of i individual ray intensities:

$$L_{bld} = 10 \log\left[\sum_i \frac{L_e \lambda^2}{4\pi d_i^2}\left[\prod_n |\Gamma_n|^2 \prod_m |T_m|^2\right]\right] \tag{7.29}$$

where d_i is the distance connecting the ith ray between the transmitter and receiver. That ray undergoes n reflections Γ_n from walls and m transmissions T_m through walls. In addition, the ith ray is subjected to a distance-dependent loss L_e in excess of free space loss due to diffraction from clutter near the floor and in the plenum. The diffraction loss factor is unity for small distances d_i and approaches an inverse 9.5 power with distance behavior for large distances in an office building. The break point between the extreme behaviors of L_e is dependent, via successive aperture diffractions, on the clear space parameter W_c. The break point, as seen in Figure 7.13, is near 30m at 900 MHz in an office environment where W_c ranges between 1.5m and 2m. The excess factor can be modeled by $-G_{rake}$ in (7.28) of the SBY model. With d_t = 30 and u = 9.5, it is indistinguishable from the diffraction loss for the ray-based model, as shown in Figure 7.13.

7.5.3 An In-Room Deterministic Propagation Model

One short-coming of statistically based models, whether theoretical or curve-fit to data, is that they do not provide a way of studying the effect of motion by the radio devices. An in-room deterministic model was designed to study UWB propagation on the personal-area network scale [34]. The model is based on ray-tracing principal propagation paths within a single room. The model captures a line-of-sight (LOS) in-room component A in Figure 7.14, and multipath is derived from the 13 primary reflections of the modeled room. These include a ground reflection B, four principal reflections from the walls (two are shown, C

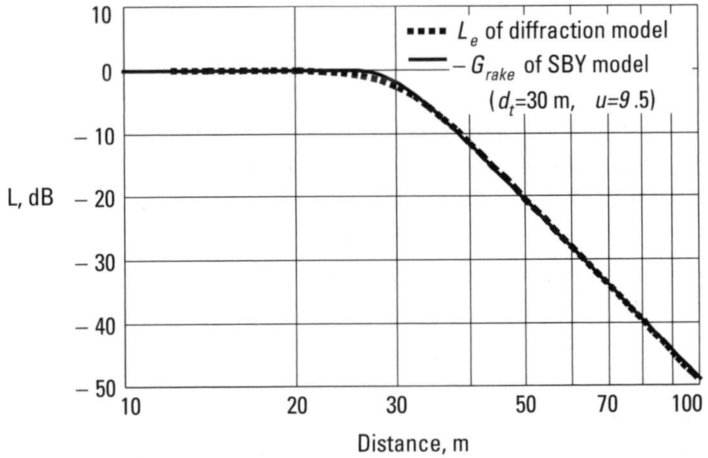

Figure 7.13 Excess propagation factors in an office environment.

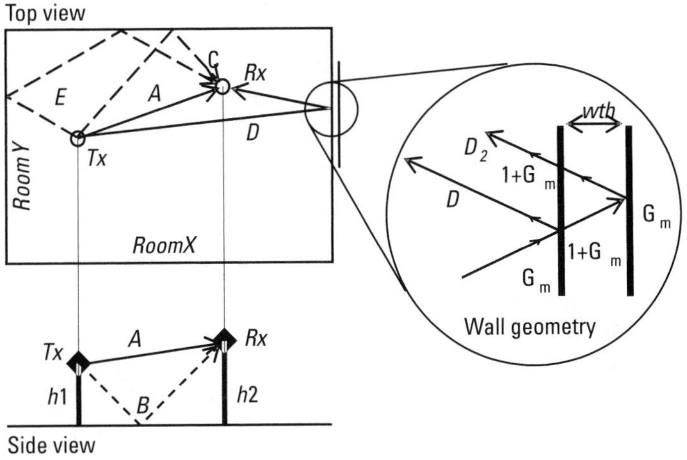

Figure 7.14 Ray paths for an in-room propagation model.

and *D*), four corner reflections (one is shown, *E*), and four secondary wall reflections capturing the internal wall-reflection components (one is shown, D_2). Objects in the room are not taken into account since the object of the model is simplicity in gaining a fundamental understanding of in-room propagation. An average reflection coefficient is used, and the result is valid for frequencies at which wall and floor reflections appear specular, that is, for very high frequencies, ultra high frequencies, and above.

The model accounts for total energy in the room. That is, the energy transferred between the transmitter and receiver can exceed that of a free space direct path because multiple paths are considered. A channel impulse response is provided as an output, which can then be convolved with the transmitted signal to predict the received signal at the receiver. Initially developed for UWB pulses (see [34]), the model is general and can be used with harmonic waves as well.

A Mathcad source template, <IN-ROOM.mcd>, is included on the textbook CD-ROM. The primary input parameters are the room length and width, *RoomX* and *RoomY*; the two antenna heights, *h*1 and *h*2; the minimum permitted separation between the antennas and any wall, *wt*; and the wall thickness, *dt*. Since the model is completely deterministic, it is possible to study the impulse response as a specific function of the relative position of the two antennas in the room; thus, the effects of motion on multipath can be easily studied. The antenna positions are defined in by variable $(X1r_i, Y1r_i, H1_i)$ for the first antenna and $(X2r_i, Y2r_i, H2_i)$ for the second antenna by equation (12) in <IN-ROOM.mcd>. As supplied, these variables are randomly chosen from a uniform distribution. An output file, <Channel-model-data-output.txt>, of 100 realizations of the positions and the multipath components is generated when the template is exercised. This data file can be used for further analysis of the channel impulse responses.

Figure 7.15 shows the impulse responses for 100 random antenna locations in a room represented by the dots in the figure inset. The inset also shows a pair of bold dots representing a specific location for which the specific multipath components are identified. There is a nearly exponential distribution for the 100

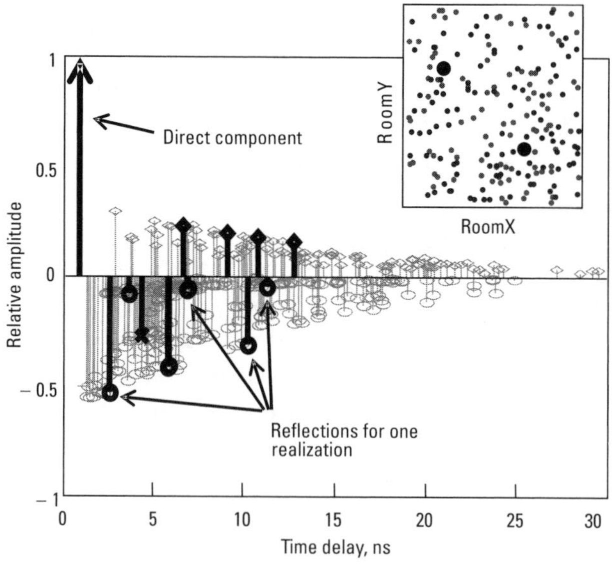

Figure 7.15 Multiple and a specific impulse realization.

cases, with definite structure relating to the various reflections (walls, corners, floor). The theoretical distributions based on diffusion theory [35], as will be shown in Chapter 8, predicts an exponential distribution for propagation confined to a single room.

[IN-ROOM.mcd] Exercise the in-room model for various room dimensions *RoomX* and *RoomY*. Compare the RMS delay spread, τ_{rms}, with $0.2D/c$, where $D = \sqrt{(RoomX - dt)^2 + (RoomY - dt)^2 + (h2 - h1)^2}$ is the maximum distance that can separate the antennas in the room, and c is the speed of light. Are τ_{rms} and $0.2D/c$ linearly related?

7.5.4 Propagation near Buildings

Personal communications systems are often microcellular systems. A characteristic of these systems is that performance is limited, not by the extreme range because of propagation losses but by interference from nearby cells that reuse the same frequency. For short distances, there is a break point beyond which the propagation law has increased losses with distance, as predicted by the SBY propagation model. The phenomenon can be exploited in system designs to help improve the signal-to-interference performance of interference-limited systems. Figure 7.16 shows an indoor application of the SBY model for a UWB personal-area network operating between 7 and 9 GHz. A break point at d_t = 3m is seen to separate the near-in free space propagation to inverse third power later.

Multiple-ray models have been developed that show the same break-point behavior for more realistic representations of the personal communications

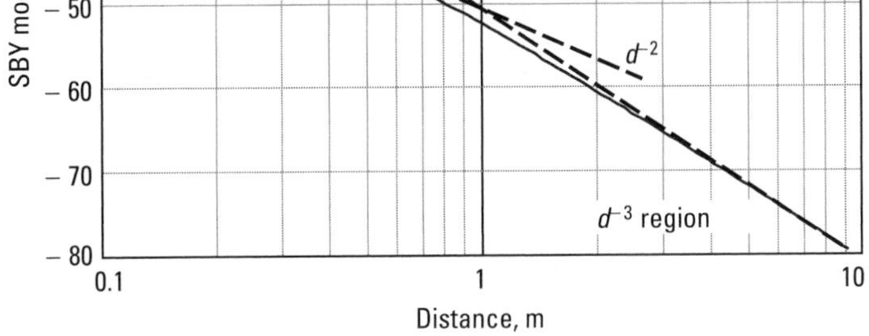

Figure 7.16 Indoor personal-area network application using the SBY model.

propagation paths, similar to the behavior shown by (7.27) for propagation in an office environment [36]. Another study concludes that microcellular propagation characteristics depend heavily on antenna heights and on local geometry, so simple propagation formulas such as the Hata formulas are inadequate for microcellular applications [37]. Propagation within a small microcellular region can be studied using expressions similar to (7.27) or (7.29).

Measurements in a dense urban downtown [38] area of propagation between a 12m-high fixed-site antenna and a mobile on the street reveal that the path attenuation breaks from the free space behavior within tens of meters of the fixed site into a fairly high inverse power law with distance, as seen in Figure 7.17. The figure presents average propagation loss over many measurements taken for distances from about 10m to 1,000m in several directions and under several conditions; measurements include those taken down single streets aligned with the antennas, as well as those taken over completely obscured paths. The data was first averaged over a distance of several wavelengths to remove mutipath variation, then averaged over multiple measurements in various directions at like distances to arrive at the averaged behavior. Note the significant departure of the measured data from a simple two-ray simulation. The average behavior can be curve-fit at distances greater than about 50m or 60m using

$$L_{low} = -71.2 - 52.9\log(d_{km}) - 20\log(f_{MHz}) \tag{7.30}$$

where d_{km} is the range in kilometers and f_{MHz} is the frequency in megahertz. Expression (7.30) should be taken as an average example of the kind of

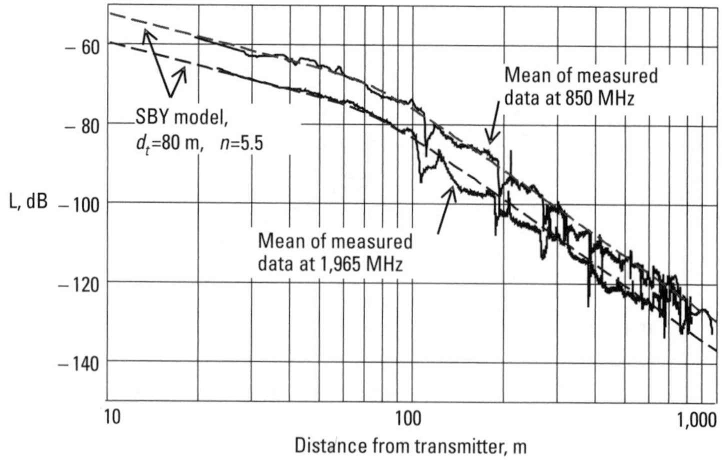

Figure 7.17 Urban path attenuation when antennas are below roof level.

propagation losses that occur with this type of geometry. Results for specific geometries may differ significantly.

7.5.5 Propagation into Buildings

Waves penetrate buildings with a frequency-dependent loss that also depends on the type of building, its construction, and the floor level. Since penetration loss decreases by about 1.5 dB per floor with increasing floor level [39], the conservative design specifies coverage to the ground-level floor of a building. Note that the 1.5 dB per floor figure is consistent with the increase in street-level signal, as shown, for example, by Figure 7.5. Figure 7.18 shows the median building penetration loss for buildings classified as "urban," "medium," and "residential," according to the definitions in Table 7.5

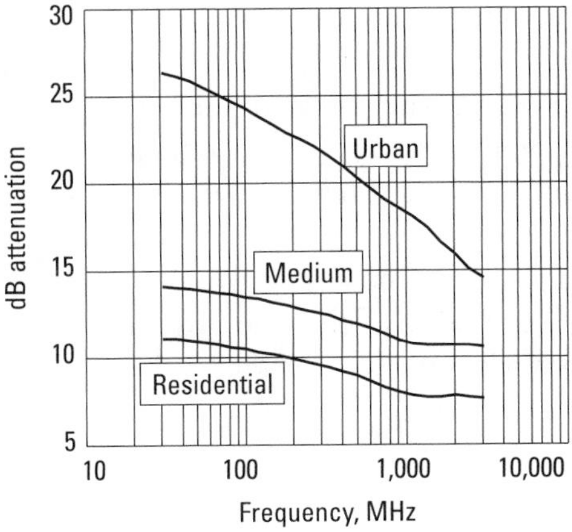

Figure 7.18 Building attenuation at ground level.

Table 7.5
Building Type Definitions

Building Type	Description
Urban	Typically large downtown office and commercial buildings, including enclosed shopping malls
Medium	Medium-sized office buildings, factories, and small apartment buildings
Residential	One- and two-level residential buildings, small commercial and office buildings

There is a steady decrease in building attenuation as frequency increases. The companion description to loss is the standard deviation of the loss shown in Figure 7.19 for the three building types. It is important to note that the building loss and standard deviation do not apply to a specific building but form a median value over a large group of similar structures. In fact, the standard deviation shown in Figure 7.19 is indicative of the variation. The building attenuations are averaged from several sources and agree with the determinations in [39].

7.6 Polarization Effects

The theoretical models for urban and suburban propagation presented in this chapter, including the knife-edge diffraction analysis, are independent of polarization. That is, the approximations used absorbing screen models of diffraction, which do not distinguish between polarizations. The empirical models presented earlier were derived from measurements involving only vertically polarized fixed-site antennas and vertically polarized portables and mobiles. None of these models can therefore be used to predict (a) any differences in the urban-area propagation of vertically or horizontally polarized waves or (b) coupling between the polarizations. The microcellular propagation expressions like (7.29) can account for polarization effects if the transmission and reflection coefficients are written for each polarization, as was shown in the two-ray example in Figure 7.16, but models of that sort require a rather detailed geometric description of the local vicinity. W. C. Y. Lee and Y. S. Yeh [40] presented measurements in a suburban environment showing that vertical polarized signals

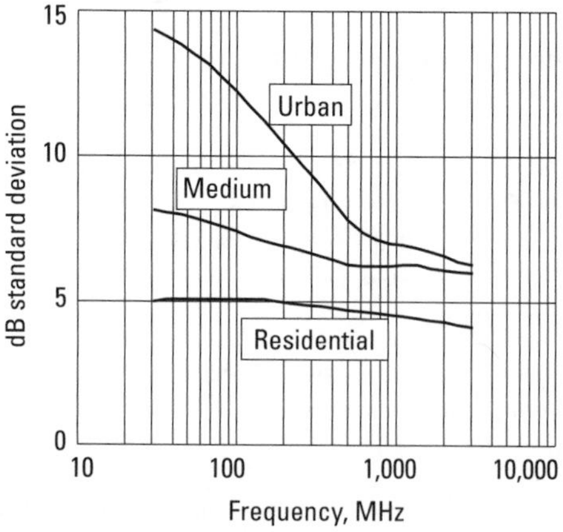

Figure 7.19 Standard deviation of building attenuation.

and horizontally polarized signals behave very similarly. Their local means are highly correlated and are within ±3 dB nearly 90% of the time. They, and later W. C. Jakes [41], also present evidence and analysis that demonstrates the feasibility of providing two diversity branches at UHF by polarization diversity.

7.6.1 Polarization Cross-Coupling Model Using Diffraction

The diffraction coefficients from geometric theory of diffraction (GTD) for scattering by a conducting wedge are well known (see [42]). In the shadow region, $(\gamma_n - \gamma) < (\pi/2 - \gamma')$, shown in Figure 7.20, the GTD diffraction coefficients $G^{\pm}$ reduce to

$$G^{\pm} = \frac{e^{-j\pi/[4\,\sin(\pi^2/\gamma_n)]}}{\frac{\gamma_n}{\pi}\sqrt{2\pi k}}\left[\frac{1}{\cos[\pi^2/\gamma_n] - \cos[\pi(\gamma-\gamma')/\gamma_n]} \pm \frac{1}{\cos[\pi^2/\gamma_n] - \cos[\pi(\gamma+\gamma')/\gamma_n]}\right] \quad (7.31)$$

where k is the wave number, γ_n is the exterior wedge angle, and γ' and γ are the angles of incidence and diffraction as identified in Figure 7.20.

The upper (+) sign in (7.31) is used for the polarization component parallel to the plane of incidence, and the lower (−) sign is used for polarization perpendicular to the plane of incidence. The rooftop edge vertex is perpendicular to the plane of incidence and is aligned with the horizontal polarization component. The building corner edge is aligned with the vertical polarization

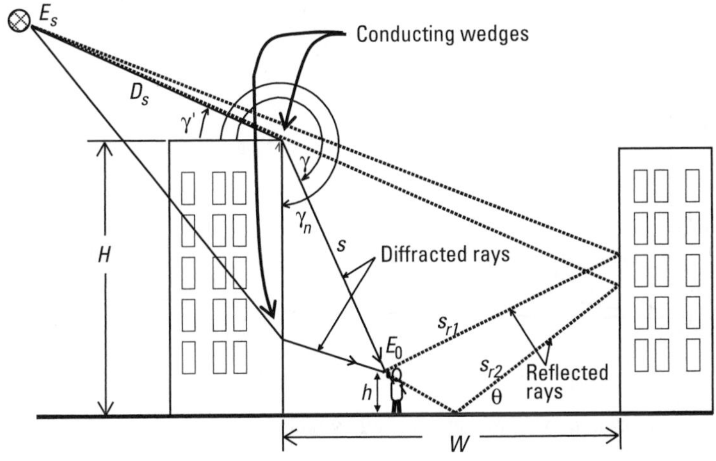

Figure 7.20 Geometry for urban diffraction and reflection.

component. The electric field at the field point that is not close to a shadow or a reflection boundary is then

$$\frac{E_0}{E_s} = \frac{e^{-jk(D_s+s)}}{D_s}\sqrt{\frac{D_s}{s(D_s+s)}}G^{\pm} \tag{7.32}$$

where E_s is the spherical wave source amplitude, and E_0 is the field point amplitude, both in volts per meter. The distance D_s depicted in Figure 7.20 represents the free space distance to a satellite or corresponds to the range parameter. Distance s is from the diffracting edge to the field point. Since $G^{\pm}$ is polarization dependent, then the diffracted field E_0 at the field point is also polarization dependent. More accurate expressions for the diffracted fields are available (e.g., see [43]), but (7.32) with (7.31) suffices for an investigation of polarization randomization due to the diffracted wave polarization.

Two diffraction scenarios are illustrated in Figure 7.20, a rooftop edge diffraction (heavy solid line) and a building corner edge diffraction. Additionally, reflected rays are portrayed. Figure 7.21 shows the polarization axial ratio P in the shadow region calculated using (7.31) and (7.32) for circular polarization. As the incident angle γ' decreases, the polarization aligned with the wedge increasingly attenuates; thus, a circular polarization incident on a rooftop edge

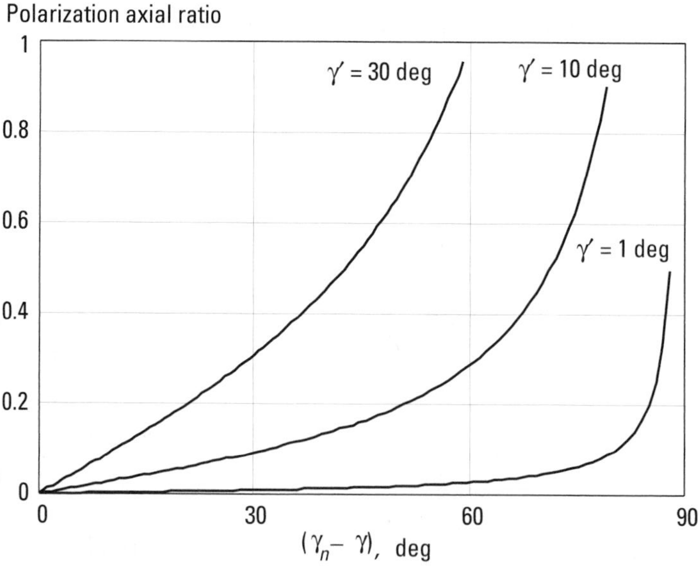

Figure 7.21 Polarization axial ratio in the shadow region.

tends toward linear polarization perpendicular to the roof edge. Incidence on a building corner edge results in polarization perpendicular to that edge. In general, both ray paths exist, so the polarization in the shadow of a building tends to be random.

Building corner edges behave like polarization "filters," in the diffraction shadow region. This precludes, for example, the use of orthogonal polarizations as a degree of freedom in reuse strategies of multiple access telecommunications systems. It does, however, support the use of orthogonal polarizations for diversity gain.

7.6.2 An Urban Model of Polarization Cross-Coupling

A homogeneous urban environment can be generalized by an average street width *W* and an average building height *H*, as seen in Figure 7.20. This suggests a model for urban cross-polarization coupling as reported in [44]. The fields in the "urban canyon" between buildings are due to diffraction from the rooftops and building edges, along with reflections from buildings and streets. Polarization couples by diffraction from an incident vertical polarization on the roof edge to a $\boldsymbol{\theta}$ polarization, where the angle $\theta = \gamma - \pi/2$ in Figure 7.20. The $\boldsymbol{\theta}$ polarization is then resolved into vertical (*z*-directed) and horizontal (ρ-directed) components. The relative diffracted power in the vertical polarization is proportional to the average over *x* (see Figure 7.22) along the street width of the diffraction coefficient $G^{+}(\theta)$ resolved into the vertical (*z*-directed) polarization,

$$Z_{pol}(W/H) = \int_0^1 \left|G^{+}(\theta_1)\right|^2 \sin(\theta_1)^2 + R\left|G^{+}(\theta_2)\right|^2 \sin(\theta_2)^2\, dx \quad (7.33)$$

while the horizontal polarization diffracted power is

$$\rho_{pol}(W/H) = \int_0^1 \left|G^{+}(\theta_1)\right|^2 \cos(\theta_1)^2 + R\left|G^{+}(\theta_2)\right|^2 \cos(\theta_2)^2\, dx \quad (7.34)$$

where the angles $\theta_1 = \pi\text{-atan}[xW/(H-h)]$ and $\theta_2 = \pi - \text{atan}[(2-x)W/(H-h)]$ are functions of *x*, *W* and *H* are the separation between buildings and the building height, as seen in Figure 7.22. The average power reflection coefficient from the building across the street from the diffracting building is *R*, and *h* is the height of the field point above the street level. Here, $H \cong H - h$, approximately. An additional component *R* is the denominator ray-3, the reflected power in the vertical polarization which is not due to diffraction. The polarization cross-coupling based on the three ray paths is

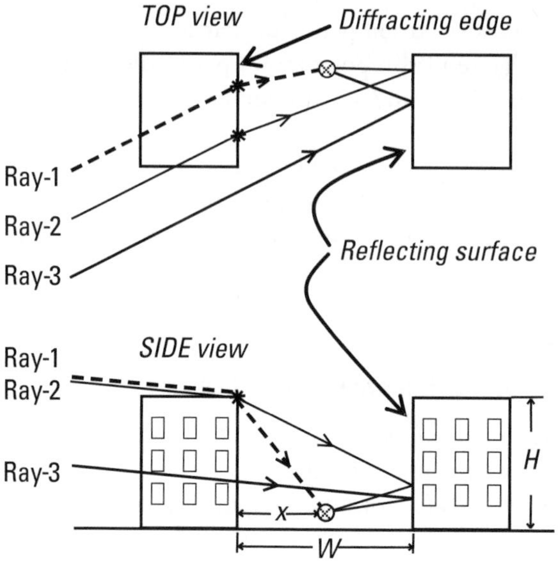

Figure 7.22 Rays diffracted and reflected by buildings.

$$X_{couple}(W/H) = 10\log\left[\frac{\rho_{pol\ of(W/H)}}{Z_{pol\ of(W/H)+B}}\frac{(W/H)}{(W/H)+B}\right] \tag{7.35}$$

Both ρ_{pol} and Z_{pol} decrease with the square root of frequency because of the wave-number term in (7.31); however, R is relatively independent of frequency.

Figure 7.23 shows the predicted polarization cross-coupling at 800 to 900 MHz for several values of reflection coefficient R. W/H ratios in the range 0.3 to about 6 represent typical heavy to light suburban ranges of urbanization.

Inspection of Figure 7.23 reveals that X_{couple} can be closely approximated by

$$X_{couple} = -3.33\log\left[\left[\frac{W}{H}\right]^3 + 1.25\right] - 5.1 \tag{7.36}$$

in the 800–900-MHz range for R = 0.25; hence, the diffraction polarization coupling mechanism relates simply to a homogeneous urban environment generalized by the ratio of building width W to building height H. Circles denote measurements made in Tokyo by T. Taga [45], and the squares were each averaged over four measurements by L. A. Ponce de Leon in Boynton Beach, Florida, and reported in [44].

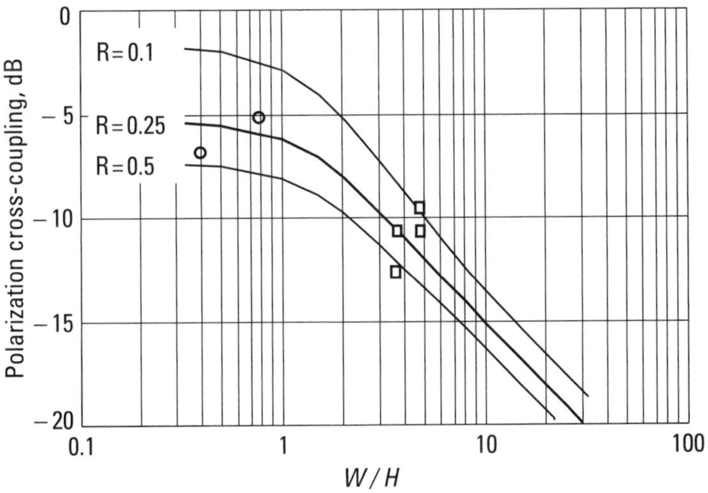

Figure 7.23 Polarization cross-coupling.

7.6.3 Polarization Cross-Coupling Measurements

Waves propagating in a suburban or urban environment are subjected to significant scattering, and because the scatterers are randomly oriented, there is cross-coupling of polarizations. One simple example of polarization cross-coupling was studied in Chapter 3, where the polarization filter was analyzed. Cross-polarization coupling in the urban and suburban environments occurs by a similar mechanism: scattering by objects that are skewed with respect to the horizontal and vertical axes. Consequently, when a vertically polarized signal is transmitted from a fixed site, there is significant horizontally polarized energy available in the scattered local vicinity of the receiver.

D. C. Cox and his associates [46] reported measurements of cross-coupling from the transmitted vertical-to-horizontal polarizations in residential houses and in commercial buildings. They defined the cross-polarized component in terms of measurements made using a coaxial dipole antenna as XPOL = $(L_h - L_v) + 3$ dB, where L_h is the average of median signals in decibels recorded with the dipole end-on and broadside to the fixed site. L_v is the median level in decibels with the dipole held vertically. The factor of 3 dB adjusts for the average gain difference of a dipole in its omnidirectional plane compared with average gain in the plane of the dipole and assumes that the signal angle of arrival is exclusively from the horizontal plane. Restating the cross polarization as $L_h - L_v$, the measured values of cross-polarization coupling are shown in Table 7.6 and compared to other similar measurements. It was further found that when locations were sorted with the copolarized signal level as a criterion,

Table 7.6
Polarization Cross-Coupling Values, $L_h - L_v$, in Different Environments

Location	Mean (dB)	Median (dB)	Standard deviation (dB)
Inside office area [48]	–0.5 to –3	–	–
Inside office area, on SALTY [48]	+0.5	–	–
Outside houses [46]	–7.1	–6.7	3.2
Inside houses [46]	–5.6	–5.5	2.4
Crawford Hill [46]	–2.8	–2.1	3.5
Holmdel [46]	–0.9	–1.2	3.1
Urban area [40]	–	–4 to –9	–
Ningyo-cho [45]	–	–5.1	–
Kabuto-cho [45]	–	–6.8	–
Yokohama National University [47]	–3 to –9	–	–

the lower signal levels tended to have higher cross-polarization coupling. The correlation between high cross-polarization coupling and weaker signal levels was also reported for measurements at Yokohama National University [47]. The fact that $L_h - L_v$ levels measured in the commercial buildings (Crawford Hill and Holmdel) are greater than –3 dB indicates that significant propagation paths exist outside the horizontal plane. Lee [40] concludes that the cross-polarization coupling in urban areas is approximately between –4 and –9 dB, while Taga [45] measured –6.8 and –5.1 dB on two urban routes (Ningyo-cho and Kabuto-cho) in Tokyo. Recent signal measurements at 930 MHz inside a large office area indicate that the polarization cross-coupling is in the range –0.5 to –3 dB when the transmitter is well shadowed [48]. Measurements on a SALTY human phantom fixture, as discussed in Chapter 10, showed horizontally polarized signal levels actually exceeded the vertically polarized signals. This is not unexpected because, as is discussed with respected to Figure 10.15, the horizontal polarization enhancement on SALTY exceeds the vertical polarization value at 930 MHz. Of further interest, the correlation coefficient between the cross-polarized signals increased to about 0.3 for the cases where L_h nearly equaled L_v, perhaps indicating the presence of nearby planar reflectors.

Taga reports that the mean equivalent gain (MEG) for a dipole inclined at 55° form the vertical is –3 dBi no matter what the cross-polarization coupling is and that when the cross-polarization coupling is –2 dB, the MEG = –3 dBi, regardless of the antenna inclination [45]. The MEG = G_e is defined as

$$G_e = \int_0^{2\pi}\int_0^{\pi}\left[\frac{X}{1+X}G_\theta(\theta,\phi)P_\theta(\theta,\phi) + \frac{1}{1+X}G_\phi(\theta,\phi)P_\phi(\theta,\phi)\right]\sin(\theta)d\theta d\phi \tag{7.37}$$

where the power ratio X is horizontally to vertically polarized signal power, G_θ and G_ϕ are the θ and ϕ components of the antenna power gain patterns, and P_θ and P_ϕ are the angular density functions of the incoming plane waves (described in the next section).

7.6.4 A Three-Dimensional Model of Incident Waves

Taga in [45] provides a model for dealing with statistical propagation properties in both the horizontal and the vertical planes by adopting angular density functions in the azimuth and elevation planes. A uniform distribution was used in the azimuth plane, while a Gaussian distribution was used in the elevation plane [40, 45]. Separate functions P_θ and P_ϕ were written for each of the two orthogonal polarizations.

Taga found that for a vertically polarized, 900-MHz, fixed-site transmitter, the signals on the street out of view of the fixed site had Gaussian distribution in the elevation angle. For the vertically polarized signals on the Ningyo-cho route, the mean elevation angle was 19° with a standard deviation of 20°, while on the Kabuto-cho route, the mean was 20° with 42° standard deviation. The cross-coupled polarization signals had means of 32° and 50° and standard deviations of 64° and 90°, respectively, for the same two routes. The copolarized signal can be seen to have a fairly well-defined elevation angle of arrival, while the cross-coupled polarization component is much more diffuse.

7.7 Summary

Urban and suburban propagation were studied by introducing and analyzing a theoretical model of a relatively homogeneous area of buildings modeled as absorbing screens of uniform height. The model allows for transmitter heights that are below as well as above rooftop level. The theoretical problem of diffraction by a single knife-edge was introduced as a way of taking into account a single prominent diffracting feature in the propagation path. Empirical models for urban and suburban propagation were introduced to present the methodology of curve-fitting measured data for parameters that can be derived from urban land-use maps. A universal propagation chart was presented for use in the urban environment. The behavior of propagation beyond the horizon was modeled empirically.

Propagation within, around, and into buildings was investigated. A propagation model, intended for UWB signals, was based on the shedding of energy into multipath and provides a physical basis for propagation laws greater than free space law. Another model took into account the diffraction from clutter in a building plenum and near the floor. A deterministic in-room ray path model was also introduced, as were measurements of UHF signals. An analytical approach to in-building propagation was presented. Finally, the theoretical models and the statistical properties of polarization cross-coupling in the urban environment and in buildings was investigated.

Problems

7.1 Path attenuation expressions like $L = A + B\log(d)$ typically express the loss between isotropically radiating antennas. Derive an expression for the field strength involving L and effective radiated power P_{eirp} dBm.

Ans:

$E^2 = (4\pi\eta_0/\lambda^2)10^{0.1L}\,P_T$, so $E_{dB\mu V/m} = 77.2160 + 20\log(f_{MHz}) + L + P_{eirp}$.

7.2 Derive an expression for $E_{dB\mu V/m}$ in terms of f_{MHz}, P_{eirp} dBm, and free space path loss for range d_{km}.

Ans:

$E_{dB\mu V/m} = 44.768 - 20\log(d_{km}) + P_{eirp}$. The result is independent of frequency.

7.3 Derive (7.8) for distance to the radio horizon.

7.4 On a normal refraction day, a seaside apartment dweller at a height of 16m above sea level sees the deck of a ship just on the horizon. Estimate the distance to the ship if the deck is 9m above the water line.

Ans:

$$d = 4.124\left(\sqrt{16} + \sqrt{9}\right) = 28.9\text{km}$$

7.5 Plot the signal strength as a function of mobile antenna height in an urban area for the Hata model at 900 MHz and compare it with the two-ray grazing incidence result.

7.6 Plot the signal strength as a function of fixed-site antenna height for the Hata model at a distance of 1 km and compare it with the two-ray grazing incidence result.

7.7 Find the exponent of the power law with distance for the Hata model with fixed-site antenna heights of 30m, 100m, and 300m.

7.8 A vertically polarized antenna at 915 MHz radiates 10 mW of effective radiated power from a height of 7m toward a 7m-tall chain link fence 10m away. Estimate the field strength 3m above ground and 3m away from the other side of the fence.

7.9 Determine the exponent of the power law for Ibraham's London propagation model at 170, 470, and 915 MHz.

7.10 Plot the signal strength as a function of fixed-site antenna height for the Ibraham London model and compare it with the two-ray grazing incidence result.

7.11 Plot the signal strength as a function of mobile antenna height for the Ibraham London model and compare it with the two-ray grazing incidence result.

7.12 Find the coefficients A and B for the $L = A + B\log(d_{km})$ form of the CCIR propagation formula for use at 1.7 GHz in a large city for fixed-site antenna heights of 30m, 100m, and 300m and for mobile antenna heights of 1m and 2m.

7.13 A portable user of a low-antenna, urban, 1.5-GHz microcellular PCS experiences an average range of 900m. What would the range be from a 100m-high fixed site in a suburban area that can be described by the diffracting screens with b = 15m, s = 40m, and w = 20m? Assuming a symmetric uplink and downlink, under what conditions, if any, could the portable communicate with an Earth-orbiting satellite? *Hint:* see (7.1), (7.2), and (7.30).

7.14 Assuming an office environment at 900 MHz is characterized by Figure 7.13, find the field strength 50m inside the building if the field strength outside the office building is 60 dBμV/m and the wall penetration loss is 6 dB.

Ans:

Source is plane wave–like: 60 – 6 – 21 = 33 dBμV/m.

7.15 A pair of 915-MHz portable radios operate at a system gain, including antennas, of 110 dB. Estimate their range within an office environment characterized by Figure 7.13.

Ans:

Total path loss is free space along with result from Figure 7.13: at 80m, F = 69.9 dB and L_e = 40, so d = 80m.

7.16 A pair of 915-MHz portable radios operate at a system gain, including antennas, of 110 dB. Estimate the range at which their call success rate is 90% within an office environment characterized by Figure 7.13 and having a 6-dB standard deviation.

Ans:

Available system gain is 110 – 6(1.28) = 102.3 dB: at d = 70m, F = 68.4 and L_e = 34.5, so range is just under 70m.

7.17 A system with a 56-dBm EIRP transmitter operating at 902 MHz requires 50 dBμV/m median signal strength. Use Figure 7.9 to estimated the range if the transmitter height is 100m.

Ans:

Chart field strength is 50 + 62.15 – 56 = 56.15 dBμV/m, so range is 10.1 km.

7.18 The median signal measured on a suburban street at 20-km range from a 100m-high transmitting antenna is 29.8 dBμV/m. Find the EiRP from Figure 7.9.

Ans:

47 dBm.

7.19 Using Figure 7.9, find the path loss at 300, 600, and 1,200 MHz from an antenna 200m high.

7.20 A personal communications device transmits 0.25W EIRP to a 10-dBi fixed-site antenna 200m high. Using Figure 7.9, find the range if the fixed-site receiver at the antenna requires a –119-dBm signal strength.

7.21 It is often suggested that system capacity in an urban setting can be doubled by simply overlaying another system with an orthogonal polarization. Demonstrate whether this supposition is true or false.

Ans:

There is enough coupling between orthogonal polarizations that simply overlaying another system in an orthogonal polarization will not double capacity. Another independent variable must be present to two such systems (see Chapter 8).

References

[1] Longley, Anita G., and P. L. Rice, "Prediction of Tropospheric Radio Transmission Loss over Irregular Terrain—A Computer Method," *ESSA Technical Report ERL 79-IOTS 67* [NTIS access number AD-676-874], 1968.

[2] Hufford, George A., Anita G. Longley, and William A. Kissick, "A Guide to the Use of the ITS Irregular Terrain Model in the Area Prediction Mode," *NTIA Report 82-100* [NTIS access number PB 82-217977], U.S. Department of Commerce, Boulder, CO, April 1982.

[3] "Microcomputer Spectrum Analysis Models (MSAM)," NTIA, 1990 [NTIS access number PB 91-0100669].

[4] Bullington K., "Radio Propagation Fundamentals," *Bell System Technical Journal*, Vol. 36, May 1957, pp. 593–626.

[5] Bullington, K., "Radio Propagation for Vehicular Communications," *IEEE Transactions on Vehicular Technology*, Vol. VT-26, No. 4, November 1977, pp. 295–308.

[6] Fujimoto, K., and J. R. James, (eds.), *Mobile Antenna Systems Handbook*, Norwood, MA: Artech House, 1994.

[7] Lee, W. C. Y., *Mobile Cellular Telecommunications Systems*, New York: McGraw-Hill, 1989.

[8] Lee, W. C. Y., *Mobile Communications Engineering*, New York: McGraw-Hill, 1982.

[9] Walfisch, J., "UHF/Microwave Propagation in Urban Environments," Ph.D. Dissertation, Polytechnic University, Brooklyn, NY, 1986.

[10] Walfisch, J., and H. L. Bertoni, "A Theoretical Model of UHF Propagation in Urban Environments," *IEEE Transactions on Antennas and Propagation*, Vol. 36, December 1988, pp. 1788–1796.

[11] Maciel, L. R., H. L. Bertoni, and H. H. Xia, "Unified Approach to Prediction of Propagation over Buildings for all Ranges of Base Station Antenna Height," *IEEE Transactions on Vehicular Technology*, Vol. VT-42, No. 1, February 1993, pp. 41–45.

[12] Vogler, L. E., "The Attenuation of Electromagnetic Waves by Multiple Knife-Edge Diffraction," *NTIA Report 81–86*, Nat. Telecommunications Inf. Admin., Boulder, CO, 1981.

[13] Furutsu, K., "On the Theory of Radio Propagation over Inhomogeneous Earth," *J. Res. NBS*, Vol. 67D, No. 1, 1963, pp. 39–62.

[14] Saunders, S. R., and F. R. Bonar, "Prediction of Mobile Radio Wave Propagation over Buildings of Irregular Heights and Spacings," *IEEE Transactions on Antennas and Propagation*, Vol. AP-42, No. 2, February 1994, pp. 137–143.

[15] Ikegami, F., et al., "Propagation Factors Controlling Mean Field Strength on Urban Streets," *IEEE Transactions on Antennas and Propagation*, Vol. AP-32, December 1984, pp. 822–829.

[16] Löw, K., "Comparison of Urban Propagation Models with CW-Measurements," *Vehicular Technology Society 42nd VTS Conference*, IEEE Cat. No. 92CH3159-1, Vol. 2, 1992, pp. 936–942.

[17] Vogler, L. E., "The Attenuation Function for Multiple Knife-Edge Diffraction," *Radio Science*, Vol. 17, No. 6, 1982, pp. 1541–1546.

[18] Anderson, L. J., and L. G. Trolese, "Simplified Method for Comparing Knife-Edge Diffraction as the Shadow Region," *IEEE Transaction on Antennas and Propagation*, July 1958, pp. 281–286.

[19] Jordan, E. C., and K. G. Balmain, *Electromagnetic Waves and Radiating Systems*, 2nd ed., Upper Saddle River, NJ: Prentice Hall, 1968.

[20] Collin, R. E., *Antennas and Radiowave Propagation*, New York: McGraw-Hill, 1985.

[21] Abramowitz, M., and I. Stegun, (eds.), *Handbook of Mathematical Functions*, New York: Dover Publications, 1972.

[22] *Mathcad User's Guide* (Versions 2.0–7.0 Plus), Cambridge, MA: MathSoft, Inc., 1988–1997.

[23] More, J. J., B. S. Garbow, and K. E. Hillstrom, *User's Guide to Minpack I*, Argonne National Laboratory, Pub. ANL-80-74, 1980.

[24] Luebbers, R. J., "Finite Conductivity Uniform GTD versus Knife Edge Diffraction in Prediction of Propagation Path Loss," *IEEE Transactions on Antennas and Propagation*, Vol. AP-32, No. 1, January 1984, pp. 70–76.

[25] Parsons, J. D., and J. G. Gardiner, *Mobile Communication Systems*, New York: John Wiley and Sons, 1989.

[26] Parsons, J. D., *The Mobile Radio Propagation Channel*, New York: John Wiley and Sons, 1992.

[27] Okumura, Y., et al., "Field Strength and Its Variability in VHF and UHF Land-Mobile Radio Service," *Rev. Elec. Commun. Lab.*, Vol. 16, September–October 1968, pp. 825–873.

[28] *CCIR Rep. 567-3*, International Radio Consultative Committee, International Telecommunications Union, Geneva, 1986.

[29] Hata, M., "Empirical Formula for Propagation Loss in Land Mobile Radio Services," *IEEE Transactions on Vehicular Technology*, Vol. VT-29, No. 3, August 1980, pp. 317–325.

[30] Siwiak, K., H. Bertoni, and S. Yano, "Relation between Multipath and Wave Propagation Attenuation," *Electronic Letters*, Vol. 39, No. 1, January 9, 2003, pp. 142–143.

[31] Yano, S. M., "Investigating the Ultra-wideband Indoor Wireless Channel," *Proc. IEEE VTC2002 Spring Conf.*, Birmingham, AL, May 7–9, 2002, Vol. 3, pp. 1200–1204.

[32] Honcharenko, W., et al., "Mechanisms Governing Propagation on Single Floors in Modern Office Buildings, " *IEEE Transactions on Vehicular Technology*, Vol. VT-41, No. 4, November 1992.

[33] Honcharenko, W., "Modeling UHF Radio Propagation in Buildings," Ph.D. Dissertation, Polytechnic University, Brooklyn, NY, 1993.

[34] "UWB Channel Modeling for under 1 GHz" *IEEE P802.15 Working Group for WPANs*, IEEE document P802.15-04/505-TG3a, December 2002.

[35] Nemati, M. A., and R. A. Scholtz, "A Diffusion Model for UWB Indoor Propagation," MILCOM 2004, Monterey CA, October 31–November 3, 2004.

[36] Xia, H. H., et al., "Radio Propagation Measurements and Modelling for Line-of-Sight Microcellular Systems," *Vehicular Technology Society 42nd VTS Conference*, IEEE Cat. No. 92CH3159-1, Vol. 1, 1992, pp. 349–320.

[37] Xia, H. H., et al., "Microcellular Propagation Characteristics for Personal Communications in Urban and Suburban Environments," *IEEE Transactions on Vehicular Technology*, Vol. VT-43, No. 3, August 1994.

[38] Siwiak, K., "Propagation Issues in Urban Areas for PCS," Workshop WMHI, *1996 IEEE MTT-S International Microwave Symposium*, San Francisco, CA, June 17, 1996.

[39] Turkmani, A. M. D., and A. F. de Toledo, "Modeling Radio Transmissions into and within Multistorey Buildings at 900, 1800, and 2300 MHz," *IEE Proceedings-I*, Vol. 140, No. 6, December 1993, pp. 462–470.

[40] Lee, W. C. Y., and Y. S. Yeh, "Polarization Diversity System for Mobile Radio," *IEEE Transactions Communications*, COM-20, No. 5, October 1972, pp. 912–923.

[41] Jakes, W. C., *Microwave Mobile Communications*, American Telephone and Telegraph Co., 1974 [reprinted by IEEE Press, New Jersey, 1993].

[42] Stutzman, W. L., and G. A. Thiele, *Antenna Theory and Design*, New York: John Wiley and Sons, 1981.

[43] Demetrescu, C., C. C. Constantinou, and M. J. Mehler, "Corner and Rooftop Diffraction in Radiowave Propagation Prediction Tools: A Review," 48th Annual International Vehicular Technology Conference, Ottawa, Canada, May 18–21, 1998, pp. 515–519.

[44] Siwiak, K., and L. A. Ponce de Leon, "Simulation Model of Urban Polarization Cross Coupling," *Electronic Letters*, Vol. 34, No. 22, October 29, 1998, pp. 2168–2169.

[45] Taga, T., "Analysis for Mean Effective Gain of Mobile Antennas in Land Mobile Radio Environments," *IEEE Transactions on Vehicular Technology*, Vol. VT-39, No. 2, May 1990, pp. 117–131.

[46] Cox, D. C., et al., "Cross-Polarization Coupling Measured for 800 MHz Radio Transmission in and around Houses and Large Buildings, " *IEEE Transactions on Antennas and Propagation*, Vol. AP-34, No. 1, January 1986, pp. 83–87.

[47] Arai, H., N. Igi, and H. Hanaoka, "Antenna-Gain Measurements of Handheld Terminals at 900 MHz," *IEEE Transactions on Vehicular Technology*, Vol. 46, No. 3, August 1997, pp. 537–543.

[48] Siwiak, K., "Radiowave Propagation Channel Engineering for Personal Communications," *Course TOO-310 Notes*, Johns Hopkins University, Whiting School of Engineering, Orgaizational Effectiveness Institute, on-line: http://www.oei-edu.com/t310.htm, last accessed January 22, 2007.

8

Signals in Multipath Propagation

8.1 Introduction

The propagation models investigated in Chapters 6 and 7 generally provide the median value of the average power in a local geographical region. Except for knife-edge diffraction, all of the described models rely on a homogeneous geometry along the ray path. This means that unless the model parameters are varied as a function of azimuthal angle measured from the fixed antenna site, the path attenuation is independent of azimuth. Terrain and building heights do vary in the suburban and urban environments, and the mean/median path attenuation values obtained from the simple models will be inaccurate because of shadowing caused by the terrain and varying urbanization. Furthermore, in the local vicinity of the mobile or portable unit, localized multipath scattering, as transmitted from antenna A depicted in Figure 8.1, causes the harmonic wave signal strength to vary from peak levels a few decibels above the median to tens of decibels below in deep fades. The shedding of energy into multipath results in an increased value of the propagation law exponent. The prediction of the exact field in a particular spot is an intractable electromagnetics problem because the boundary conditions are generally too cumbersome, and some are transitory. We rely, therefore, on a statistical description of the signal levels. That is, signals are described in terms of a median level, a distribution of levels, and a measure of temporal and frequency spreading. While multipath propagation disperses signals into time, and in the case of harmonic waves produces serious destructive interference, it also affords us the opportunity to recover from signal and/or capacity losses using time and space diversity techniques. These techniques include rake receivers, multiple-input, multiple-output (MIMO), single-input,

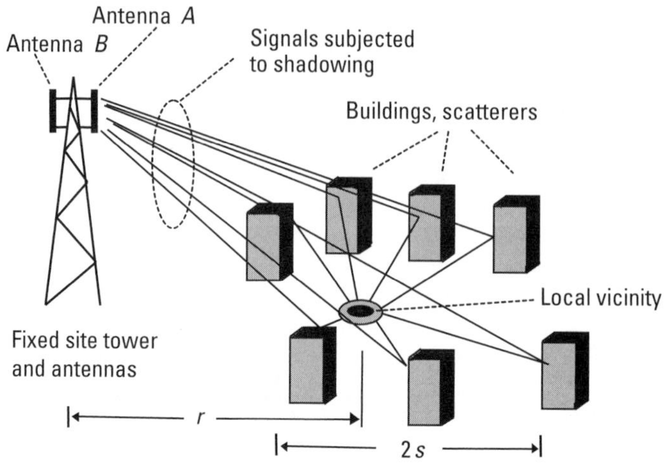

Figure 8.1 Multipath transmission to a local vicinity in an urban setting.

multiple-output (SIMO), and multiple-input, single-output (MISO) systems, which can be further enhanced by space-time coding techniques.

In this chapter, the behavior of waves in the local vicinity of a suburban or urban personal communications radio user will be studied, primarily for narrowband signals. We will also explore in-building settings for wireless local-area networks, including UWB technology. Two major influences on signal strength will be identified, shadowing and local multipath scattering. Each will be shown to have a different statistical behavior, and a manner of accounting for these effects in PCS designs will be presented. A similar multipath problem was described in Chapter 5 when coverage from satellites was examined. There are differences in the statistics of terrestrial systems because (1) in terrestrial systems the fixed-site antenna tends to appear just above the horizon, whereas satellites can appear anywhere in the sky, and (2) in satellite systems the satellites can move quickly enough relative to the mobile for Doppler shift to exceed the signal bandwidth, whereas in terrestrially based systems, the geometry between the fixed site and the mobile or subscriber unit is, by comparison, essentially stationary, especially for short messages.

We will also explore techniques of statistically improving the signal reception probability by using various transmitting and receiving diversity techniques. These include the use of repeated transmissions, transmissions sent simultaneously from multiple sites, and receiving multiple copies of signals in uncorrelated ways to statistically improve the probability of reception. The later method gives rise to MIMO, SIMO, and MISO techniques, where energy dispersed to mutipath can be recovered to improve the link and/or provide additional link capacity.

8.2 Urban Propagation: Understanding Signal Behavior

In an urban or suburban environment, signals travel from a fixed-site antenna over buildings, structures, and terrain that may vary in height. Finally, by diffraction from the last obstruction in the path, the signal arrives at the mobile or portable communications devices. Usually, multiple local paths are involved at the receiving location as depicted in Figure 8.1. The median value of the signal strength, as typified by the "flat terrain" curve of Figure 8.2, is a monotonically decreasing signal strength with distance, which can be predicted from the propagation models of Chapters 6 and 7. The basic power law is influenced by how much energy is shed into time in the form of multipath; thus, it depends on the transmitting and receiving antenna heights, on the average terrain and environment, and on additional loss parameters.

The median level is, however, subject to two major variations. One is a small-scale variation (on the order of tens of wavelengths) due to the vector combination of multiple rays arriving at the local vicinity of the portable communications antenna. At any local vicinity tens of wavelengths in diameter, the

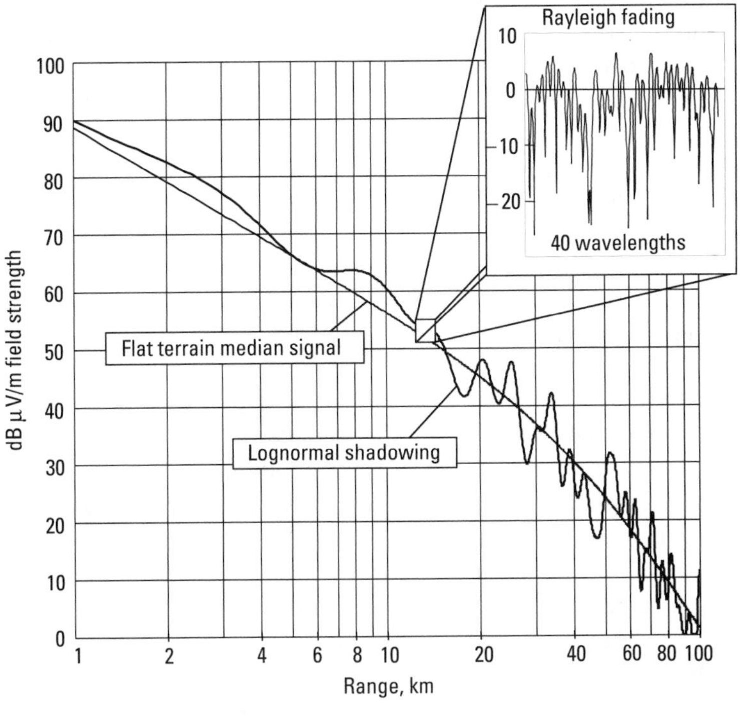

Figure 8.2 Signal behavior in a suburban region showing shadowing and multipath fading. (After [1].)

effect of multipath scattering from this mechanism, pictured in Figure 8.1, will cause deep fades for time-harmonic signals with fade distances that are about a half-wavelength apart. The small-scale variation for narrowband signals is typified by signal levels that are distributed according to Rayleigh statistics. The second variation is revealed by averaging over the small-scale variations at multiple locations separated by hundreds to thousands of wavelengths. The collection of large-scale signal variations has been found to be normally distributed when the averaged signal levels are stated in decibels; hence this variation is referred to as the "lognormal" variation. The lognormal variation of the local average signal levels is due to the multiplicative effect of the shadowing and diffraction of signals along rooftops and undulating terrain, along with a final diffraction to the portable communications device. Although the lognormal shadowing component is shown as a function of distance in Figure 8.2, the same behavior causes the signal to vary in azimuth as measured from the fixed-site antenna.

Signal behavior in a shadowed and multipath environment is described in the language of statistics because exact electromagnetic solutions are generally not available on the small, or even the medium, scale. Since we design communications systems for a desired probability (or reliability) of successful communication, the statistical description of waves is appropriate and useful. Excellent reviews and derivations of the probability concepts that apply to waves propagating in the urban and suburban environments are available in [2, 3], and simplified descriptions are in [4]. The derivations of the applicability of various statistical functions to signal distributions will not be repeated here, but the use of these functions in communications system design will be reviewed. The computational details of statistical distributions and density functions are available in [5–7].

8.3 Statistical Descriptions of Signals

Several distinct phenomena cause signal variations in a local region. The median signal level is governed primarily by the median propagation path attenuation. The larger-scale terrain variations and the building shadowing losses not included in the propagation model are the primary contributors to a lognormal standard deviation of loss. Finally, the local multipath, due to multiple reflections of signals in the immediate local vicinity, cause a standing wavelike pattern of deep nulls on the order of a half-wavelength apart. In addition to the amplitude variations, there are temporal effects due to the difference in arrival time of the local multipath. This temporal effect is known as delay spread and can place an upper limit on data transmission rates due to intersymbol interference.

The waves in a local area that are not directly visible to rays from the fixed site can be modeled very well using Rayleigh statistics, as shown in [3]. When a significant direct ray component (more than about half the total power) is present, a Rician statistical model can be used. The statistical distribution, after averaging the local Rayleigh behavior, on a larger scale is Gaussian in decibels. There are descriptions, as will be shown later, that combine the large-scale statistics with the Rayleigh behavior. A good statistical description of the signal, along with the median propagation path loss, allows us to calculate the required fixed-site transmitter power to provide a desired probability of coverage in the design coverage area.

8.3.1 Multipath and Fading: Local Variations

Harmonic waves in the vicinity of the receiver are characterized by a mean signal level and a variation about that level which can be described by Rayleigh statistics. The envelope of a sinusoidal signal affected by multipath fading can be described by

$$R_N = \sqrt{\frac{1}{\sum_i |A_i|^2}} \left| \sum_i A_i \exp\left\{ j\left[\phi_i + kr\left[\cos\left[\Phi_i - Nk\Delta d\right] + \sin\left[\Phi_i - Nk\Delta d\right]\right]\right]\right\} \right| \quad (8.1)$$

An observer moves in distance increments Δd along a semicircle of radius kr. The Nth signal point is described in terms of i signal paths with amplitudes A_i uniformly distributed in the interval [0, 1] and phases ϕ_i and Φ_i uniformly distributed in the interval $[0, 2\pi]$. The envelope is normalized to the signal amplitudes. In the physical picture of multipath, as in Figure 8.1, there are no "phase" terms. Instead, there are time delays related to the ray path distances. Those time-distance delays modulo one wavelength can then be related to a phase delay in the case of sinusoidal transmissions. This description, of course, loses the information regarding the actual time delays of the multiple paths and restricts the model to narrow bandwidths. A second parameter, delay spread, is then used to capture the time-distance delays that were suppressed by choosing the time-harmonic phase representation here. Thus, a description such as (8.1) is ill suited to the consideration of wideband or UWB signals, as will to be described later.

> 🖫 [8-3.mcd] Use (8.1) to generate the envelope of a signal subjected to multipath fading. Find the standard deviation. How does it compare to the Rayleigh standard deviation of 5.57 dB?

The probability density function (PDF) of a Rayleigh-distributed random variable is given by

$$f(s) = \frac{s}{\sigma^2}\exp\left(-\frac{s^2}{2\sigma^2}\right) \tag{8.2}$$

where $f(s)$ is the envelope of the voltage distribution of the signal, and σ^2 represents the average signal power. The notation and functional forms of G. Hess [2] are adopted and used here. Since $f(s)$ represents the envelope of the magnitude of the signal distribution, the value of $f(s)$ cannot, of course, be less than zero. The corresponding cumulative distribution function (CDF), which is the integral of (8.2) from 0 to x, is

$$F(x) = 1 - \exp\left(-\frac{x^2}{2\sigma^2}\right) \tag{8.3}$$

The true standard deviation of the Rayleigh PDF can be found from the expression $\sigma_R^2 = E\{x_{dB}^2\}E^2\{x_{dB}\} = [10/\ln(10)]^2\ (\pi^2/6) = 5.57^2$, where $E\{u\}$ is the expected value of the random variable u [2]. This is not the Rayleigh standard deviation that is appropriate in system designs. As will be shown when combined statistics are considered in the following paragraphs, a value of $\sigma_R = 7.5$ dB is usually used when the Rayleigh CDF is approximated by a Gaussian CDF.

As an example of what might be expected in a faded environment, a signal envelope can be calculated using (8.1) with 25 scatterers, sampled every 1/10 of a wavelength, around a semicircular path 50 wavelengths long. The signal, shown in Figure 8.3, has a median level 1.5 dB below the RMS value, and the calculated standard deviation is 5.5 dB below RMS. Rayleigh statistics give the median level as 1.592 dB and the standard deviation as 5.57 dB below the RMS

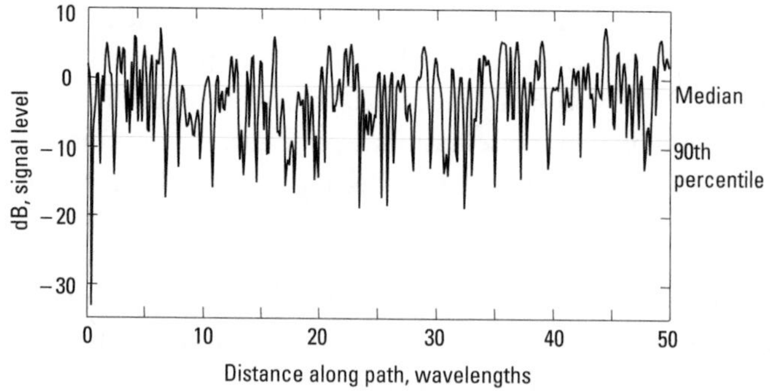

Figure 8.3 A signal subjected to multipath scattering.

value. There is a level identified as "90th percentile" in Figure 8.3 representing the level for which 90% of the signal is above the required signal level. A PCS design goal is to specify a median signal strength at the edge of a coverage local area with enough margin relative to median level so that the signal does not go below the required signal level more than 10% of the time.

The modeled signal envelope shown in Figure 8.3, like similar signal strengths that can be collected experimentally, can be analyzed by counting the number of levels that occur within an interval. This is then compared to the exact Rayleigh PDF as shown in Figure 8.4. The total sample size was 500 points, and the sample interval was 0.8 wavelengths. The cumulative distribution is shown in Figure 8.5 plotted versus the Rayleigh CDF. The model signal shows a generally good fit to the distribution. Measured signal distributions in a local area are typically this close to the Rayleigh curve.

The curves in Figure 8.4 illustrate a particular signal behavior based on a model of uniformly distributed scatterers in a local neighborhood. The Rayleigh behavior noted is typical of the signal statistics encountered in many urban and suburban environments. Most often, signals are measured on a logarithmic scale in decibels.

In cases where there is a predominant direct-signal component added in with diffused scattered signal, the signal envelope distribution is called Rician, and the density function is

$$r(s) = \frac{s}{\sigma^2} \exp\left(-\frac{s^2 + c^2}{2\sigma^2}\right) I_0\left(\frac{sc}{\sigma^2}\right) \tag{8.4}$$

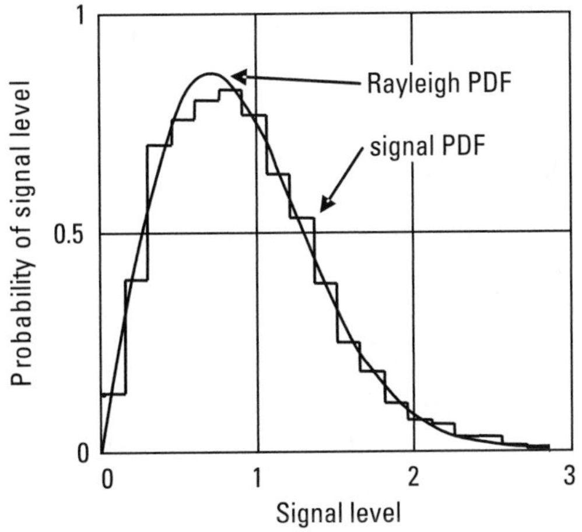

Figure 8.4 Signal statistics compared with a Rayleigh PDF.

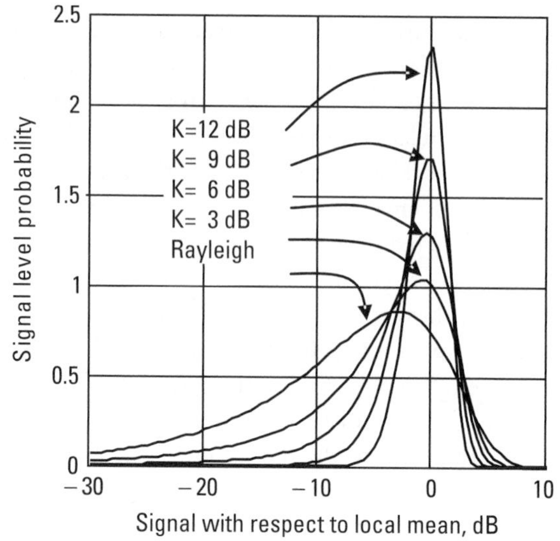

Figure 8.5 Rician PDF normalized to the local mean.

where c^2 represents the signal power in the direct line-of-sight component, and I_0 a is modified Bessel function of the first kind and order zero. The Rician *K*-factor equal to $c^2/2s^2$ is the relationship between LOS signal power and diffuse power. As the LOS power diminishes to zero, (8.4) becomes the same as (8.2), and the density function becomes Rayleigh. Figure 8.5 shows the Rician probability density function for Rician *K*-factors of 3, 6, 9, and 12 dB, compared with the Rayleigh PDF as a function of signal level in decibels. In Chapter 6, we compared a 0.1% bandwidth signal with a 35% bandwidth UWB pulse signal. We can now state that the 0.1% bandwidth PDF resembles a Rayleigh distribution, while the 35% bandwidth signal looks Rician with a *K*-factor of about 14 dB.

8.3.2 Large-Scale Signal Variations

The Rayleigh statistics describe the nature of small-scale signal variations very well over distances of tens to hundreds of wavelengths. Larger-scale variations, which are due to signal shadowing, terrain variations, and perhaps fixed-site antenna pattern undulations, take on Gaussian statistics when the signal is expressed in decibels. The character of slow fading statistics, or variations in the small-area average, arise because propagation past undulations in terrain and rows of buildings of random height, which imparts one random variation to the amplitude, is followed by a second random amplitude variation associated with the last diffraction down to the personal communications terminal. The two random events acting in multiplicative sequence produce the lognormal variation of slow fading. Multiplying random variables is equivalent to adding their logarithms, and by the Central Limit Theorem, the sum goes into a normal distribution, hence exhibits Gaussian statistics. This larger-scale variation is characterized by a lognormal standard deviation typically in the range of 5 to 12 dB.

The Gaussian, or normal, density function takes the form

$$g(w) = \frac{1}{\sqrt{2\pi\sigma^2}} \exp\left[-\frac{(w-\mu)^2}{2\sigma^2}\right] \tag{8.5}$$

When applied to the large-scale signal distribution envelope, w is the signal level expressed in decibels, σ is the standard deviation in decibels of the signal distribution, and μ is the mean signal level stated in decibels. The Gaussian density function is two-sided and symmetric about the mean value. The CDF corresponding to (8.5) is

$$G(w) = \frac{1}{2} + \frac{1}{2}\operatorname{erf}\left[\frac{w-\mu}{\sigma\sqrt{2}}\right] \tag{8.6}$$

8.3.3 Combining Cumulative Distribution Functions

The Rayleigh CDF applies to a local region measured in tens of wavelengths, while the normal CDF applies over the large scale. The Rayleigh CDF, (8.3), and the normal CDF can be combined to give a composite Rayleigh-lognormal distribution function. The composite Rayleigh and lognormal CDF, where σ and μ are the standard deviation and mean in decibels, is

$$P(x \le X) = 1 - \frac{1}{\sqrt{2\pi\sigma}} \int_{-\infty}^{\infty} \exp\left[\frac{(\beta-\mu)^2}{2\sigma^2} - 10^{-(\beta-x)0.1}\right] d\beta \tag{8.7a}$$

Expression (8.7a) can be evaluated numerically; however, an approximate formula is given by Hess [2] as

$$P(x \le X) \approx 1 - \frac{1}{\sqrt{u+1}} \exp\left[-\frac{u(u+2)}{2\sigma^2}\right] \tag{8.7b}$$

where $u = \gamma_0 - \mu$, and γ_0 is from the transcendental relationship $\gamma_0 = \mu + X\sigma^2 \exp(-\gamma_0)$.

8.3.4 Normal Approximation to Composite CDF

Although (8.7a) is closer to the true physical behavior of wave propagation, the composite distribution is often approximated by a normal distribution having a standard deviation σ_{total} given by

$$\sigma_{total} = \sqrt{\sigma_R^2 + \sigma_{LN}^2} \tag{8.8}$$

where σ_{LN} is the lognormal standard deviation in decibels, and $\sigma_R = 7.5$ dB is a popular choice for curve-fitting the Rayleigh CDF to a Gaussian CDF, as can be inferred from Y. Okumura and colleagues [8]. As shown in [2], the true Rayleigh standard deviation is 5.57 dB; however, the curve-fit value of 7.5 dB is approximately correct for $P\{x \leq X\} < 0.5$. The form of (8.8), as will be shown later, is useful in combining the effects of other signal standard deviations, which can be approximated by a Gaussian distribution using the method of G. Hagn [9]. Figure 8.6 shows the combined Rayleigh and lognormal distributions for several lognormal standard deviations. In Figure 8.6, the lognormal standard deviation σ_{LN} is identified as $\sigma_{Lognormal}$. For comparison, the lognormal distribution with Rayleigh and lognormal standard deviations combined according to (8.8) are also shown.

The approximation of (8.8) gives useful results when $P < 0.5$. The mean of the normal distribution is μ and also equals the median. The median of the Rayleigh distribution is, however, 1.592 dB below the mean value. The median value for the composite distribution depends on σ_{LN} and is presented in Table 8.1 for several values of lognormal standard deviations

Table 8.2 compares the composite distribution having lognormal standard deviation $\sigma_{LN} = 10$ dB with the normal approximation having Rayleigh standard deviation $\sigma_R = 7.5$ dB, $\sigma_{LN} = 10$ dB, and $\mu = 0$ dB.

Table 8.3 extends the comparison to a composite distribution having lognormal standard deviation $\sigma_{LN} = 6$ dB, with the normal approximation having Rayleigh standard deviation $\sigma_R = 7.5$ dB, $\sigma_{LN} = 10$ dB, and $\mu = 0$ dB.

Tables 8.2 and 8.3 show that the normal approximation is close (within 1.2 dB for $P \leq 0.1$) to the results obtained with a composite distribution.

[8-3b.mcd] Calculate the combined Rayleigh and lognormal probabilities using (8.7a) for $\sigma_{LN} = 2$ and 8, and compare with the lognormal approximation (8.8) with −1.5 dB mean.

8.3.5 Small-Scale Signal Variations and Delay Spread

Signals that are subjected to multipath scattering travel by paths of varying lengths from the fixed site to the local urban site. As was noted earlier, the traditional descriptions of multipath assume a time-harmonic signal and thus use ray

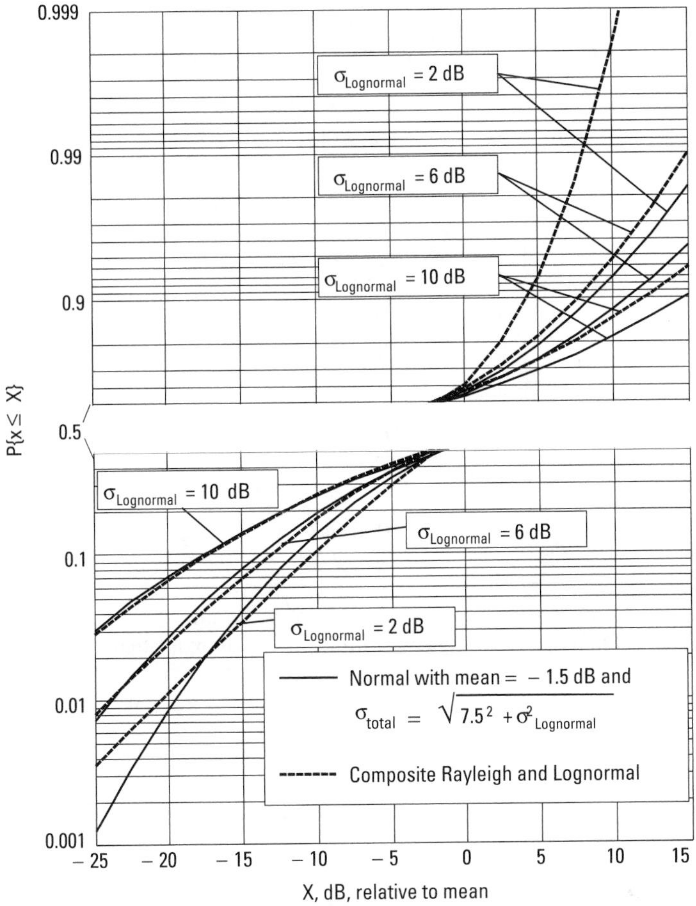

Figure 8.6 Combined Rayleigh and lognormal signal distributions.

path distances that are simplified by modulo one wavelength. The missing time-delay parameter is captured in the delay-spread formulation. If a transmitter sends an impulse $\delta(t)$, the signal at the receive point will be

$$s(t) = \sum_n a_n \delta(t - t_n) \tag{8.9}$$

where t_n is the arrival time of the nth signal and has amplitude a_n. The channel impulse response (CIR) $s(t)$ has no bandwidth restriction until the component amplitudes are specified and the CIR is convolved with the emitted signal. In conventional systems, the transmitted signal is a sine wave; in UWB, it may be a short-duration pulse. We let the power-delay profile $P_n = |a_n|^2$ and time delay of arrival $d_n = t_n - t_0$ of each ray relative to the arrival time t_0 of the first ray. The

Table 8.1
Median Relative Mean for Several Values of Lognormal Standard Deviations

σ_{LN}	**Median Relative to Mean (dB)**
0	–1.592
2	–1.71
4	–1.93
6	–2.10
8	–2.22
10	–2.29
20	–2.51

Table 8.2
Composite Distribution and Normal Approximation for σ_{LN} = 10 dB

P{x ≤ X}	**Composite X (dB)**	**Normal Approximation**
0.01	–30.4	–29.1
0.05	–21.7	–20.6
0.1	–17.2	–16.0
0.5	–2.288	0
0.9	12.0	16.0
0.95	16.0	20.6
0.99	23.3	29.1

Table 8.3
Composite Distribution and Normal Approximation for σ_{LN} = 6 dB

P{x ≤ X}	**Composite X (dB)**	**Normal Approximation**
0.01	–24.0	–22.3
0.05	–16.5	–15.8
0.1	–13.0	–12.3
0.5	–2.099	0
0.9	7.6	12.3
0.95	10.2	15.8
0.99	15.0	22.3

RMS delay spread τ is defined as the square root of the second central moment of the power-delay profile. The delay spread can be calculated from [10]

$$\tau = \sqrt{\frac{\sum P_n d_n^2}{\sum P_n} - \left[\frac{\sum P_n d_n}{\sum P_n}\right]^2} \tag{8.10}$$

When (8.10) is evaluated for two rays of equal amplitude, the delay spread is one half the time delay between the two rays. We refer to twice the delay spread, 2τ, as the signal-weighted differential delay.

Table 8.4 contains examples of typical delay spreads measured in various environments. The distance parameter $\Delta d = c\tau$, where c is the velocity of light, is essentially a measure of the RMS distance to the field of scatterers.

8.3.6 Multipath with UWB Pulses

With UWB pulse transmissions, we can use (8.9) to model the received signal directly by convolving $s(t)$ with a copy of the transmitted pulse, as introduced in Chapter 6. There is no phase to keep track of, only time delays t_n. An interpretation of some measured data [3] indicates that the power-delay profile of (8.9) is exponentially distributed, so a CIR model can be developed in which

$$s(t) = \frac{1}{\sqrt{W}} \sum_n h_n e^{-\Delta t(n+r)/2\tau} \delta[t - \Delta t(n+r)] \tag{8.11}$$

where r is a random variable uniformly distributed in [0, 1], and h_n takes on positive and negative real values. The random variable r takes the place of the random phase variable in the time-harmonic formulation and guarantees that an impulse can occur anywhere in time. The signal is normalized by the square root of the pulse energy:

Table 8.4
Delay Spread in Different Environments

Environment	Delay Spread τ	RMS Distance to Scatterers Δd (m)
In-building	4–25 *ns*	1.2–7.5
Rural	0.2 μs	60
Suburban	0.2–2 μs	60–600
Urban	1–10 μs	300–3,000

$$W = \sum_n h_n^2 \tag{8.12}$$

The h_n^2 values are normally distributed with unity standard deviation. The impulse arrival interval Δt is equal to the lesser of half the UWB pulse duration or $\tau/10$. Figure 8.7 shows one realization of (8.11) with $\tau = 4$ ns and 0.75-ns UWB pulse at 4 GHz. The maximum value of n is $5t/\Delta t$.

A multipath distorted UWB signal $P(t)$ is obtained by convolving $s(t)$ with a UWB pulse signal. Unlike a narrowband signal, $P(t)$ has an extremely small probability of fading more than a few decibels, as was noted in Chapter 6. In fact, the PDF of $P(t)$ is approximated by a Rician distribution having a K-factor of around 14 dB. It is also quite possible to use the CIR of (8.11) with time-harmonic waves by convolving $s(t)$ with a sine wave.

8.3.7 Relation Between Multipath and Propagation Law

As UWB pulses propagate through a building, they are scattered by the many encountered obstacles. Creation of these multipath echoes removes energy from the primary pulse. The echoes of these pulses arrive at a receiver spread out in time, thereby creating the received time-delay profile. Integrating the received energy over the entire delay profile gives the total received energy. The time resolution of UWB technology allows the separation of the direct pulse, which is the first arrival and often the strongest pulse, from the rest of the received pulses.

Because of energy removed from the direct pulse by scattering, it is to be expected that the intensity of the direct pulse will decrease more rapidly with

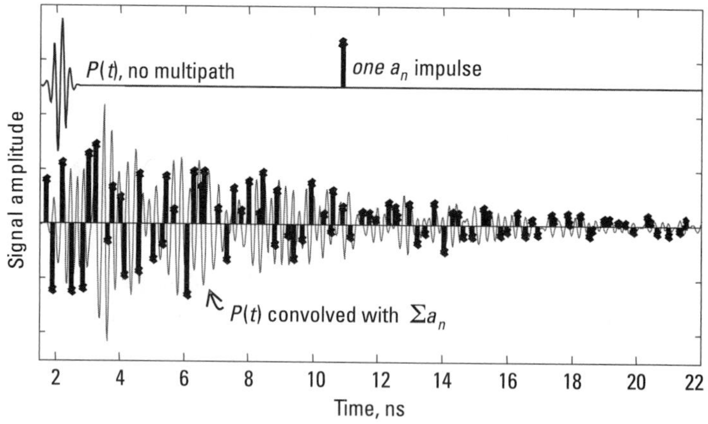

Figure 8.7 A channel impulse response and a UWB signal in multipath.

distance than the $1/d^2$ dependence in free space. At least for moderate distances d, the total energy flux will not be diminished by conversion to heat in the material encountered, and energy is conserved. Thus, in a full three-dimensional scattering environment, such as a large building, conservation of the energy flux leads to a $1/d^2$ propagation dependence.

Energy delay profiles resulting from propagation inside buildings are typically a sequence of impulses arriving at discrete times $t = n\Delta t$ at an average interval $\Delta t << \tau$. They have the classic exponential profile of the average realization in (8.11). The average normalizing energy is found by integrating the summation in $s(t)$ over all time, which leads to the relation

$$W(d) = \left[1 - \exp(-\Delta t / \tau(d))\right] \tag{8.13}$$

for the strongest (leading) arrival energy of the average case in (8.11). The value $\Delta t/\tau(d)$ represents a break-point distance d_t in the range propagation index. Generally, the delay spread increases with distance by some power law u. Thus, combining the free space law and u gives n, so $u = n - 2$ and

$$\tau(d) = \tau_0 d^{\,n-2} \tag{8.14}$$

Consideration of (8.13) leads to the SBY propagation model [11] encountered in Chapter 7. For $\Delta t/\tau(d) < 1$, the term in brackets can replaced by the negative of the argument of the exponential in (8.13) and with $\Delta t/\tau(d) = d_t/d^{n-2}$

$$W(d) = d_t / d^{\,n-2} \tag{8.15}$$

One set of measurements in buildings using UWB pulses shows that $\tau(d) = \tau_C d$ [11, 12]. Thus, for indoor propagation, $W(d) = d_t/d$, which multiples the free space factor to give a net $1/d^3$ factor for $d > d_t$. For propagation of narrowband signals in urban environments from base station antennas to mobiles, the average delay spread is thought to vary as $\tau(d) = \tau_C d^{0.5}$. Distance d is in kilometers, and τ_C, which is the average delay spread at 1 km, has a value in the range 0.4 to 1.0 μs [13]. For such radio links, the scattering environment is essentially a two-dimensional layer of finite thickness, so power scattered upward is lost. As a result, the average of the total power density in the scattering layer exhibits a distance dependence $1/d^n$ with $n > 2$. Equations (8.13) and (8.14) suggests that the strongest arriving multipath component will have distance dependence $1/d^{2.5}$. The difference in range index for the total power and strongest pulse power does not appear to have been measured directly or discussed in the literature.

The exponential distribution of the delay spreads is a good approximation in many cases, but other distributions are often used. One distribution, proposed by A. A. M. Saleh and R. A. Valenzuela [14], comprises a series of delayed exponential distributions and captures the effects on multipath of several major reflectors within a cluttered environment. This model has found application for indoor propagation models for UWB systems. Another model, by M. A. Nemati and R. A. Scholtz [15], proposed a spatially averaged diffusion mechanism. They model the propagated signal energy distribution in time by

$$S(t) = K_1 e^{m_1 t} + K_2 e^{m_2 t} \tag{8.16}$$

where coefficients m_1 and m_2 are negative. For same-room propagation, $K_1 + K_2 = 1$, while for propagation into other rooms, $K_1 - K_2 = 0$. The model approximates exponential distribution in same-room propagation and is consistent with the deterministic in-room model described in Chapter 7.

8.4 Signal Strength Required for Communications

We define a communications coverage area in terms of the geographical regions where there is a given probability P_s that a communication event will be successful. The design value of probability P_s is typically 0.90 on the perimeter or edge of the coverage area for personal communications services. Note that, as can be inferred from [3] and as [16] shows, there is a distinction between the probability of coverage on the edge of the coverage area and probability of coverage averaged over the total area. There are many "standard" definitions for the establishment of a successful communication link. Table 8.5 list some definitions that are applicable to personal communications systems; there are other definitions of successful, or desired, communications events. In Table 8.5, an 80% paging call success rate (CSR) is tied to a specific message, such as an alert, or the correct reception of a 40-character message, or even "the reception of a 55-character alphanumeric message with no more than two consecutive characters in error," as defined in [17] for the Enhanced (formerly European) Radio Messaging System (ERMES). The CSR results in the specification of a required field strength for the paging device worn on a standard body at a specified height, as will be studied in Chapters 9 and 10. Successful communications events in the land-mobile services are tied to a receiver input sensitivity level, which, with the addition of an antenna/body or antenna/vehicle factor, can be related to a required field strength at a specific height. In digital cellular telephony service, for example in IS-95 CDMA systems, the measure of a successful

Table 8.5
Examples of Personal Communications Events

Event	Definition
99% paging call rate	99% probability of receiving a message correctly
80% paging call rate	50% probability of three correct calls in a row
1% bit error rate (BER)	Reception of data with an average BER $\leq$ 0.01
1% WER	Reception of data with an average WER $\leq$ 0.01 (used with IS-95 signaling)
0.1% WER	Reception of data with an average WER $\leq$ 0.001 (used with paging signaling)
10^{-9} BER or 10^{-3} packet error rate	Wireless video decoders
CM (circuit merit)	Audio quality measure based on perceived merit. CM=3 is typical level in land-mobile service.
12-dB SINAD level	Reception of 1 kHz tone 12 dB above the combined signal, noise, and distortion level
20-dB quieting	Reception of an unmodulated signal at a level that reduces the no-signal nosie level by 20 dB

event is defined by the 1% word error rate (WER). In fact, in that system, the portable or mobile power is adjusted to target a 1% WER.

8.4.1 Signal Call Success Probability

Once a required signal strength based on the desired communications event is determined, signal margin is added to result in the desired performance under shadowing and multipath conditions, as is described in [18]. Hagn [9] discusses a method for including many factors that affect communications success probability. Assuming all of the factors are independent Gaussian random variables, mean values and associated standard deviations can be assigned. The mean values L_i of the i various loss factors add directly to the margin, while the standard deviations are combined as the sum of their variances to obtain a single total standard deviation, σ_{total}, in the manner of (8.8). The expression for margin M takes the form

$$M = \sum_i L_i + z\sigma_{total} \tag{8.17}$$

The probability P that a Gaussian random variable z is less than or equal to an arbitrary value Z is found from

$$P\{z \leq Z\} = 1 - P_s = \frac{1}{2} - \frac{1}{2}\mathrm{erf}\left[\frac{z}{\sqrt{2}}\right] \tag{8.18}$$

where P_s is the desired communication call success probability, while erf(u) is the error function. Typically, we might desire $P_s = 0.90$. Expression (8.10) is readily solved numerically, and Table 8.6 lists values of z for some useful call success probabilities P_s.

[8-4.mcd] Calculate z for the values of P_s in Table 8.6 by solving for z in (8.18).

Hagn lists many factors to which he assigns standard deviations, such as the standard deviation of transmitter power, transmission line losses, antenna gain, basic transmission loss, antenna circuit loss, long-term fading, location variability, vegetation loss, and urban loss. We will concern ourselves here only with the total standard deviation of basic transmission loss, which is expressed as

$$\sigma_{total} = \sqrt{\sigma_R^2 + \sigma_{LN}^2 + \sigma_{veg}^2 + \sigma_B^2 + \ldots} \tag{8.19}$$

where

σ_R = 7.5 dB, which is the normal approximation to the Rayleigh standard deviation;

σ_{LN} = the lognormal standard deviation, typically between 5 and 12 dB;

Table 8.6
Call Success Probability Versus z Factor

P_s	z
0.50	0
0.60	0.253
0.70	0.524
0.80	0.842
0.90	1.282
0.95	1.645
0.99	2.326

σ_{veg} = standard deviation of vegetation loss, typically 2.5 dB when present;

σ_B = standard deviation of building losses, between 4 and 14 dB when applicable;

σ_V = standard deviation of vehicle losses, typically 5 dB when applicable.

Not all of the factors are always present; for example, building losses and vehicle losses are typically exclusive. The ellipsis in (8.19) signifies other standard deviations that may be included where applicable, such as variations due to the fixed-site antenna pattern, as was discussed in Section 2.9.4. Putting it all together now, once a performance or signal call success reliability level S is specified, as, for example, resulting in one of the communications events of Table 8.5, and the local environment is characterized in terms of additional loss factors L_i and a combined standard deviation σ_{total}, the margin M required above the median field strength can be computed from (8.17).

8.4.2 Determining the Fixed Station Power

The required median signal strength S_r in terms of a performance field strength S and margin M is

$$S_r = S + M \tag{8.20}$$

The required field strength S_r must now be stated in terms of the transmitted power, the propagation losses, and the fixed-site gain-loss budget. The field strength S_r, stated as decibels in ratio to 1 μV/m (abbreviated dBμV/m), is

$$S_r = [P_{TX} + L_{TL} + G_{TX}] + L_{prop} + [77.2159 + 20\log(f)] \tag{8.21}$$

where the first set of brackets represents the effective isotropically radiated power (EIRP) from the fixed site with P_{TX} dBm transmitter power, L_{TL} transmission line losses (a negative value) in decibels, and G_{TX} dBi antenna gain. The second term is decibel path attenuation (a negative value), and the last set of brackets contains frequency f in megahertz and contains the units conversion constant. The median propagation loss is L_{prop} and can be calculated using the models given in Chapter 7. S_r is a function of distance since L_{prop} is a function of distance.

Given a geographic area for which communications with call success probability greater than P_s is specified, the factor z is found from (8.18), and the various environmental factors are combined to get M in (8.17), from which the required field strength S_r is found from (8.20). That field strength relates to a fixed-site EIRP through (8.21) via the median propagation loss.

8.5 Diversity Techniques

Signal reception can be improved whenever there is a probability of receiving the signal in at least two independent ways. The degree of independence determines the degree of improvement. Repeating transmissions is one method of improving the probability of signal reception. Another way is by using MIMO and MISO techniques, especially in combination with time-space coding techniques. Simultaneous transmission, also called simulcast and quasisynchronous transmissions, can be used to flood an area with signal and provide opportunities for receiving independently faded signals from multiple sources. Fixed receive sites may employ diversity in the form of multiple receive antennas separated horizontally by several meters or by collocated antennas that receive two orthogonal polarizations. At the portable or mobile end, the sampling of the signal in multiple ways can be achieved by considering multiple electromagnetic field components, for instance, by using polarization diversity or "energy density" antennas. Energy density in this context refers to the reception of an electric and a magnetic field component associated with the same polarization. Finally, systems employing very short channel symbol times, that is, symbol or CDMA chip times commensurate with and shorter than the multipath differential delay, can employ rake receivers [19] to detect and combine individual multipath components. CDMA, as defined in the IS-95 specification, requires rake receivers with at least three signal branches.

8.5.1 Diversity Improvement by Repeated Transmission

We have shown that margin is required with respect to the median signal to achieve a specified level of performance. In fact, examination of Figure 8.6 shows that when the shadowing component of the standard deviation is 10 dB (equivalent total standard deviation is 12.5 dB), getting from probability of a missed signal of, for example, 0.316 to a miss probability of 0.1 requires about 10 dB. This is equivalent to transmitting ten times as much energy to get the required transmission reliability for the message. In cases where real-time transmission is not a factor, the same improvement may be achieved by simple repeating the transmission twice, as long as the two transmissions are not correlated. In general, the call probability P_N after N uncorrelated transmissions is given in terms of the single call probabilities, $P_{s,n}$, by

$$P_N = 1 - \prod_{n=1}^{N} \left(1 - P_{s,n}\right) \tag{8.22}$$

Hence, two equally probable uncorrelated transmissions at P_s = (1 – 0.316) give P_N = 0.9. If the repeat strategy is carefully chosen so that the repeat is not correlated with the original transmission, the total energy required for a given level of performance can be optimized. In the example given, two transmissions requires twice the energy of a single transmission, but simply increasing the power requires one transmission using ten times the energy. The net "diversity gain" here is 10 dB – 3 dB = 7 dB. In systems where energy is costly, like satellite-based communications, the economic benefits of a repeat strategy can be enormous. The relative benefit depends to a large degree on the standard deviation. For example, if in the normal case the standard deviation were only 3 dB, the difference in signal levels between P_s = 1 – 0.316 and P_s = 1 – 0.1 is only 2.4 dB. The net diversity gain in this case is 2.4 dB – 3.0 dB = –0.6 dB, a net loss. It can be shown that for P_n = 0.9 with a Gaussian distribution, the threshold for diversity improvement from a single repeat occurs when the standard deviation is 3.75 dB. Figure 8.8 shows improvement in the calling probability, or CSR, as a function of the number of uncorrelated transmissions with initial call probability as a parameter. Keeping multiple transmissions uncorrelated becomes increasingly difficult to accomplish in practical systems, and there is a diminishing return after just a few such transmissions.

8.5.2 Simultaneous Transmissions in Radio Communications

Simultaneous transmission, where the same information is transmitted with the same modulation on essentially the same frequency with geographic over-

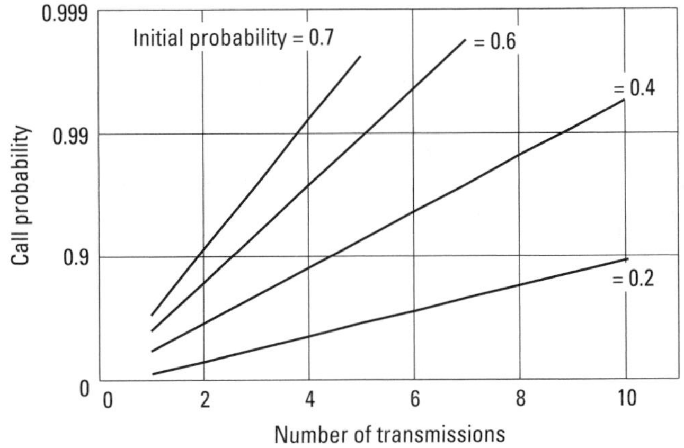

Figure 8.8 Call probability for multiple transmissions.

lap in coverage, is sometimes referred to as simulcast [2] or as *quasisynchronous transmission* [17]. In some wide-area radio communications systems, several synchronized transmitters are used to deliver a signal to a large geographic area and to "flood" that area with power. Wide-area coverage is available from multiple transmitters, and large increases in user capacity are possible over sequential transmission. The principles of diversity embodied by (8.22) apply to simultaneous transmissions. For example, an area covered with 70% probability from each of two transmitters will have a net coverage probability equal to $(1 - (1 - 0.7)^2) = 0.91$, or 91%. Figure 8.9 shows the improvement in the calling probability in the overlap region between two transmitters. One transmitter is located at the 0-km point in Figure 8.9, and the second simulcasting transmitter is at 40 km. At the midpoint in coverage, each individual transmitter provides enough signal for a calling success probability of about 0.7. The overlap of two such transmitter patterns raises that probability to 0.91 in the absence of any simulcasting distortion effects.

Simulcast systems need close coordination of transmitter frequency errors, transmission time delays, and propagation overlap between such simultaneous transmissions. The radiowave propagation issues in simultaneous transmissions are wave interference in the signal overlap area and relative propagation delay of the signals. Severe intersymbol interference is possible if the differential delay between two signals is too great and the received power levels are nearly equal.

Simulcast transmitting systems require close attention to the transmitting frequencies of the multiple sites. In digital paging applications, for example, frequencies are often offset from each other to mitigate the effects of standing wave interference patterns, which could otherwise cause localized areas of poor coverage. Table 8.7 shows the offset frequency increments for digital messaging systems in noninterleaved signal formats having symbol rates up to about 1,200

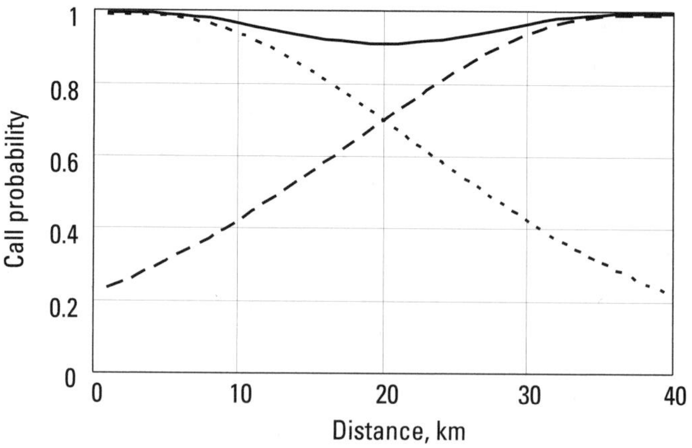

Figure 8.9 Calling probability from two transmitters.

Table 8.7
Frequency Offset Increments for Overlapping Simulcasting Transmitters for Noninterleaved Signal Formats

Number of Overlapping Transmitters	Frequency Offset Increment (Hz)
2–3	450
4–5	300
6–7	200
8–9	150
10–13	100

symbols per second for different numbers of transmitters in the overlap region. The maximum offset of the carrier frequency is chosen never to exceed ±600 Hz. For interleaved signal formats like FLEX™, frequency offsets must be carefully selected to avoid multiple signal phase cancellations within an interleave block. Optimum offset choices for FLEX™ are 15, 30, 35, 45, 60, 70, and 90 Hz. Offsets below 10 Hz and multiples of 20 and 25 Hz should be avoided.

Signals arriving from multiple transmitters are subject to different delays due to different individual propagation path lengths. The combined effects of local multipath scattering ensure that the signal levels from different sites have wide ranging amplitudes. When the delay differential in the digital transmission system is more than about 0.15 to 0.25 of a symbol time, and the amplitudes are nearly the same, severe intersymbol interference is possible, resulting in loss of data. This is true even in strong signal conditions [20], as shown in Figure 8.10, where the performance of a 3,200-baud (3,200-bps FSK and 6,400 4-FSK) and a 1,600-baud (1,600-bps FSK and 3,200 4-FSK) receiver is depicted. Word error rates begin to increase when either the signal-to-noise ratio is too small or when nearly equal simulcast signal levels are excessively differentially delayed, as shown in Figure 8.10(a). For a fixed WER of, for example 0.1%, the required CNR (defined there as the ratio of the sum of the desired signal power to the noise in decibels) increases with differential delay between two unequal-power signals. The power difference in Figure 8.10(a) is 10 dB. *The objective of a simulcasting paging system design is to provide signals that appear above and to the left of the curves in Figure 8.10(a).* Performance at low CNR is limited by noise and by "zero-beating" between the simulcasting signals. At high differential delays, even for strong signals, performance is limited by the receiver's inability to "capture" one of the signals. This behavior is seen also in the capture curves of Figure 8.10(b), where the power difference ΔP in decibels is shown plotted versus delay time difference for CNR equal to 10, 20, and 30 dB at the 0.1% WER

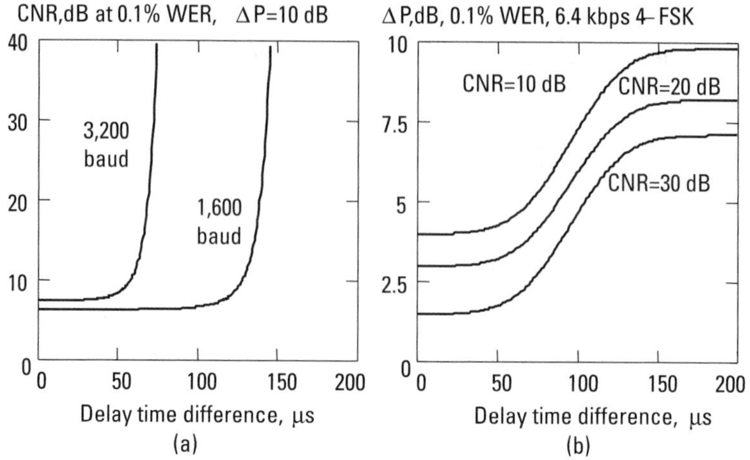

Figure 8.10 Differential delay performance of 3,200- and 1,600-baud FM signaling.

performance level. The performance shown in Figure 8.10 is generally illustrative of 3,200-baud FM receivers; however, specific performance of particular devices must be measured in the laboratory.

Simulcast systems are designed so that the differential delay is smaller than about 0.15 to 0.25 symbol times. A multispeed signaling format such as FLEX allows system operators to transition gracefully from lower performance signaling of POCSAG 1,200 by permitting operation at 1,600 bps with essentially no infrastructure changes. As capacity requirements demand, the signaling speed can be increased and system infrastructure upgraded according to the performance demands shown in Figure 8.10. Performance in a simulcasting environment can also be improved by using signal-diversity techniques, including time diversity [21]. Table 8.8 shows the maximum permissible delay times based on a differential delay of between 0.15 and 0.25 symbol times and the corresponding differential distances for digital paging and messaging systems operating at various symbol rates. In effect, it is easier to space transmitters farther apart when the symbol rates are slower. The values in Table 8.8 are approximate and will vary depending on the actual modulation and on the particular forward-error-correcting code used.

Table 8.8 shows, for example, that a simulcasting paging system designed to operate using POCSAG signaling at 1,200 symbols per second (1,200 bps) will also typically operate with minimal simulcast differential delay degradation using FLEX signaling at 1,600 symbols per second (1,600-bps FSK or 3,200 bps 4-FSK). Paging network operators exercising that path can then upgrade their networks to 3,200-baud (3,200-bps FSK and 6,400-bps 4-FSK) FLEX operation only in the geographical service regions where demands on network capacity warrant the network infrastructure enhancements and investments.

Table 8.8
Data Rates and Maximum Differential Delay and Distances Based on 0.15 to 0.25 Symbol Time

Data Rate (symbols/s)	Maximum Permitted Differential Delay (μsec)	Maximum Permitted Differential Distance (km)
512	293–488	88–146
1,200	125–208	37–62
1,600	94–156	28–48
3,200	47–78	14–23

8.5.3 Diversity Reception by Multiple Antennas

Signals transmitted by personal communications devices (PCDs) are subjected to the same multiple paths as received signals. From the point of view of a fixed receive site, the signal appears to emanate from a distributed source (see Figure 8.1), which encompasses the many scattering obstacles. The resulting "transmit pattern" of one such distributed source is shown in Figure 8.11 and affords us an opportunity to study fixed-site receiver diversity. The signal across several meters width at the fixed site, for example antennas A and B in Figure 8.1, is just a small angular extent of that pattern but exhibits significant signal variation, as seen in Figure 8.11. Thus, two antennas separated by just a few meters will, on average, encounter signal levels that are several decibels different.

The problem can be studied using Monte Carlo methods and also analytically. The distributed source is modeled approximately by multiple sources of random amplitude and phase that are distributed randomly in azimuth at a scattering radius from the portable. The radius is chosen from a knowledge of the delay profile of signals received at that location. The pattern is sampled by the two physically separated antennas. Samples are collected for various points on the pattern, and the average decibel magnitude difference is reported. The circle symbols in Figure 8.12 show the results of one such Monte Carlo simulation, specifically, the average decibel difference for signals received from a portable or mobile at a distance of r = 15 km, illuminating an s = 300m radius of scatterers, with wave number k evaluated at 930 MHz. The result scales with scattering radius, frequency, and the inverse of the range to the PCD, so normalized antenna separation is shown. For the studied range and scattering radius, the average signal variation increases from 0 dB for coincident antennas to a limiting value of 6.02 dB at normalized separations beyond $kds/r = 2$, or about d = 5m. That signal variation is the opportunity for diversity improvement of the radio link. A similar result can be found from the correlation coefficient of signals received by two fixed-site antennas. W. C. Jakes [3] gives that correlation

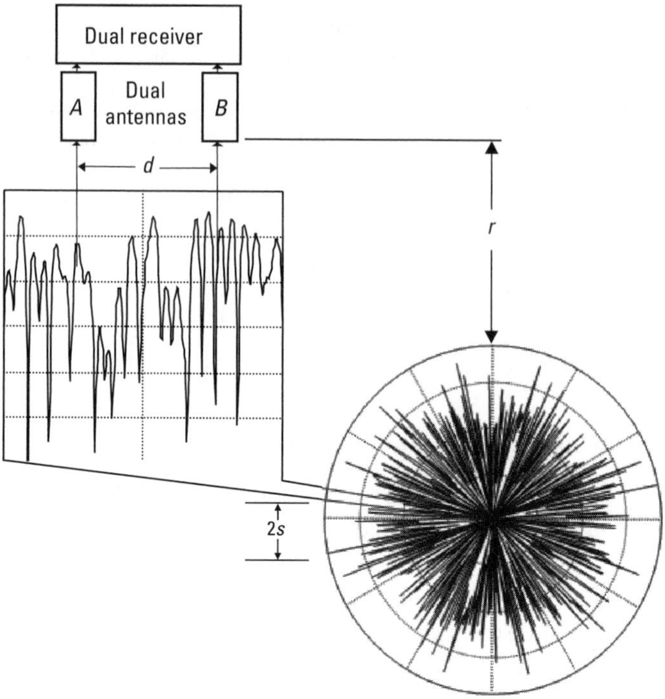

Figure 8.11 The signal at a fixed site from a multipath environment.

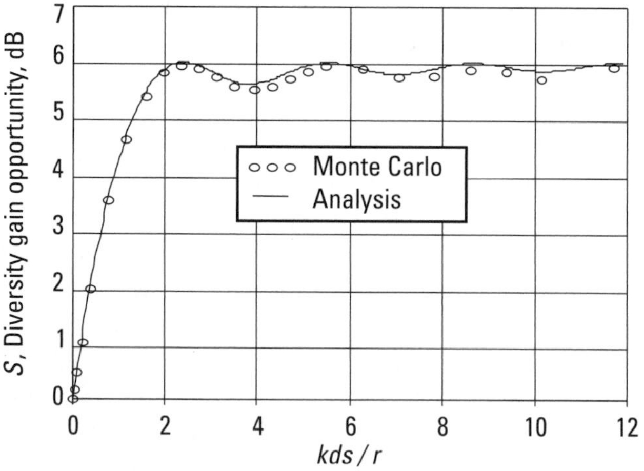

Figure 8.12 Diversity gain opportunity from two fixed-site receiver antennas versus normalized antenna separation.

function ρ_e between *voltage* samples from two fixed-site antennas separated by d as

$$\rho_e = J_0^2\left[kd\,\frac{s}{r}\sin(a)\right]J_0^2\left[\frac{kd}{2}\left(\frac{s}{r}\right)^2\sqrt{1-\frac{3}{4}\cos(a)^2}\right] \tag{8.23}$$

where a is the angle between the axis of the fixed-site antennas and the PCD. Since ρ_e represents the signal power corresponding to correlated voltage signal components, then $1 - \rho_e$ is the uncorrelated signal power. The average signal variation, or diversity gain, from signals received by the two antennas can then be expressed as

$$S = 20\log\left[1+\sqrt{1-\rho_e}\right] \tag{8.24}$$

which is shown as the solid trace in Figure 8.12. For discrete samples, the correlation coefficient between sample sets u_i and v_i is defined by

$$\rho_{uv} = \frac{\sum_i \frac{u_i v_i}{N} - \langle u\rangle\langle v\rangle}{\sqrt{\sum_i \frac{u_i^2}{N} - \langle u\rangle^2}\sqrt{\sum_i \frac{v_i^2}{n} - \langle v\rangle^2}} \tag{8.25}$$

where $\langle u\rangle$ and $\langle v\rangle$ are the mean values of the N samples u_i and v_i, respectively.

[8-5a.mcd] Simulate a 900-MHz transmitting PCD that illuminates scatterers at 300m radius and is at 15-km range. Find the average decibel magnitude, $|10\log(X/Y)|$, of two signal power levels X and Y received by two antennas separated horizontally by 1m, 5m, and 10m. Compare with (8.24).

Having multiple copies of the same signal obtained from spatially separated samples of a single field component of a signal is called *space diversity*. If the signal statistics are Rayleigh distributed, as is likely with a portable or mobile PCD, the correlation coefficient ρ between samples spaced a distance d given in [3] is

$$\rho = J_0^2(kd) \tag{8.26}$$

where $k = 2\pi/\lambda$ is the wave number, and $J_0(x)$ is the Bessel function. A correlation coefficient of less than 0.5 is generally deemed useful; hence, from (8.26), the samples must be separated by at least 0.2 wavelengths.

Multiple copies of the same signal can also be obtained from the electric and magnetic field of the same polarization; having multiple copies is called *field diversity* or *energy density diversity*. If the signal statistics are Rayleigh distributed, the cross-correlation coefficient ρ_X between the two electromagnetic fields is [3]

$$\rho_x = J_1^2(kd) \tag{8.27}$$

where $J_1(x)$ is the Bessel function. Note that from (8.27), collocated electric and magnetic antennas have zero cross correlation.

8.5.4 Diversity Reception of Lognormally Distributed Signals

The same signals may be received at two or more fixed sites and exploited to improve the probability of reception. This form of reception is sometimes called *macrodiversity* (although that term also implies a particular signal combining method). We can study the opportunity for this form of diversity by characterizing the signals received at two sites from a portable or mobile PCD in terms of the signal mean value and a lognormal standard deviation. The signals are assumed to travel to each of the two fixed sites along independent paths and to be subjected to independent shadowing variation. Although the composite CDF of (8.7a) is closer to the true physical behavior, the lognormal approximation leads to a more tractable solution. We write the combined distribution of two lognormally distributed signals having standard deviations σ_a and σ_b dB and differing in average power by m dB as

$$F(z) = \frac{1}{2\pi\sigma_a\sigma_b}\int_{-\infty}^{\infty}\exp\left[-\frac{(w+z)^2}{2\sigma_a^2}\right]\exp\left[-\frac{(w-m)^2}{2\sigma_b^2}\right]dw \tag{8.28}$$

which integrates to

$$F(z) = \frac{\exp\left(-\dfrac{(z-m)^2}{2\sigma_e^2}\right)}{\sqrt{2\pi\sigma_e^2}} \tag{8.29}$$

We make the assignment

$$\alpha_e = \sqrt{\alpha_a^2 + \alpha_b^2 - 2\rho\alpha_a\alpha_b} \tag{8.30}$$

where ρ represents the correlation between the signals from sites a and b. The magnitude average of (8.29) is found by integrating over z. This has the closed-form solution

$$g_L = \sqrt{\frac{2}{\pi}}\sigma_e \exp\left(-\frac{m^2}{2\sigma_\varepsilon^2}\right) + m\,\mathrm{erf}\left(\frac{m^2}{\sqrt{2}\sigma_\varepsilon}\right) \qquad (8.31)$$

Expression (8.31) gives the decibel magnitude of the average signal difference received at two sites *a* and *b*, where the lognormal standard deviations for the paths from the portable PCD to the sites are σ_a and σ_b dB, and the mean difference in signal strength is *m* dB. For the case when the signals are of equal strength, that is $m = 0$, (8.31) reduces to

$$g_0 = \sqrt{\frac{2}{\pi}}\sigma_e \qquad (8.32)$$

Thus, the stronger of the two signals received at either site *a* or site *b* may be improved, on average, by

$$S_L = g_L - m \qquad (8.33)$$

when the signal powers differ by *m* dB. Figure 8.13 shows the average signal improvement in decibels by considering the stronger of the signals received at two sites as a function of the mean power ratio expressed in decibels. The lognormal standard deviation of the signal to each site is 8 dB in the example. The same result may be obtained using a Monte Carlo simulation.

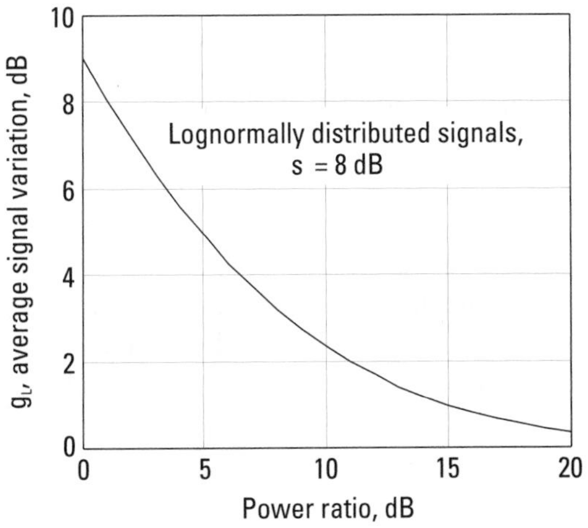

Figure 8.13 Diversity gain opportunity from lognormally distributed signals received at two fixed sites.

[8-5b.mcd] Find the decibel magnitude of the average signal difference received at two sites *a* and *b* and the diversity opportunity if the lognormal standard deviations for the paths from the portable PCD to the sites are σ_a and σ_b dB and the mean difference in signal strength is *m* dB by Monte Carlo simulation. Use $m = 0$, 1, and 3 dB, and $\sigma_a = \sigma_b =$ 0.1, 2, 4, and 8 dB.

8.5.5 Diversity Reception of Rayleigh-Distributed Signals

Portable devices in personal communications systems are often very small and afford little opportunity for spatial sampling of signals subjected to a multipath environment. Likewise, large physical separation of antennas may not always be feasible at a fixed site. It is reasonable to expect the average power measured in two orthogonal polarizations at a fixed-site point to be nearly equal. PCDs are rarely held exactly vertically, and the multipath environment further contributes to polarization randomization. The scattering environment also randomizes the wave impedance to some extent. That, therefore, provides an opportunity to exploit the reception of the signal in different ways to improve the link margin statistically. If two Rayleigh-distributed fields are available for sampling, as, for example, two orthogonally polarized waves or two field components, such as the electric and the magnetic fields, then we are interested in knowing and exploiting the absolute decibel difference between these sample pairs.

The problem can be solved using Monte Carlo methods. From the Rayleigh cumulative distribution function $F(s)$ of (8.3), we solve for the power random variable $X = s^2$ in a distribution of average power *a*,

$$X = -a\ln(1 - F_X) \tag{8.34}$$

and a second independent power variable Y in a distribution of average power *b*,

$$Y = -b\ln(1 - F_Y) \tag{8.35}$$

The CDFs F_X and F_Y are uniformly distributed between 0 and 1. The average magnitude decibel difference $|10\log(X/Y)|$ is found to be 6.02 dB for a sufficiently large number of samples X and Y. When the average power of distribution Y is A dB below the average power of the X distribution, the opportunity for diversity improvement is

$$g_R = |10\log(X/Y)| - A \tag{8.36}$$

[8-5c.mcd] Find the average magnitude decibel difference $|10\log(X/Y)|$ for two equal-power, independent, Rayleigh-distributed, random power variables X and Y by Monte Carlo simulation and show that the result is 6.02 dB.

The analytical solution was suggested by Hess [22] and extended here for distributions of unequal average power. We let $Z = X/Y$, where X and Y are independent, identically distributed, exponential random variables of the form

$$f_W(w) = (1/a)\exp(-w/a); \quad w \geq 0 \tag{8.37}$$

where a is the average power, and W equals X or Y. The density function for Z is given [23] by

$$\begin{aligned} f_Z &= z^{-2}\int_0^{-\infty} f_X(x) f_Y(y)\, x dx \\ &= \frac{1}{abz^2}\int_0^{-\infty} \exp(-x/a)\exp(-x/(bz))\, x dx \\ &= 1/(z + a/b)^2 \end{aligned} \tag{8.38}$$

where the signal distributions are in a power ratio of $a/b \geq 1$. The average value g_R of the decibel difference between two signals is the expectation of a decibel magnitude of z, which from [24] can be evaluated as

$$\begin{aligned} g_R = \int_0^{-\infty} \frac{10|\log(z)|}{(a/b+z)^2} dz &= \frac{10}{\ln(10)}\left[\int_1^{\infty} \frac{\ln(z)}{(a/b+z)^2} dz - \int_0^1 \frac{\ln(z)}{(a/b+z)^2} dz\right] \\ &= 10\left[2\frac{\log(1+a/b)}{a/b} - \frac{\log(a/b)}{a/b}\right]; \quad a/b \geq 1 \end{aligned} \tag{8.39}$$

Thus, on the average, two equal-power ($a = b$), independent, Rayleigh-distributed signals the will differ by 6.02 dB. This, then, is the opportunity for diversity reception in the ideal case of equal-power, independent distributions. When the two signal distributions are correlated or the average power is unequal, then the magnitude of the average signal difference, hence the opportunity for diversity reception, is reduced. For example, when the two power distributions are in a ratio of 3 dB, then from (8.39), the opportunity for diversity is g_R = 3.27 dB. Figure 8.14 shows that the diversity opportunity decreases rapidly as the power ratio between the distributions increases.

Sometimes the Rayleigh-distributed signals are slightly correlated, that is, $\rho > 0$. The maximum available diversity gain is, in that case, governed by (8.24). An estimate of the available diversity gain in decibels for the case of two

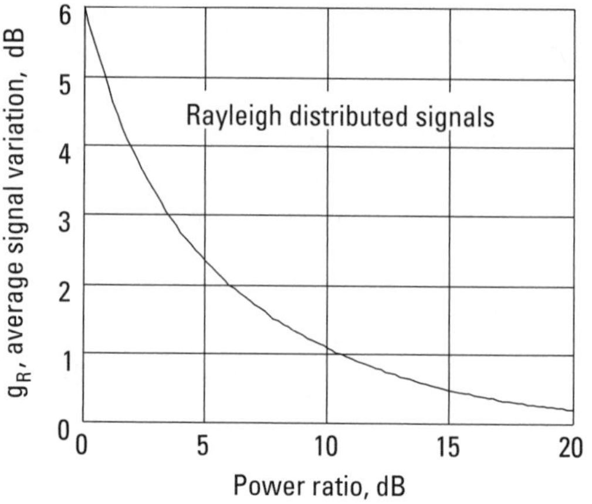

Figure 8.14 Diversity gain opportunity with unequal-power, independent, Rayleigh-distributed signals.

unequal-power, Rayleigh-distributed signals with correlation coefficient ρ can be written as

$$G_R = 20\log\left[1+\left[10^{g_R/20}-1\right]\sqrt{1-\rho}\right] \tag{8.40}$$

[8-5d.mcd] Find the diversity gain opportunity for a dual-polarized receiver system operating in an environment where the horizontally and vertically polarized signal powers vary over a range of 0 to 10 dB and the correlation coefficient is between 0 and 0.8.

8.5.6 Mitigation of Multipath Effects

CDMA spread spectrum systems use high chip rates, which often have chip durations that are shorter than the multipath differential delays. A rake receiver [19] is used to detect, align, and combine these individual multipath components. The IS-95 specification for CDMA cellular telephony requires rake receivers with at least three rake fingers or signal branches. The multipath components due to unequal propagation path lengths of scattered signals result in multiple copies of the same signal delayed in time. A rake receiver uses multiple correlators to separately detect several of the multipath components. The outputs of the correlators are weighted and combined to provide an optimized estimate of the received signal, which is better than any single signal component estimate. The rake receiver can also be set to combine the signals received from multiple codes originating from multiple transmitters, providing a unique solution to the intercell handoff problem in cellular telephony, as well as an opportunity for multisite diversity improvement.

8.5.7 Maximum Rake Gain for UWB Pulses in Multipath

The maximum available rake gain is defined by the ratio of total energy density to single-pulse energy density. When expressed in decibels, using (8.11), on average,

$$G_{\max} = -10\log\left[1-\exp\left(-d_t / d^{n-2}\right)\right] \tag{8.41}$$

Based on the indoor measurements and for $d > 1$, $G_{max} = 10\log(d/d_t)$, so G_{max} increases with $d > d_t$ as $10\log(d)$. However, in the case of cellular systems in the urban environment, the dependence of τ is $d^{0.5}$, which will result in G_{max} increasing only as $5\log(d)$. The negative of G_{max} is the modifier to free space propagation in the SBY model of Chapter 7.

8.6 Multiple-Input, Multiple-Output Systems

MIMO systems exploit multiple independent propagation paths between a transmitter array and a receiver array to increase the capacity and/or the link margin performance of a wireless system. In the general case, MIMO also includes single-input, multiple-output (SIMO) and multiple-input, single- output (MISO) systems. The latter case was combined with space-time coding by S. M. Alamouti [25] to improve the performance of MIMO channels. The capacity of a MIMO system depends on the coupling between paths in the propagation channels; hence, antenna coupling, as well as propagation path correlation, will affect the outcome. A MIMO system can be modeled using the approach in [26], based on receive and transmit correlation matrices. Coupling among the transmitter and receiver antennas occurs due to antenna mutual coupling and due to propagation channel path cross correlation.

8.6.1 A MIMO System Reference Model

In general, the MIMO system comprises N transmitter antennas and M receiver antennas, and a corresponding $M \times N$ channel matrix. Thus, there is an $N \times N$ transmitter mutual-coupling matrix, and an $M \times M$ receiver mutual-coupling matrix. A system reference model is depicted in Figure 8.15. The propagation channel may include both line of sight (LOS) and nonline-of-sight (NLOS) components.

The MIMO system model is given by

$$Y = C_{RX} H_{chan} C_{TX} X + V \tag{8.42}$$

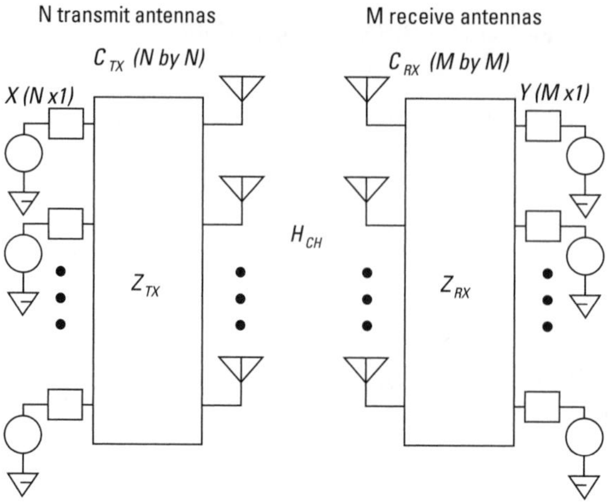

Figure 8.15 The MIMO system reference model. (After [27]).

where Y represents the receiver system output vector, and X represents the transmitter input vector. C_{RX} and C_{TX} are the receiver and transmitter system antenna mutual-coupling matrices, H_{chan} is the channel matrix, and V is the complex white noise vector. The MIMO channel matrix H_{chan} can be separated into a LOS path matrix and a diffuse path, or NLOS, matrix [27, 28]. An example channel matrix employing four transmit and four receive antennas is

$$
\begin{aligned}
H_{chan} &= \left(\sqrt{K_R}H_{LOS} + \sqrt{1-K_R}H_{NLOS}\right) \\
&= \left(\sqrt{K_R}\begin{bmatrix} e^{j\phi_{11}} & e^{j\phi_{12}} & e^{j\phi_{13}} & e^{j\phi_{14}} \\ e^{j\phi_{21}} & e^{j\phi_{22}} & e^{j\phi_{23}} & e^{j\phi_{24}} \\ e^{j\phi_{31}} & e^{j\phi_{32}} & e^{j\phi_{33}} & e^{j\phi_{34}} \\ e^{j\phi_{41}} & e^{j\phi_{42}} & e^{j\phi_{43}} & e^{j\phi_{44}} \end{bmatrix} \right. \\
&\qquad \left. + \sqrt{1-K_R}\begin{bmatrix} R_{11} & R_{12} & R_{13} & R_{14} \\ R_{21} & R_{22} & R_{23} & R_{24} \\ R_{31} & R_{32} & R_{33} & R_{34} \\ R_{41} & R_{42} & R_{43} & R_{44} \end{bmatrix}\right)
\end{aligned}
\tag{8.43}
$$

where $K_R = \kappa/(\kappa+1)$, and κ is the Rician K-factor: a ratio of the LOS power to scattered signal power. The diffuse matrix H_{Ray} is a stochastic matrix representing the scattered or diffuse NLOS components, and the deterministic matrix H_{LOS} represents the LOS elements $\exp(j\phi_{ij})$.

Equation (8.42) identifies three coupling mechanisms that impact the capacity between an array of receiving signals Y and an array of transmitting signals X. The three mechanisms are the receiver antenna mutual-coupling matrix C_{RX}, the transmitting antenna mutual-coupling matrix C_{TX}, and the propagation channel matrix H_{chan}. Each of these deteriorates the channel capacity when the correlation coefficients increase.

8.6.2 MIMO System Capacity

A MIMO system capacity evaluation can be demonstrated for the case of base station receiving antennas that are in the clear and spaced in the horizontal plane. When the pair of receiving antennas is totally uncorrelated, the problem reduces to one of finding the correlation between propagation paths, as shown in Figure 8.11.

MIMO capacity in bits per second per hertz is bounded by the Shannon-Hartley capacity formula:

$$C = \log_2\left\{\det\left(1 + \left(\sigma_2 / \sigma_N^2\right)\left(K_R H_{LOS} + \left(1 - K_R\right)\right)\right) H_{NLOS}\right\} \quad (8.44)$$

In this case, only the NLOS components are present and

$$R_{ij} = E\left\{H_{NLOS} H^*_{NLOS}\right\} \quad (8.45)$$

The capacity is directly related to the correlation $R_{ij} = \rho_e$ of (8.23) for the case of two base station antennas. In this two-by-two antenna system, the ergodic capacity based on this geometry is

$$C = 1 + \sqrt{1 - \rho_e} \quad (8.46)$$

which, for $\rho_e \to 0$, limits to $C = 2$. Thus, the two-antenna base station system studied earlier for its diversity reception gain is seen to be suitable for a potential doubling of MIMO system capacity.

8.6.3 MIMO System Capacity with a LOS Component

In the LOS condition, the ergodic capacity with M antenna separated >> $\lambda/2$ reduces from (8.44) in terms of the signal covariance to (see [27])

$$C_{LOS} = \log_2\left(1 + \left(\sigma^2 / \sigma_N^2\right) M^2\right) \quad (8.47)$$

Thus, the capacity for the LOS case is affected by the increase in the signal-to-noise ratio, hence increases with the logarithm of the number of paths, as seen in Figure 8.16(a).

For NLOS channels, with antenna separation >> $\lambda/2$, (8.44) reduces to

$$C_{NLOS} = M \log_2\left(1 + \sigma^2 / \sigma_N^2\right) \tag{8.48}$$

In the NLOS (Rayleigh-fading) case, the limit of capacity is proportional to the number of independent channels, as in Figure 8.16(b). A mixture of LOS and NLOS will give results between the two extremes.

8.7 Summary

Signals in the local vicinity of the portable or mobile personal communications device were described in terms of the signal statistical properties. The basic loss is cast as a median level found from the propagation models investigated in Chapters 6 and 7. Terrain and building heights vary in the suburban and urban environments, and a shadowing standard deviation was added to the mean/median values of path attenuation to account for the varying terrain and urbanization not accounted for by the propagation model. Lognormal statistics were used to characterize the added large-scale losses. In the local vicinity of the mobile or portable unit, localized multipath scattering causes the signal strength to vary from peak levels a few decibels above the median to tens of decibels below in deep fades. Rayleigh statistics were used to describe that fading phenomenon. Additional loss factors and their standard deviations were added, and

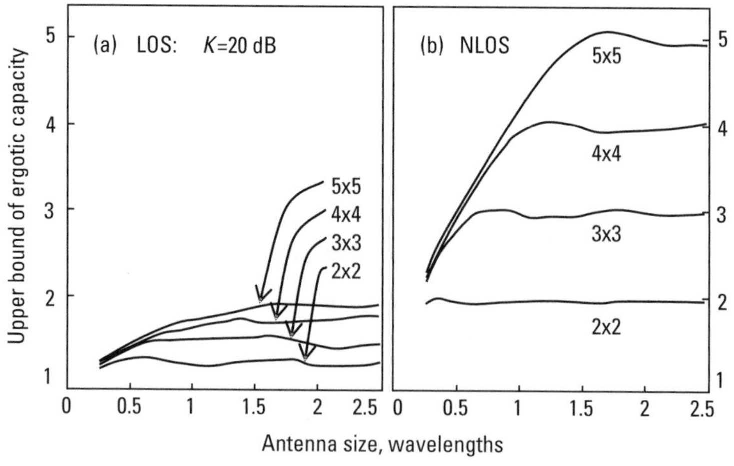

Figure 8.16 MIMO capacity: (a) LOS and (b) NLOS cases. (After [27].)

composite statistical models were used to arrive at the signal strength levels required for a desired level of portable communications call success probability.

The unique properties of UWB pulse propagation in multipath were shown, and the relationship between the multipath delay spread and propagation power law was seen to lead to the SBY propagation model.

Signal reception probability can be improved by various transmitting and receiving diversity techniques. Transmissions can be repeated or sent simultaneously from multiple sites. The design of simulcasting systems is constrained by signal strength, as well as by differential delays from the various transmitting sites. Receivers and antennas may be configured to "sample" signals subjected to multipath propagation in ways that are uncorrelated to improve the probability of reception statistically. Multipath leads to diversity reception opportunities, as well as to a potential increase in the channel capacity by using MIMO techniques.

Problems

8.1 What is the probability that a sample measurement from a Rayleigh- distributed signal is 10 dB below the RMS value?

8.2 A cellular communications system requires that an interfering signal be 12 or more decibels below the RMS value of the desired signal, so a distant same-frequency cell is designed to have an RMS level 12 below the desired signal at a particular location. If the signal levels are Rayleigh distributed, find the probability of interference at that location.

8.3 The signal along a propagation path exhibits shadowing variations that are described by a normal distribution having a standard deviation of 6 dB. Find the probability that the signal (a) exceeds the median value, and (b) is lower than the median value by10 dB.

8.4 An engineer uses the normal approximation to Rayleigh statistics to calculated the interference probability of a signal subjected to multipath and shadowing with a standard deviation of 6 dB. Quantify and describe the nature of the error committed.

8.5 Find the total standard deviation, including the normal approximation for the Rayleigh distribution when the shadowing, vegetation, and in-vehicle standard deviations are 5.5, 2.5, and 5 dB.

8.6 Find the signal margin required to ensure that in 95% of the locations, the RMS value exceeds a threshold value if the total standard deviation is 12 dB.

8.7 A portable receiver requires a field strength 27 dB above 1 mV/m to operate with 0.99 calling probability on an receiver-sensitivity test range. A 10W transmitter delivers an RMS signal level of 27 dB above 1 mV/m into an urban location. What is the calling probability at that location, and what must the transmitter power level be to provide a 0.90 calling probability if the signal environment is characterized by a lognormal distribution with a 10-dB standard deviation.

Ans:

$P_{call} = (0.5)(0.99) = 0.495$

$P = 0.9 = P_s/0.99$; $P_s = 0.909$, from $1 - P_s = 0.5 - 0.5\text{erf}(z/0.7071)$; $z = 1.335$. So, transmitter power needs to be 13.35 dB more than 10W, or 216.36W.

8.8 In an overlap area between two transmitters, the probabilities of successfully decoding a message from each of the two transmitters individually are 0.80 and 0.50. Find the success probability if both transmitters operate in simulcast fashion, and simulcast distortion is not a factor. If the standard deviation of the signal variation is 10 dB, estimate the power increase required at each transmitter if the transmitters are operated as independent systems.

Ans:

$P = [1 - (1 - 0.5)(1 - 0.8)] = 0.9$

$[z(0.9) - z(0.5)]10 = 12.82$ dB; $[z(0.9) - z(0.8)]10 = 4.40$ dB

8.9 A communications system uses a strategy of repeated transmissions to improve message reception reliability. Find the probability of reception if four uncorrelated tries are used, each with a success probability of 0.4, 0.5, 0.6, and 0.3.

Ans:

$P = [1 - (1 - 0.4)(1 - 0.5)(1 - 0.6)(1 - 0.3)] = 0.916$

8.10 Two simulcasting transmitters are d km apart. Write an expression describing the locus points of equal differential propagation delay from the two transmitters.

8.11 On a planar area representing the coverage from a two-transmitter simulcasting system, locate the regions of highest differential delays.

8.12 A receiver fixed site requires an mean signal ratio of at least 5 dB between signals on a two-antenna diversity system. Referring to Figure 8.11, find the minimum antenna separation if the scattering radius in the service area at a 5-km range is 200m.

Ans:

From Figure 8.11, separation = 0.25(200/5) = 10m.

8.13 Two independent signal distributions are available for diversity exploitation. The median signal levels are in a ratio of A decibels. Devise a Monte Carlo strategy to estimate the diversity opportunity, and find the diversity opportunity when A is 3 dB.

Ans:

Use (8.23) to (8.26).

8.14 In a highly scattered environment, the average ratio of horizontally to vertically polarized signals is 2 dB. (a) Determine the average decibel difference between signals received on horizontally and vertically polarized antennas of equal efficiency. Now, suppose additionally that the correlation between the signals is 0.3. (b) Find the diversity gain.

Ans:

(a) From (8.38), $g_R(10^{2/10})$ = 3.94 dB; (b) From (8.39), G_R = 3.41 dB.

8.15 Estimate the average decibel difference between signals received on horizontally and vertically polarized antennas of equal efficiency for the environments in Table 7.6.

8.16 Two vertically polarized quarter-wave whip antennas operating at 860 MHz are mounted on the roof of a van. The van is located at the receive location of an open-air test side and oriented so that the separation between the antennas is perpendicular to the site transmitter. What is the correlation coefficient between the signals received from the site transmitter if the antenna separation is 15 cm?

Ans:

$\rho = 1$

8.17 Find the correlation coefficient in a multipath environment between two van-mounted, vertically polarized quarter-wave whip antennas separated by a quarter-wavelength.

8.18 Find the EIRP of a 930-MHz transmitter servicing an urban area characterized by a total lognormal standard deviation of 10 dB if the propagation path loss is 120 dB and field strength is above 50 dBμV/m 90% of the time.

Ans:

From (8.21), EIRP (dBm) = 50 + 1.28(10) + 120 – 77.22 – 20log(930) = 46.21 dBm.

8.19 Find the diversity gain opportunity for a dual-polarized system in an environment where the power levels of the two polarizations are in a ratio of 0, 3, and 10 dB and the correlation between the signals is 0 and 0.5.

Ans:

🖫 [8-5d.mcd] in MCAD60 directory.

8.20 In Chapter 7 problem 7.21, it was suggested that system capacity in an urban setting can be doubled by simply overlaying another system with orthogonal polarization. Discuss what might be necessary to make such a system work.

Ans:

If the correlation, including antennas and multipath reflections, for the two paths (one in each polarization) is sufficiently randomized, MIMO techniques might be employed to improve the channel capacity up to twofold.

References

[1] Kuznicki, W. J., and K. Siwiak, "The World of Paging," *USTTI Course Notes*, Motorola Paging Products Group, Boynton Beach, FL, 1993.

[2] Hess, G., *Land-Mobile Radio System Engineering*, Norwood, MA: Artech House, 1993.

[3] Jakes, W. C., *Microwave Mobile Communications*, American Telephone and Telegraph Co., 1974 [reprinted by IEEE Press, New Jersey, 1993].

[4] Lee, W. C. Y., *Mobile Communications Design Fundamentals*, Indianapolis, IN: Howard W. Sams and Co., 1986.

[5] Abramowitz, M., and I. Stegun, (eds.), *Handbook of Mathematical Functions*, New York: Dover Publications, 1972.

[6] *Mathcad User's Guide*, Cambridge, MA: MathSoft, Inc., 1988–2006.

[7] More, J. J., B. S. Garbow, and K. E. Hillstrom, *User's Guide to Minpack I*, Argonne National Laboratory, Pub. ANL-80-74, 1980.

[8] Okumura, Y., et al., "Field Strength and Its Variability in VHF and UHF Land-Mobile Radio Service," *Rev. Elec. Commun. Lab.*, Vol. 16, September–October 1968, pp. 825–873.

[9] Hagn, G., "VHF Radio System Performance Model for Predicting Communications Operational Ranges in Irregular Terrain," *IEEE Transactions on Communications*, Vol. COM-28, No. 9, September 1980, pp. 1637–1644.

[10] Cox, D. C., "Correlation Bandwidth and Delay Spread Multipath Propagation Statistics for 910-MHz Urban Mobile Radio Channels, *IEEE Transactions on Communications*, Vol. COM-23, November 1975, pp. 1271–1280.

[11] Siwiak, K., H. Bertoni, and S. Yano, "Relation between Multipath and Wave Propagation Attenuation," *Electronic Letters*, Vol. 39, No. 1, January 9, 2003, pp. 142–143.

[12] Yano, S. M.: "*Investigating the Ultra-wideband Indoor Wireless Channel*," *Proc. IEEE VTC2002 Spring Conf.*, Birmingham, AL, May 7–9, 2002, Vol. 3, pp. 1200–1204.

[13] Greenstein, L.J., et al., "*A New Path-Gain/Delay-Spread Propagation Model for Digital Cellular Channels*," *IEEE Trans. on Vehicular Technology*, Vol. 46, No. 2, May 1997.

[14] Saleh, A. A. M., and R. A. Valenzuela, "A Statistical Model for Indoor Multipath Propagation," *IEEE J. Select. Areas Commun.*, Vol. 5, No. 2, February 1987, pp. 128–137.

[15] Nemati, M. A., and R. A. Scholtz, "A Diffusion Model for UWB Indoor Propagation," MILCOM 2004, 2004, Monterey CA, October 31–November 3.

[16] Hill, C., and B. Olsen, "A Statistical Analysis of Radio System Coverage Acceptance Testing," *IEEE Vehicular Technology Society News*, February 1994, pp. 4–13.

[17] "Paging Systems; European Radio Message System (ERMES) Part 5: Receiver Conformance Specification," *ETS 300 133-5*, ETSI, Valbonne, France, July 1992, ammended (A1), January 1994.

[18] Fujimoto, K., and J. R. James, (eds.), *Mobile Antenna Systems Handbook*, Norwood, MA: Artech House, 1994.

[19] Price, R., and P. E. Green. "A Communication Technique for Multipath Channel," *Proceedings of the IRE*, March 1958, pp. 555–570.

[20] Souissi, S. "Wide-Area Coverage for Next Generation Paging Networks: Surmounting the Constraints of Simulcast," *Optimising Paging Network Design and Management—Technical Conference*, London, UK, September 24, 1997.

[21] "FLEX-TD Radio Paging System," *RCR STD 43A*, 1996, Association of Radio Industries and Businesses, Japan.

[22] Hess, G., "Absolute Power Difference of Independent Rayleigh Samples," private communication, June 25, 1997.

[23] Beckmann, P., *Probability in Communication Engineering*, New York: Harcourt, Brace and World, 1967.

[24] Gradshteyn, I. S., and I. M. Ryzhik, *Table of Integrals, Series, and Products*, New York: Academic Press, 1965.

[25] Alamouti, S. M., "A Simple Transmit Diversity Technique for Wireless Communications," *IEEE Journal on Selected Areas in Communications*, Vol. 16, No. 8, October 1998, pp. 1451–1458.

[26] Kermoal, J. P., et al., "Experimental Investigation of Correlation Properties of MIMO Radio Channels for Indoor Picocell Scenario," *Proc. IEEE Veh. Technol. Conf.*, Boston, MA, September 2000, Vol. 1, pp. 14–21.

[27] Skafidas, E., et al., "Analysis of MIMO Capacity for Ricean Channels with Antenna Effects," 2004 Wireless Networking Symposium, WCNG 2004, Austin, TX, October 20–22, 2004.

[28] Farrokhi, F. R., et al., "Linkoptimal Space-Time Processing with Multiple Transmit and Receive Antennas," *IEEE Communications Letters*, Vol. 5, No. 3, March 2001, pp. 85–87.

9

Receiver Sensitivity and Transmitted Fields

9.1 Introduction

Our path between the antenna of a fixed site and the small antenna of the personal communications device proximate to the body involves field-strength measurements in the vicinity of the body and the measurement of signals transmitted by the personal communications device (PCD). The basics of receiver-sensitivity measurements are introduced in this chapter. Receiver field-strength sensitivity is measured using statistically based methods for determining a specified calling rate for a data receiver. The details of this testing are presented in this chapter. Since the types of communications devices that we portray here often have integral antennas and therefore do not have accessible receiver input terminals, the performance is reported as "field-strength sensitivity" rather than "receiver input sensitivity" or antenna gain or loss. The relationship between field strength and power delivered to a receiver is derived. We present the techniques for measuring field-strength sensitivity, including methods of reporting sensitivity averaged over azimuth angles. The characteristics of open-field test sites, where receiver-sensitivity measurements are often performed, which we studied analytically in Chapter 6, are presented here in terms of field-strength sensitivity measurement accuracy, reliability, and repeatability.

Open-field antenna test ranges are often specified for use in measurements of the sensitivity of pagers and data receivers in the very-high- and ultra-high-frequency ranges (30 MHz to 3 GHz) [1–4]. Receiver-sensitivity measurements are actually signal-to-noise measurements that involve statistical methods of arriving at the transmitted power required to evoke a particular receiver response (successful message reception). As such, errors arise from uncertainties in range calibration, path loss, transmitter power levels, gain standards, geometric placement of test devices, ambient temperature, temperature changes, interference, and malfunctioning equipment, including cables. Both systematic and random errors are encountered, and both absolute and relative measurements are affected. We investigate the calibration methods for test ranges configured to measure the EIRP of transmitting PCDs. Finally, we examine the effects of the human body on PCD performance, as well as compliance with radio frequency (RF) exposure standards and regulations.

9.2 Field-Strength Sensitivity of Receivers

Unlike transmitters, receivers do not have a convenient "radiated output power" with which to measure antenna patterns and antenna directivity. Consequently, statistical methods are used to establish receiver field-strength sensitivity. The statistical methods have been developed [1–3] for reliable and repeatable measurements. For best accuracy, the sensitivity is measured at the 80% call success rate using the simplest possible message. Correlation to other calling success rates more suitable to system design is done in a test fixture. Since selective call receivers are often worn on the body, the sensitivity measurement is most often complemented with a simulated human device, which will be detailed in the following chapter. The characteristics of the testing procedure, measurement reliability, and repeatability will be detailed here.

9.2.1 Statistical Method for Measuring Field-Strength Sensitivity

The objective in receiver-sensitivity measurements is to find a field-strength level that results in a specified response at the receiver. One such response is a 20-dB quieting level or a 12-dB SINAD level measured in the audio circuits of a voice receiver. Other often-used responses include the 1% bit error rate and 8% packet error rate in various test scenarios involving digital data receivers. In digital data and selective call, or paging, receivers, the common measurement response is the 80% calling success rate. This measurement method was designed to determine the field strength required to evoke the 50% probability of responding correctly to three calls in a row. Mathematically, this probability, P_{call}, is written

$$(P_{call})^3 = 0.5 \tag{9.1}$$

Hence,

$$P_{call} = (0.5)^{1/3} \approx 0.8 \quad (9.2)$$

A statistically based testing procedure for arriving at the 80% call success signal level is called the *20-call method.* Several variations of the 20 call testing method have been described in various specifications documents [1, 2].

Pager sensitivity is reported as the field strength (usually expressed as the equivalent electric field in decibels with respect to 1 μV/m) required for the 80% probability of responding correctly to a sent message.

9.2.2 Determining the 80% Calling Response Rate

A typical 20-call method described here is detailed in [2]. An 80% calling rate is determined by adjusting the attenuator settings of a signal source modulated with a preselected message and transmitted to a message receiver. The sensitivity is determined by a sequence of events involving the "hits" (correctly received messages) and "misses" (messages not received correctly), according to the following rules:

1. Starting at a low signal level, transmit a call up to three times, terminating the sequence if the receiver misses a response.
2. If there is a miss, decrease the transmitter attenuator by 2 dB and go back to step 1.
3. If three hits in a row are measured, increase the attenuator, record that value, and begin counting calls.
4. Transmit a call up to three times.
5. If there is a miss, decrease the attenuator by 1 dB, record that value, and go back to step 4.
6. If 3 hits occur in a row, increase the attenuator by 1 dB, record that value, and continue with step 4 until 10 values have been recorded.
7. Average the recorded attenuator values.

Figure 9.1 shows a record of one such 20-call test in which the determined attenuation level for the sensitivity is 15.1 dB, the average of the 10 recorded values. If the signal source is a signal generator connected to a transmitting antenna at a receiver-sensitivity test site, the attenuation value is added to the signal generator level, and the result is compared with a standard signal generator and attenuator level setting used to generate a calibrated field strength to arrive at the receiver field-strength sensitivity. The generation of standard field-strength levels and the calibration of open-field receiver-sensitivity test sites was described generally in Chapter 6 and will be revisited in Section 9.4. There

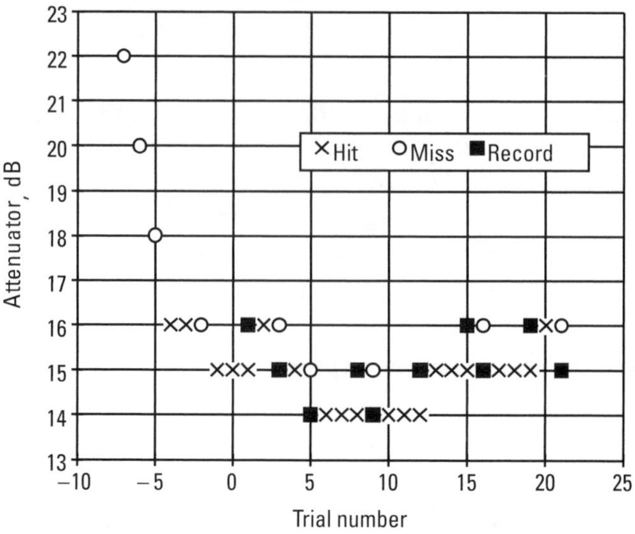

Figure 9.1 A statistical determination resulting in the 80% calling response rate level. (*Source:* [5].)

are many variations of the basic statistical testing methods. The outlined method is based on [2].

9.2.3 Accuracy of the 20-Call Test

The 20-call method results in a slightly conservative approximation of the 80% calling success rate. Figure 9.2 shows the results of an analysis [1] of the 20-call test compared with the true calling probability. The 20-call result is shifted slightly to the lower attenuator settings, which corresponds to reporting a conservative (worse) sensitivity. The width of the probability curve depends on the coding protocol and on the message length. Longer messages will result in a wider curve, hence a less reliable sensitivity determination.

The standard deviation, σ_{call}, of a single 20-call sensitivity determination has been found to be in the vicinity of σ_{call} = 0.3 dB, based on a large number of measurements [4, 6]. The 20-call method is time-consuming: about 30 calls are transmitted to get about 20 responses, of which 10 are recorded and averaged. As a result, usually no more than eight positions, equally spaced in azimuth, are measured to estimate the directivity of a body-mounted receiver-antenna combination. The eight positions are averaged using a gain-average formula, which reports a good estimate of the average sensitivity value, as will be shown in Chapter 10. The resulting average sensitivity value is relevant in a Rayleigh fading environment because it reports the effect of average antenna directivity.

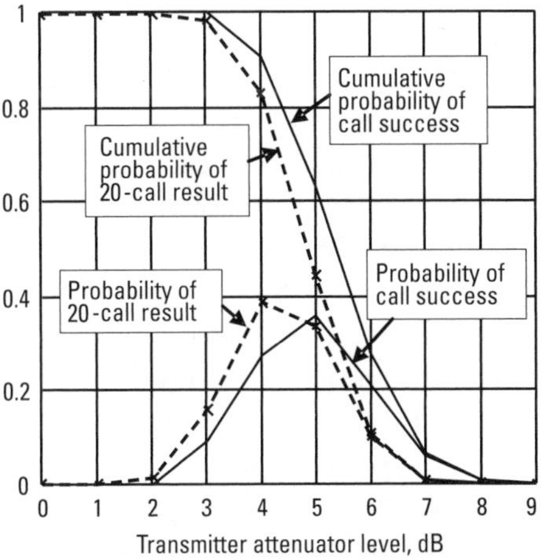

Figure 9.2 Accuracy of the 20-call method. (*Source:* [5].)

9.2.4 A Simplified Three-of-Three Method

A shortened test, called the *three-of-three test*, may sometimes be used to save time in receiver-sensitivity determinations. When eight azimuth positions of a sensitivity pattern are required, and the eight positions are reported as a single average sensitivity value, the following test procedure may be used. For the "best" sensitivity direction, usually the front position in body-mounted receiver testing, use the full 20-call test as outlined above. For the remaining seven azimuth positions, use the following shortened procedure:

1. Starting at a low signal level, transmit a call up to three times, terminating sequence if receiver misses a response.
2. If there is a miss, decrease the transmitter attenuator by 1 dB and go to step 1.
3. If three hits in a row are measured, increase the attenuator by 1 dB and record that value.
4. The rest of the 20-call procedure is truncated.

The attenuator is initially varied here in 1-dB steps, compared with 2-dB steps in the full procedure. The inherent averaging of measurement errors in the application of the gain-average formula results in an acceptable level of performance for the shortened procedure.

The result of applying the shortened "three-of-three" test is, on the average, a 0.3-dB more pessimistic determination of receiver average field-strength sensitivity than is determined by the full 20-call method. The repeatability and accuracy are also slightly worse than for the full-length test. The shortened test is most useful when eight equally spaced azimuthal positions are tested, and the results are gain-averaged. Since the front position (best sensitivity) is measured using the full 20-call method, this more accurately determined value tends to dominate the averaging procedure. Gain-averaging smoothes out some of the measurement inaccuracy introduced by the shortened procedure.

The three-of-three method has found application in measuring cellular/mobile phone receiver sensitivities in radiated tests [7]. The usual method for testing mobile phone receiver sensitivity [8] involves measurements conducted at the mobile phone test port. In order to assess sensitivity in the presence of an interferer, however, the sensitivity in a radiated field is required. Since bit error or packet error rates are not always available, a method based on the three-of-three test was adopted. The radiation-based tests revealed significant radiated self-interference from the mobile phone digital circuitry that did not manifest itself in the conducted sensitivity tests.

9.3 Relating Field Strength to Received Power

The field strength at the receiver antenna can be related to the received power by observing that the received power is the antenna effective aperture area multiplied by the wave power density. In receiver-sensitivity measurements, it is customary to report the field strength E as the RMS quantity, not as the peak amplitude of a sine wave, as is customary in electromagnetics and antenna analysis. Consequently, expressions involving power density and field strength differ by a factor of 2 from the corresponding expressions that are typical of other antenna and electromagnetics texts.

The power P_r in a wave in terms of the equivalent aperture A_e of an antenna and the RMS electric field E, using the method of Section 3.3.1, is

$$P_r = A_e \frac{E^2}{\eta_0} = G \frac{\lambda^2}{4\pi} \frac{E^2}{\eta_0} \tag{9.3}$$

The gain factor G is the product of efficiency and directivity and equals 1.5 for a lossless, electrically small dipole or loop antenna. The wavelength is λ, and η_0 is the intrinsic impedance of free space. We then observe that the power received at the terminals of the receiver can be written in terms of the receiver terminal voltage, V_r, across the input resistance R_r.

$$P_r = \frac{V_r^2}{R_r} \tag{9.4}$$

Hence, equating (9.3) with (9.4) and solving for E, the receiver RMS input voltage V_r, in terms of the electric field strength E, is

$$E = \frac{V_r}{\lambda}\sqrt{\frac{4\pi\eta_0}{GR_r}} \tag{9.5}$$

which, in an $R_r = 50\Omega$ system, reduces to

$$E = \frac{V_r}{\lambda}\sqrt{\frac{94.68}{G}} \tag{9.6}$$

so that for a directive gain of $G = 1.5$ appropriate for small dipoles and loops,

$$V_r = \frac{37.7}{f_{MHz}} E \tag{9.7}$$

When (9.6) is stated in decibels, the field strength, antenna gain, and received power are related by

$$E(dB\mu V/m) = P_r(dBm) - G(dBi) + 77.226 + 20\log(f_{MHz}) \tag{9.8}$$

where P_r is the power delivered to the receiver input.

The ratio of field strength E to voltage across the receiver input terminals V_r is often stated in decibels and is called the *antenna factor* (AF)

$$AF = 20\log(E/V_r) \tag{9.9}$$

The antenna factor AF appears as a calibration value with standard-gain dipoles. The value includes the antenna losses and the directive gain.

9.3.1 Pattern Gain Averaging

A formula often applied to average the eight-position receiver-sensitivity measurements can be shown to report the sensitivity for the average directive gain of the receiver-antenna-body combination. Gain averaging is appropriate for multipath and scattered fields. The gain-average formula is derived by considering first the expression for received power, P_r, in terms of the incident field strength, E.

$$P_r = A_e \frac{E^2}{\eta_0} \tag{9.10}$$

In an eight-position test, the field-strength sensitivity, E_i, is recorded for eight equally spaced azimuth angles. The relationship between the antenna effective aperture and the directive gain G, as a function of the azimuth angles θ_i, is

$$A_e(\theta_i) = G(\theta_i)\frac{\lambda^2}{4\pi} \tag{9.11}$$

Averaging the directive gain values for the eight azimuth positions gives

$$G_{avg} = \frac{1}{8}\sum_i G(\theta_i) = \frac{1}{8}\sum_i A_e(\theta_i)\frac{4\pi}{\lambda^2} \tag{9.12}$$

The average gain can now be restated in terms of average field-strength sensitivity using (9.10) and (9.11):

$$G_{avg} = P_r \eta_0 \frac{4\pi}{\lambda^2}\frac{1}{8}\sum_i \frac{1}{E_i^2} \tag{9.13}$$

Finally, using (9.3) for the averages, then using (9.10) and (9.13), the average field-strength sensitivity in terms of the measured field-strength values is written as

$$E_{avg} = \sqrt{\frac{P_r \eta_0}{A_{eavg}}} = \sqrt{\frac{P_r \eta_o}{\left[P_r \eta_0 \frac{4\pi}{\lambda^2}\frac{1}{8}\sum_i \frac{1}{E_i^2}\right]\frac{\lambda^2}{4\pi}}}$$

which simplifies to the commonly used eight-position gain-averaging formula,

$$E_{avg} = \frac{1}{\sqrt{\frac{1}{8}\sum_{i=1}^{8}\frac{1}{E_i^2}}} \tag{9.14}$$

The gain average of (9.14) is the appropriate value for system designs in areas of multipath signals that can be described with Rayleigh statistics.

9.3.2 Averaging Methods for Mobile Phone Testing

Mobile phone total radiated power (*TRP*) and total isotropic sensitivity (*TIS*) are measured three-dimensionally on an great circle grid as described in the CTIA test plan [8]. The mobile phone is mounted as described in [8] on an anthropomorphic simulated head (described in Chapter 10). Data is collected for N intervals of θ and for M intervals of ϕ in the standard polar coordinate system, where ϕ is measured from the x-axis in the xy-plane and θ is measured from the z-axis. Figure 9.3 shows the measurements coordinate system and the great circle grid for 30° measurement intervals. The *TRP* is found from

$$TRP \cong \frac{\pi}{2NM}\sum_{i=1}^{N-1}\sum_{j=0}^{M-1}\left[EIRP_\theta(\theta_i,\phi_j) + EIRP_\phi(\theta_i,\phi_j)\right]\sin(\theta_i) \tag{9.15}$$

The TIS is

$$TIS \cong \frac{2NM}{\pi\sum_{i=1}^{N-1}\sum_{j=0}^{M-1}\left[\frac{1}{EIS_\theta(\theta_i,\phi_j)} + \frac{1}{EIS_\phi(\theta_i,\phi_j)}\right]\sin(\theta_i)} \tag{9.16}$$

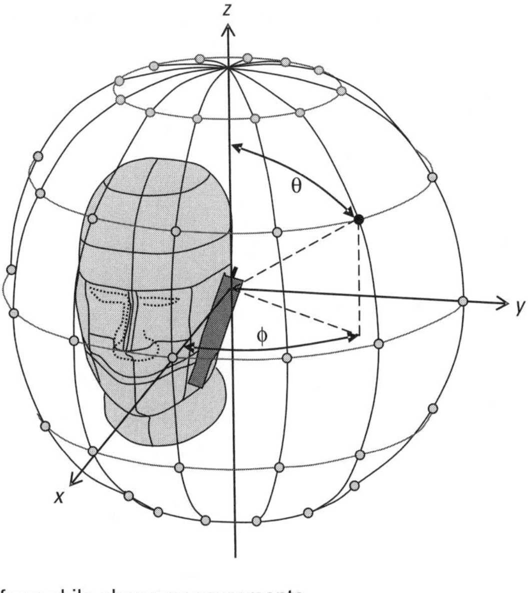

Figure 9.3 Points for mobile phone measurements.

$EIS(\theta,\phi)$ are the receiver sensitivities at the specified coordinates. Formula (9.16) reduces to (9.14) when just the $\theta = 90°$ principal plane and eight equally spaced ϕ values are chosen.

Often, the partial radiated power and the partial isotropic sensitivity are measured. For example, assuming that the total sphere is measured with $N = 12$ θ intervals and $M\phi$ intervals, the power from 45° below to 45° above the horizon is found with i in (9.15) and (9.16), ranging from 3 to 9. In measurements from 30° below to 30° above the horizon, i ranges from 4 to 8. By comparison, the body-mounted receiver sensitivity in (9.14) is eight azimuth positions on the horizon.

9.4 Test Site Field-Strength Calibration

An open-field receiver-sensitivity test site complemented with a human body test device and suitable for measuring receiver sensitivities is depicted in Figure 9.4. The field strengths on such a range are calibrated at the measurement heights, which are usually 1m above ground for belt-level measurements and, most often, 1.4m or 1.5m for chest-level measurements. The open-field receiver-sensitivity range of Figure 9.4 has a transmitting antenna height of 3.2m above ground and a range distance of 45.4m. It corresponds to the range studied analytically in Chapter 6.

The calibration factor is defined here as the field strength in ratio to 1 μV/m, expressed as decibels (dBμV/m) at the measurement position for a

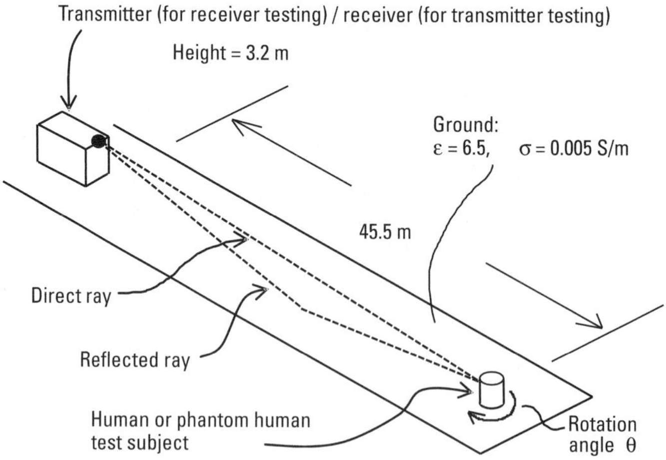

Figure 9.4 A receiver-sensitivity and PCD transmitter test site. (*Source:* [5].)

transmitter power of 1 mW. Calibration factors should be measured regularly, and the observed standard deviation in those measurements indicates the day-to-day repeatability of measurements for this type of receiver field-strength sensitivity test range. A long-term observation of receiver-sensitivity measurements and range calibration is shown in Figure 9.5(a, b), respectively. Figure 9.5(a) shows the results of sensitivity measurements of a standard correlation radio taken repeated over nearly a 1-year period. The sensitivity was determined after a careful range calibration shown in Figure 9.5(b). Receiver sensitivities during days 150 through 240 are seen here to be the same, even though calibration factors vary by almost 2 dB during the same time period. In general, there is no correlation evident between the results in Figure 9.5(a, b). Careful calibration is required for repeatable measurements. There is evidence that the calibration does indeed remove some range variability due to environmental and meteorological changes because the shapes of the curves in Figure 9.5(a, b) are different.

9.5 Reliability and Repeatability of Sensitivity Measurements

Since the 20-call and three-of-three methods are statistical in nature, they can be expected to produce a range of results in repeated trials, as suggested by the

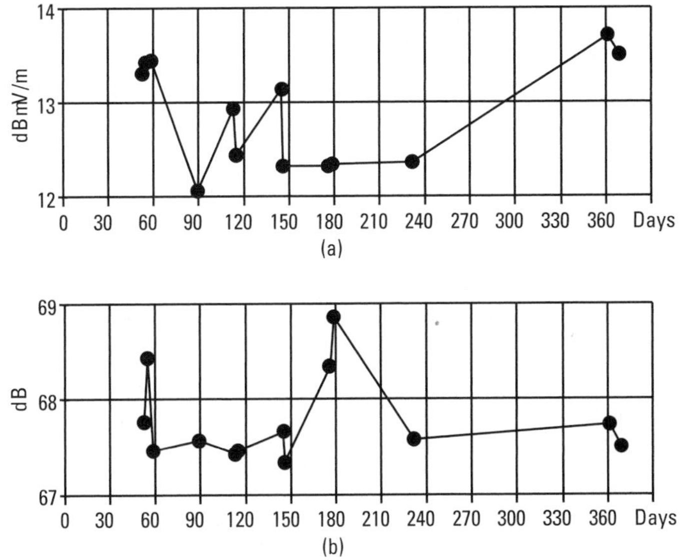

Figure 9.5 Receiver sensitivity as (a) measured over time with (b) the corresponding range calibration factors. (*Source:* [5].)

width of the call probability curve shown in Figure 9.2. The standard deviation associated with repeated measurements of a data receiver was given earlier as $\sigma_{20call} = 0.3$ dB. Additional variations give rise to additional standard deviations due to calibration factor determination and due to the accuracy of the gain standards.

9.5.1 Repeatability of Sensitivity Measurements

Antenna range calibration factors can be calculated from the physical geometry of the range and from a best-fit set of antenna range ground parameters using the methods of antenna test range analysis developed in Chapter 6. We note a close correspondence between measured values and calculated calibration factors in Table 9.1. This reinforces confidence in the antenna range analysis developed in Chapter 6 and used here to estimate calibration variations as a function of range geometry and ground parameters. Standard deviations calculated for the several calibration determinations at each calibration frequency are shown in Table 9.1, which summarizes the results shown in Table 6.2. These are combined as the mean of variances with the resulting single calibration standard deviation of $\sigma_{cal} = 0.36$ dB.

The antenna range was modeled by a direct ray from the transmitter added to a ray reflected from a layered lossy dielectric half-space representation of the ground. The best-fit ground parameters were a relative permittivity or dielectric constant of $\varepsilon = 5.2$, with a conductivity of $\sigma = 0.002$ S/m for the upper layer and a lower layer 0.66m deep having a relative dielectric constant of $\varepsilon = 11.9$ and conductivity of $\sigma = 0.002$ S/m. Similarly, good results can be obtained with a simpler model having a homogeneous dielectric half-space ($\varepsilon = 5$ and $\sigma = 0.002$ S/m) representation of the ground.

Table 9.1

Open-Field Range Calibration Factors at $h = 1$m and $h = 1.4$m Test Heights. (*From* [9].)

Frequency (MHz)	Measured Calibration Factor (dBμV/m)	Calculated Calibration Factor (dBμV/m)
150, $h = 1.0$m	66.5 ($\sigma = 0.40$)	66.4
$h = 1.4$m	68.2	68.3
170, $h = 1.0$m	68.2 ($\sigma = 0.42$)	68.1
$h = 1.4$m	69.9	70.1
470, $h = 1.0$m	73.4 ($\sigma = 0.22$)	73.1
$h = 1.4$m	75.1	75.3
929, $h = 1.0$m	76.9 ($\sigma = 0.35$)	76.8

9.5.2 Variations in the Calibration Factor Due to Ground Parameters

The permittivity of the Earth is not constant but varies with ground moisture content and soil composition. The effect of ground parameter variations on the antenna range path loss can be analyzed at 45, 167, 280, 460, and 930 MHz, the frequencies of most interest to personal communications receiver-sensitivity measurements. Calculated variations in the ground relative permittivity ε, from 5 to 25, and variations in conductivity σ, from 0 to 1 S/m, result in no more than a ±2-dB change in absolute path loss at the 1m test height using the analysis developed in Chapter 6. Receiver field-strength sensitivity tests are relative measurements; that is, calibration removes the effects of ground parameters. The absolute path loss is not important as long as there are no physical changes to the receiver-sensitivity test site during a testing sequence, such as those caused by rain or the ground's drying just after rain, and a careful calibration is performed as part of the measurement sequence. Realistically, an estimated ±0.1-dB variation can be expected during a long series of measurements.

[9.5a.mcd] The calibration factor is the field-strength value at the test point for 0 dBm at the generator. Assuming a transmitter antenna gain of 10 dB and cable and attenuator losses of 7 dB, find the field strength at a 1m height for 150, 280, 460, and 930 MHz. Use a two-ray range model with $\varepsilon = 6.5$ and $\sigma = 0.005$ S/m. Does the path loss vary much when $\varepsilon = 20$?

9.5.3 Field-Strength Variations with Height

Calculations of the electric field as a function of test point height show less than a 2-dB change in field strength over the height range of interest (about 0.7m to 1.5m) at all frequencies between 30 and 932 MHz. The analysis also shows that there are no field nulls in the 0.5m to 1.5m height range and that the fields vary smoothly with height. Figure 9.6 shows the calculated electric field as a function of height at 150 MHz and 930 MHz for the antenna range geometry shown in Figure 9.4. The results shown here as field strengths are the same as shown in Figure 6.7 in terms of path attenuation.

[9.5b.mcd] Using the same range model as used in Section 9.5.2, calculate the field variation with heights at 150, 450, and 930 MHz. What happens when the range distance is decreased from 45.4m to 10m?

For a constant-range geometry, the field-strength peak at a 1m height for 930 MHz migrates up as the frequency is lowered. At 150 MHz, a smoothly varying field that increases with height is evident. This is exactly the type of field strength with height behavior that exists at propagation distances of importance to radio systems. Height errors affect the accuracy of the calibration factor and

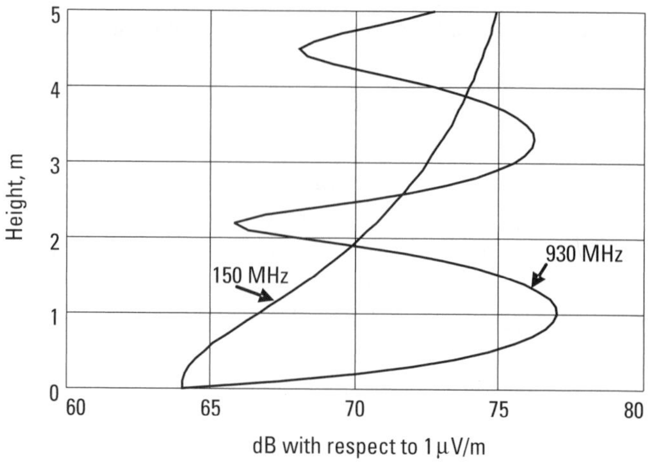

Figure 9.6 Field-strength variation with height. (*Source:* [5].)

are related to the errors in the geometric placement of calibration antennas, in the location of the "simulated human" measurement devices SALTY or SALTY-LITE, and in the placement of the pager receiver on the measurement device.

9.5.4 Accuracy of the Calibration Gain Standards

The calibration procedure averages the electric field over the length of the calibration gain standard. The field is weighted at each point by the nearly sinusoidal current distribution on the dipole when a dipole gain standard is used, and over a nearly uniform aperture of a standard-gain horn antenna, which is sometimes used at frequencies above 800 MHz. In the case of dipoles, an incorrectly designed or damaged balun will cause an asymmetric current distribution on a dipole standard, which will shift the apparent height of the field calibration point. An improperly designed balun will not be properly decoupled from the dipole feed line, and interchanging the relative orientations of the top and bottom elements by rotating the dipole will affect the accuracy of the calibration. When there is doubt, average the values of two calibrations using the calibration dipole in its standard and rotated orientations.

The most carefully constructed gain standards are typically accurate to better than 0.4 dB and are often better than 0.1 dB with respect to calculated gain performance [10–12]. Mismatches and current distortions due to mutual coupling to ground will increase the uncertainty of the standard gain by an additional 0.5-dB bias. Thus, 0.9 dB of absolute gain uncertainty exists in the use of carefully constructed gain standards to calibrate the open-field antenna range.

As shown in Figure 9.6, the electric field varies with height at the test position. In particular, at 930 MHz a vertical lobe is centered roughly at the 1m test height. A standard-gain horn antenna used in this frequency range has an aperture vertical dimension of 0.613m. When this standard is used to calibrate the field, the electric field is weighted over the 0.613m height of the horn aperture with a uniform aperture illumination, compared with cosine illumination over a dipole. Consequently, the reported factor by calibrating with a horn is 0.44 dB smaller than would be reported by a dipole calibration standard like that shown analytically in Section 6.3.5. Receivers thus measured appear to be 0.44 dB more sensitive than they would if the field were calibrated with a dipole reference.

The actual horn calibration is based on a calculated curve [13] supplied by the horn manufacturer and is typically quoted as ±1 dB, including gain ripple and measurement uncertainties.

[9.5c.mcd] Using the equations from [13], calculate the standard horn antenna gain over a range that includes 930 MHz. The horn antenna is 32.56 in. wide and 24.13 in. high, the waveguide dimensions are 9.8 in. by 4.89 in., and the flare distance is 27.69 in.

9.5.5 Intercomparison of Receiver-Sensitivity Test Sites

Measurements of communications receiver sensitivities were carried out in [14, 15] using several digital data receivers operating at a 1,200-bps data rate at three independent receiver-sensitivity test sites. In another test site intercomparison [16], an electric field-strength measurement was transferred among multiple international sites. The measurements in [16] all fell within a range of +0.75 to –0.5 dB, compared with the overall average. That kind of repeatability is comparable to the test results shown here. The intent here is to show the correlation between results measured on open-field receiver-sensitivity test sites that conform to the specifications in [2, 3]. The receiver-sensitivity intercomparison amounts to a transfer of calibration, or calibration factor, among the participating test sites. The measured sensitivity values correlate to within an average of 0.62 dB among three completely independent testing sites. This is well within the measurement uncertainty expected (standard deviation of 1 dB) based on the test range errors discussed here for such measurements. The key factors that contributed to this good correlation are:

- Gain-averaging of the sensitivity using (9.14);
- Use of a test range 30m or greater in length;

- Use of a fixed geometry to ensure a repeatable field profile over the simulated body test fixture;
- Careful experimental technique.

Five pagers, designated A1, A2, A3, B1, and B2, were measured at three different open-field receiver-sensitivity ranges; the results are summarized in Figure 9.7. The sensitivity is shown plotted versus the measured null depth. The temperature of the simulated body device, SALTY, has a significant effect on the depth of the measured pattern null, but the overall effect on averaged sensitivity is negligible. The measurements taken at site "A" are grouped around a null depth of 7 dB, while the site "B" measurements are grouped near a null depth of 11 dB, a difference of about 4 dB. Site "C" results are near a 9-dB null depth.

The difference between single receiver-sensitivity measurements among sites "A," "B," and "C" was never more than 1 dB from the average of the three sites, and the average difference was 0.62 dB. This value represents the measurement repeatability between two completely independent open-field receiver-sensitivity test ranges using the same set of pagers and is consistent with expectations of such measurements. There were significant physical differences between the two test conditions, but these did not significantly alter the outcome. Most notably, the ambient temperature was between 30°C and 35°C at site "A" and between 0°C and 5°C at site "B"; it was 18°C at site "C."

Temperature significantly affects the permittivity and conductivity of the saline solution in the SALTY simulated body test fixture, as will be studied in

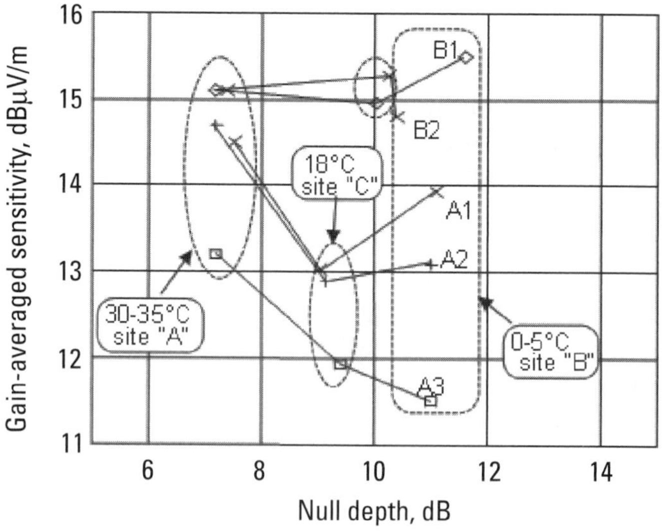

Figure 9.7 Gain-averaged sensitivity (dBμV/m) and null depth (dB) for five body-worn receivers.

detail in Chapter 10. A two-dimensional analysis, presented there, of fields around a lossy cylinder having the same cross section as SALTY reveals a significant temperature dependence in the depth of the pattern null of the body-worn receiver. The calculated null depth in decibels for an infinitely long SALTY filled with a saline solution of 1.5g NaCl per liter of water, as a function of saline-water temperature, is shown in Figure 9.8. The difference in the null depth at 30°C compared with the null depth at 5°C is very nearly 4 dB, in agreement with the measured difference encountered in Figure 9.7. For comparison, the "×" marks the calculated null depth for an infinitely long SALTY filled with homogeneous biological muscle tissue.

The gain-averaged field strength in ratio to a free space field for an infinitely long SALTY can be calculated using the analysis in Chapter 10 and the averaging formula (9.12). With the temperature-dependent model of saline water of Section 10.3.1, the gain-averaged field strength can be computed as a function of saline-water temperature, as shown in Figure 9.9. Despite the null differences due to temperature, the calculated gain-average is essentially constant over temperature.

9.5.6 Test Range Error Uncertainties

The calibration standard deviation was found earlier to be σ_{cal} = 0.36 dB and takes into account the randomness inherent in geometrically locating the calibration standard with respect to the measurement point. The pager measurement method standard deviation is σ_{meas} = 0.27 dB, which includes transmitter-generator random and repeatability errors, as well as 20 call measurement method random errors.

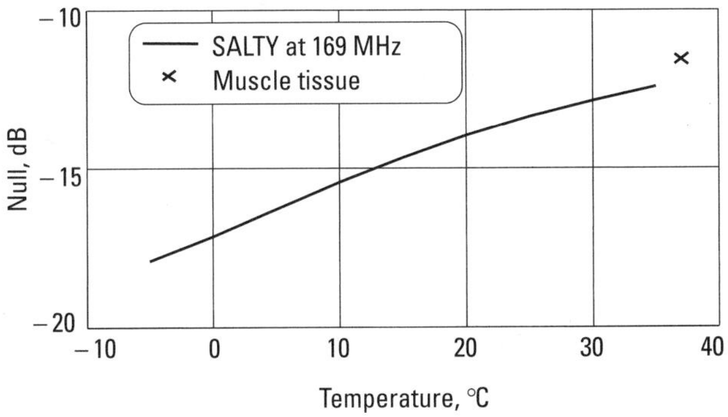

Figure 9.8 Calculated null depth in a body-worn receiver as a function of saline-water temperature.

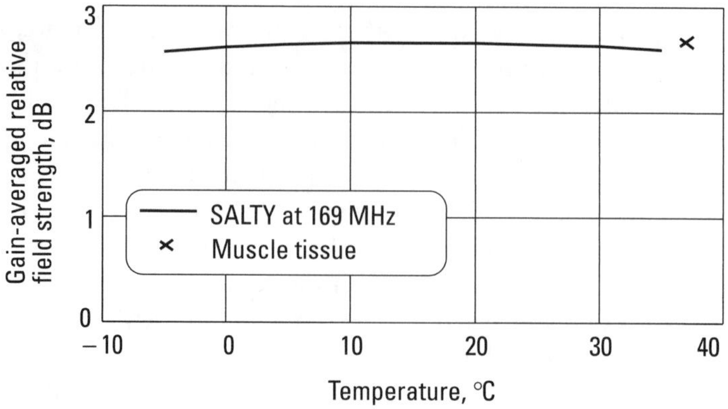

Figure 9.9 Calculated gain-averaged relative field strength as a function of saline-water temperature.

Figure 9.5(a) shows measurements of sensitivity on a single correlation receiver performed over nearly 1 year. The standard deviation of the sensitivities is σ = 0.55 dB. The range calibrations for those measurements are shown in Figure 9.5(b). Range effects are evident between days 150 and 240 that are not present in the corresponding sensitivity results. That is, the range effects have been successfully eliminated by the calibration. The calibration standard deviation for this series of measurements is σ = 0.46 dB, which is consistent with the values shown in Table 9.1. Additional variations of typically ±0.1 dB due to ground parameter changes during a measurement can be expected during extended measurements.

Systematic errors include the unknown actual gain of the dipole and horn gain standards. These are estimated here as ±0.4 dB in absolute gain and ±0.5 dB in mismatch and mutual-coupling errors. Additionally, the generator and calibration receiver or network analyzer systematic uncertainty is ±0.5 dB. All the errors are treated as uncorrelated, and the error estimates are treated like standard deviations. Then, the combined standard deviation due to random errors is of the order

$$\sigma_{random} = \sqrt{\sigma_{cal}^2 + \sigma_{meas}^2 + 0.1^2} = 0.55\ \text{dB} \tag{9.17}$$

The systematic errors total up to

$$\sigma_{system} = \sqrt{0.4^2 + 0.5^2 + 0.5^2} = 0.81\ \text{dB} \tag{9.18}$$

The systematic error represented by the systematic standard deviation will bias the sensitivity by a fixed, but unknown, amount having a standard deviation of about 0.8 dB. The random errors will cause a scattering of the results with a standard deviation of 0.6 dB about that unknown value. Said another way, we can correlate to another independent sensitivity measurement (another range of similar design, a fixture, or a factory test set) to within the random spread, but we do not know the absolute value because of the systematic errors. Any particular measurement will have an absolute accuracy to within the combined systematic and random standard deviations of about 1 dB if the measurements are carefully performed and the open-field site is carefully calibrated. The 95% confidence bound is approximately twice that value, or ±2 dB. Based on the analysis here, verified by measurements, such accuracy should be considered routinely possible.

In another experiment, the sensitivities of 15 pagers was measured three times each on three consecutive days. The standard deviation for the measurement was $\sigma_{all} = 0.45$ dB. If the range calibration errors and the pager measurement errors are treated as statistically independent, then the measurement standard deviation, σ_{meas}, can be derived by inverting (9.17) to get

$$\sigma_{meas} = \sqrt{\sigma_{all}^2 - \sigma_{cal}^2} = 0.27\,\text{dB} \tag{9.19}$$

which includes transmitter repeatability, as well as 20 call measurement method effects. This compares with $\sigma_{20call} = 0.29$ dB noted earlier. The 95% confidence interval band is approximately $U = 2\sigma_{meas}$; thus, we can report a measurement y as $Y = y \pm U$, where $U = \pm 0.54$ dB. This compares with the "maximum" ±0.5 dB and "typical" ±0.2 dB reported for the 20 call method [4].

The theory behind the uncertainty analysis used here is based on [17] and is recommended by the International Committee for Weights and Measures (CIPM). Briefly, the combined standard uncertainty of a measurement result is taken to represent the estimated standard deviation of the result. It is obtained by combining the individual standard uncertainties (and covariance as appropriate) using the usual method for combining standard deviations resulting in σ_C, as, for example, in (9.17) to (9.19).

An expanded uncertainty, U, is then obtained by multiplying the combined standard deviation of a measurement σ_C by a coverage factor k. Thus, $U = k\sigma_C$, and it can be confidently asserted that $Y = y \pm U$, where Y is the desired quantity. Consistent with current international practice, by convention, $k = 2$ with the result that the interval defined by $U = 2\sigma_C$ defines an interval having a level of confidence of better than 95%. The expansion factor k is comparable to the z-factor used in Table 8.6.

9.6 EMC and EMI Test Chamber

Test on outdoor sites is subject to radio frequency interference, particularly for tests involving radio receiver sensitivity. In many cases, clear frequencies are not available, and testing is thus driven to remote locations like that pictured in Figure 9.10, presumably to achieve a quiet radio environment. The receiver under test is mounted on the belt of the test subject, and a paging signal is transmitted from an antenna 30m away.

The site, a remote river bed located about 135 km from Beijing, China, was used for testing VHF paging receiver sensitivities. However, even this remote site surrounded by mountains was subject to in-channel interference from powerful paging transmitters in Beijing. The solution was to bring the measurements indoors, when possible, into well-constructed, shielded chambers that were also well provisioned with anechoic materials. The shielding of the example chamber in Figure 9.11 is rated for 100-dB attenuation. The chamber, located at Florida Atlantic University, is a certified EMI/EMC facility, which has a reflecting ground surface. Anechoic material was positioned on the floor between the transmitting and receiving positions to suppress the ground reflection and to provide a completely anechoic environment to emulate a free space environment. In that configuration, the facility was used to test compatibility between UWB signals and mobile phone receivers in the 1.7–1.9-GHz bands (see [7]).

The mobile phone is fixed to the phantom human head device on the

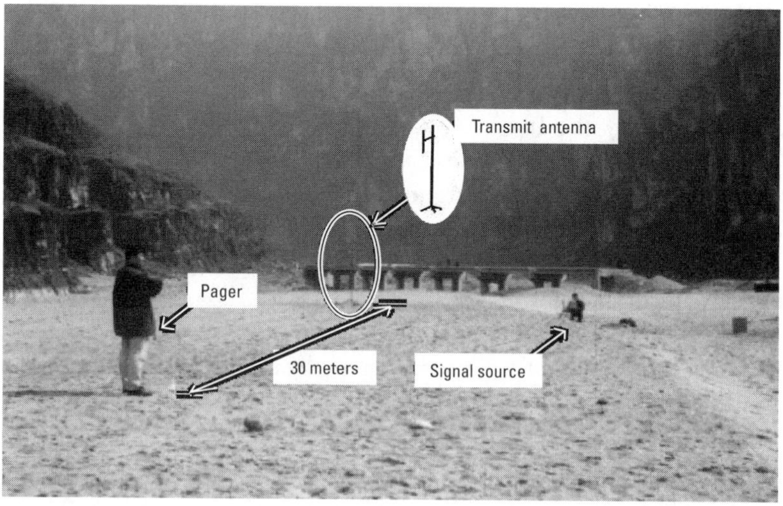

Figure 9.10 Open-air test site in a remote location.

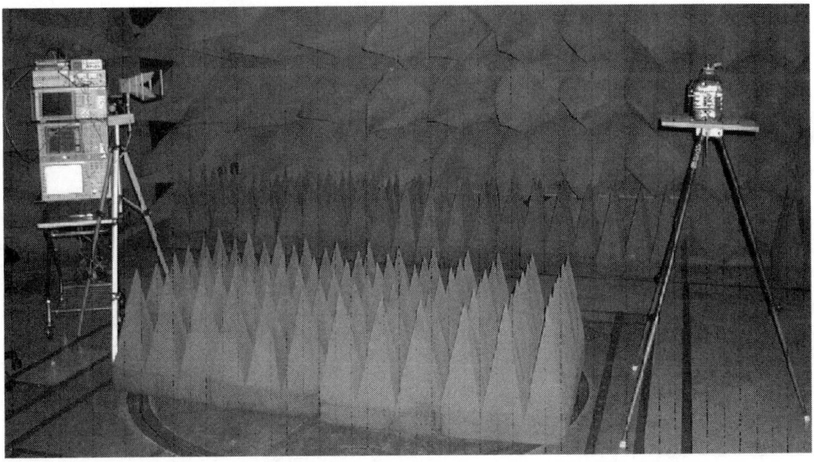

Figure 9.11 A shielded EMI/EMC facility configured as an anechoic chamber.

tripod to the right in the picture, and the test signal is sent from the dual-ridged horn antenna 3m away on the left. Use of anechoic chambers does not eliminate measurement errors, however, a well-shielded chamber enables receiver-sensitivity measurements because external sources of radio interference are greatly suppressed. Electromagnetic measurements in anechoic chambers are intrinsically more accurate than open-air test site measurements. Measurement uncertainties in open-air sites can be kept below ±1 dB, but measurement uncertainties due to anechoic chamber site effects can be in excess of ±4 dB.

9.7 Transmitter Test Sites

PCD transmitter test sites and receiver-sensitivity test sites are identical from the physical layout point of view. As with receiver test sites, there is a distinct difference between transmitter test sites and EMC/EMI compliance test sites. For testing the intentional radiation characteristics of transmitting PCDs, the range setup and calibration are simply reversed from that used in receiver testing. Referring to Figure 9.4, the test subject now performs the radiation function, and the fixed antenna receives the signal. By reciprocity, the signal received at the fixed antenna will vary, as shown both in Figure 9.6 and earlier in Figure 6.7, as the transmitting PCD height is varied at the test location. Similarly, the variation in signal as a function of height over a conducting ground screen will resemble the traces in Figure 6.8.

Calibration of a transmitter test site involves placing a calibration antenna at the test location and transmitting a known power level to the fixed antenna.

The received signal level is noted in relation to the transmitted calibration signal. The EIRP of the PCD is then determined directly in relation to the calibrated signal. The calibration should be performed at the same height above ground as the measurement. For voice devices, the appropriate height is the normal operating level in a natural standing position.

With the PCD transmitting, obviously the tedious process of statistically determining signal levels is not required, and testing times are accordingly shorter. Tests involving large numbers of human subjects are practical, and much statistical data about performance variation among human subjects has been gathered [18].

9.8 Effect of the Human Body

There are two separate electromagnetics problems with regard to the effects of the human body. The first, of primary concern here, is the influence that the human body has on the field-strength pattern of a body-mounted receiver or a transmitter. This is a question of radio system performance, and as Chapter 10 will show, rather coarse models give accurate representations of fields external to the body. The second electromagnetics problem concerns wave coupling into body tissues. This second problem is extensively covered elsewhere with respect to biological issues and dosimetry [19–23] and accurate representations of near fields [24–29], and it is the specific subject of standards [30–32]. We will explore this issue only in the context of radio performance in Chapter 10 and in terms of compliance to exposure standards and guidelines here.

9.8.1 Fields External to the Body

The fields scattered by the body and those external to it are of most concern regarding the communications problem. The fields external to the body can be determined analytically, as shown in Chapter 10 and [22, 23, 29] or experimentally using either transmitting or receiving devices. The analytical results of H.-R. Chuang [29] suggest that the radiation efficiency of a resonant 840-MHz dipole antenna proximate to the body may be reduced to as little as 29% near the head and as little as 15% at belt level, compared to the efficiency in free space conditions. Further, loop antenna efficiency at belt and pocket level near the body are degraded 60% to 62% at 152 MHz, 29% to 32% at 280 MHz, and 31% to 34% at 460 MHz by the presence of the body [33]. Furthermore, because the body is asymmetric with respect to the dipole, there is coupling of energy from the nominal polarization to the cross-polarization component. Receiver measurements involving human subjects are statistically based and are very difficult because people have difficulty remaining still for the duration of a lengthy receiver-sensitivity test. The statistical method can easily be biased by motion of the subject under test, so the receiver field-strength sensitivity test is preferably accomplished using a simulated body test device. The characteristics of such devices, SALTY and SALTY-LITE, are detailed in Chapter 10. Here, we note that the human body, or a simulated body test device, exhibits a whole-body resonance to vertically polarized incident waves. The resonance depends on body height and on coupling to ground. In the standard testing configuration, the body or body device is placed on Earth ground. The magnetic fields near the body are increased relative to the incident waves due to body resonance and to the conductivity of the body. We will show in Chapter 10 that, for fields external to the body, the body can be modeled as a lossy wire antenna in the resonance range and as an infinitely long, saline-water-filled cylinder at radio frequencies above the resonance.

9.8.2 Biological Aspects

The subject of the biological effects of nonionizing electromagnetic (EM) waves is well covered elsewhere [19–23, 30–32, 34–39]. The intent in this section is to introduce, in a very general fashion, the biological aspects of wave coupling to the human body. The electrical properties and dielectric parameters of biological materials are such that the simplified equations for skin depth δ_s, given earlier by (3.6), do not give accurate results. We must use the complete, general formula, $\delta_s = 1/\alpha_g$, where α_g is given exactly by (3.5). In the range of dielectric parameters of interest to biological studies, then, skin depth is given by

$$\delta_S = \left[\frac{k^2}{2} \left[\sqrt{\varepsilon_r^2 + \left[\sqrt{\frac{\sigma}{\omega \varepsilon_0}} \right]^2} - \varepsilon_r \right] \right]^{-1/2} \tag{9.20}$$

where k is the free space wave number $2\pi/\lambda$, the real part of the dielectric constant is ε_r, and $\sigma/\omega\varepsilon_0$ is the imaginary part of the dielectric constant having conductivity σ S/m.

The intrinsic impedance of biological material is fairly low in the range of radio frequencies, 30 MHz to 3 GHz, of primary interest in radio communications. That impedance, calculated for the biological materials (muscle) given in Table 10.1, is shown here in Figure 9.12 compared with the wave impedance magnitude, $|E_{total}|/|H_{total}|$, computed for the small dipole using (1.26) to (1.28) and for the small loop using (1.29) to (1.31). The magnitude of the intrinsic impedance of the example biological tissue is between 38Ω and 57Ω over a wide radio frequency range. The wave impedance as a function of distance

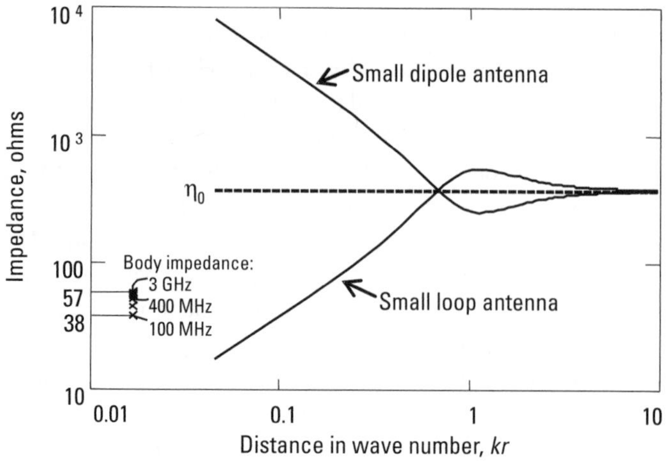

Figure 9.12 Small loop antenna and dipole antenna wave impedances in relation to body impedances.

normalized to wave number k for the small loop and small dipoles approaches the far-field value of $\eta_0 = 376.73\Omega$ for distances greater than about $5/k$.

Antennas of interest to small communications devices, especially antennas internal to radio housings, are often closer than $r = 2$ cm to the body when body worn. That is, at frequencies below 1 GHz, they are closer than $kr = 0.42$. An inspection of Figure 9.12 shows that small loops are better matched to the body than small dipoles. In fact, a dipole parallel and close to the body is effectively short-circuited by the body. For the purposes of radio performance, antennas exhibiting low wave impedance behavior in their near fields are preferred for body-worn radio applications.

In transmitting applications, the measure of the rate at which energy is absorbed by the body is called the *specific absorption rate* (SAR) and is defined as the time (t) derivative of incremental energy (dW) absorbed by an incremental mass (dm) contained in a volume element (dV) of a given density (ρ):

$$\text{SAR} = \frac{d}{dt}\frac{dW}{dm} = \frac{d}{dt}\frac{dW}{\rho dV} \tag{9.21}$$

In terms of an impressed RMS electric field E V/m on a dielectric material of conductivity σ S/m and ρ mass per volume in kilograms per meter cubed, the SAR is

$$\text{SAR} = \frac{\sigma E^2}{\rho} \tag{9.22}$$

in W/kg. The basic premise of modern standards is that the severity of an effect is directly related to the rate of RF energy absorbed, hence the introduction of the concept of SAR. Fields external to the medium are not easily related to fields within the medium, so the determination of SAR is complex and often relies on precise measurements, the details of which are beyond the scope of this book.

9.9 RF Exposure Standards

Transmitting devices, those at fixed sites and especially portable ones, expose people to radio frequency energy. New FCC actions in 1996, spurred by the U.S. Congress, have resulted in new compliance requirements regarding transmitting radio equipment [40]. The FCC compliance requirements, as contrasted with compliance guidelines elsewhere, will be used as an example to study RF exposure and methods of compliance. This section is not the definitive guide for RF electromagnetic field exposure compliance, and the reader is cautioned to refer to the latest relevant standards and regulations that apply within the jurisdiction of interest. Here, we explore the nature of some guidelines, using the IEEE/ANSI standards and the 1996 FCC regulation as particular examples.

Natural EM fields come from three main sources: the sun (130 mW/cm^2 at all frequencies and far less at RF), thunderstorm activity (quasistatic electric fields in the several volts per meter range), and the Earth's magnetic field (on the order of 40 A/m). In the last 100 years, man-made RF fields at much higher intensities and with a very different spectral distribution have altered this natural EM background in ways that are still under study. RF fields are classified as nonionizing radiation because the frequency is much too low for photon energy to ionize atoms. Still, at sufficiently high power densities, they can heat body tissue. Various standards organizations and government entities, including the American National Standards Institute (ANSI), the Comité Européen de Normalisation Electrotechnique (CENELEC), the U.S. Environmental Protection Agency (EPA), the U.S. Federal Communications Commission (FCC), the Institute of Electrical and Electronics Engineers (IEEE), the International Radiation Protection Association (IRPA), the National Council on Radiation Protection and Measurements (NCRP) chartered by the U.S. Congress, and the United Kingdom's National Radiological Protection Board (NRPB) have issued documents on RF exposure, including protection guidelines and regulations. Australia, Belgium, Canada, the former Czeckoslovakia, Finland, Norway, Poland, Russia, the former Soviet Union, Sweden, the United Kingdom, and the United States have all issued standards regarding RF exposure levels. Details of some standards and their biological aspects have been collected by O. P. Gandhi [20], S. M. Michaelson, and J. C. Lin [38, 39].

Biological tissues subjected to RF energy will absorb that energy and convert it to heat as governed by (9.21). External fields couple most efficiently to the body when the electric field is aligned with the body's axis in the whole-body half-wave resonance range. For adult humans, the resonance is between 35 MHz for a grounded person and about 70 MHz for a body isolated from the ground, as shown in Chapter 10. For small infants, that resonant range extends upwards, so special attention is paid to RF exposure in the resonant frequency region of 30 to 300 MHz. Additionally, body parts may exhibit resonant behavior. The adult head, for example, is resonant around 400 MHz, while a baby's smaller head resonates near 700 MHz. Body size thus determines the frequency at which RF energy is absorbed most efficiently [22]. As the frequency is increased above resonance, less RF heating generally occurs, and because the RF skin depth decreases with increasing frequency, the heating is increasingly confined to surface tissue. All of these factors have led to RF exposure guidelines that have varying limiting levels of power density exposure with frequency and, in some cases, different exposure limits for electric and magnetic fields. Note that SAR as defined in (9.21) and SAR limits are generally frequency independent. The coupling mechanism that results in the internal body fields, as E in (9.22), however, involves resonances and hence, is frequency dependent. Tissue heating is the primary effect of concern in the RF electromagnetic fields standards, SAR is the relevant mechanism, and the fields external to the tissue which give rise to the SAR are what we attempt to control with the relevant exposure standards. We shall concern ourselves with the external fields and with the resulting field and power density exposure criteria.

9.9.1 Radiated RF Exposure Guidelines and Regulations

A recent survey by CENELEC found more than a hundred documents worldwide relating to standards and regulations concerned with RF electromagnetic fields [41]. The CENELEC working group had as its aim the investigation of existing regulations, standards, and related documents about human safety in radio frequency electromagnetic fields. Although good agreement was found in the basic limits for whole-body SAR, specifications for local peak SARs varied by an order of magnitude. In the specification of field strength and power density, limits varied over two orders of magnitude in some cases. Furthermore, many documents were not self-consistent in derived field-strength limits as compared with specified basic restrictions.

ANSI issued RF protection guidelines in 1982, and these were adopted in part by the NCRP in the United States in 1986 [34]. The ANSI guidelines were replaced by the ANSI/IEEE C95.1-1992 guidelines [31], summarized for our study in Tables 9.2 and 9.3. We note that the protection guidelines distinguish between the controlled and uncontrolled environment. Each table further

Table 9.2
ANSI/IEEE C95.1-1992 Radio Frequency Protection Guidelines for Controlled Environments [18]

Frequency Range, f(MHz)	Electric Field Strength, E(V/m)	Magnetic Field Strength, H(A/m)	Power Density E Field; H Field S(mW/cm^2)	Averaging Time $\|E^2\|$; $\|H^2\|$; S(min.)
0.003–0.1	614	163	(100; 1,000,000)*	6
0.1–3.0	614	$16.3/f$	$(100; 10{,}000/f^2)$*	6
3.0–30	$1{,}824/f$	$16.3/f$	$(900/f^2; 10{,}000/f^2)$*	6
30–100	61.4	$16.3/f$	$(1.0; 10{,}000/f^2)$*	6
100–300	61.4	0.163	1.0	6
300–3,000	—	—	$f/300$	6
3,000–15,000	—	—	10	6
15,000–300,000	—	—	10	$616{,}000/f^{1.2}$

* Plane-wave equivalent power density. This is not appropriate for near-field conditions but is sometimes used for comparisons.

distinguishes between the effects of electric and magnetic fields at frequencies below 100 MHz. Table 9.3 additionally distinguishes between averaging times for exposure to electric and magnetic fields. This, of course, recognizes that wave impedance for waves in the presence of reflectors and scatterers is not necessarily 376.73Ω.

In 1996, the U.S. FCC, by congressional mandate, issued regulatory limits for maximum permissible exposure (MPE) for the occupational/controlled and general population/uncontrolled environments. Table 9.4 summarizes the occupational/controlled limits, and Table 9.5 summarizes the general population/uncontrolled limits. The FCC defines occupational/controlled as applying in situations in which persons are exposed as a consequence of their employment, provided those persons are fully aware of the potential for exposure and can exercise control over their exposure. Radio amateurs and their immediate households fall into this category. General population/uncontrolled exposures apply to situations in which the general public may be exposed or in which persons exposed as a consequence of their employment may not be fully aware of the potential for exposure or cannot exercise control over their exposure. The neighbors of radio amateurs also fall into this category. The effect of the FCC regulations on U.S. radio amateurs is discussed in detail in [36, 37].

Tables 9.2 to 9.5 all recognize the whole-body resonant region and apply stricter limits in the 30–300-MHz range. They also deal differently with individuals who are aware of and can control their circumstances and those who may

Table 9.3
ANSI/IEEE C95.1-1992 Radio Frequency Protection Guides for Uncontrolled Environments [30]

Frequency Range, f (MHz)	Electric Field Strength, E (V/m)	Magnetic Field Strength, H (A/m)	Power Density E Field; H Field S (mW/cm^2)	Averaging Time $\|E^2\|$; S (min.)	Averaging Time $\|H^2\|$; S (min.)
0.003–0.1	614	163	(100; 1,000,000)*	6	6
0.1–1.34	614	$16.3/f$	$(100; 10{,}000/f^2)$*	6	6
1.34–3.0	$823.8/f$	$16.3/f$	$(180/f^2; 10{,}000/f^2)$*	$f^2/0.3$	6
3.0–30	$823.8/f$	$16.3/f$	$(180/f^2; 10{,}000/f^2)$*	30	6
30–100	27.5	$158.3/f^{1.668}$	$(0.2; 940{,}000/f^{3.336})$*	30	$0.0636f^{1.337}$
100–300	27.5	0.0729	0.2	30	30
300–3,000	—	—	$f/1{,}500$	30	—
3,000–15,000	—	—	$f/1{,}500$	$90{,}000/f$	—
15,000–300,000	—	—	10	$616{,}000/f^{1.2}$	—

* Plane-wave equivalent power density. This is not appropriate for near-field conditions but is sometimes used for comparisons.

Table 9.4
1996 FCC Limits for Occupational/Controlled Environments [40]

Frequency Range, f (MHz)	Electric Field Strength, E (V/m)	Magnetic Field Strength, H (A/m)	Power Density S (mW/cm^2)	Averaging Time (min.)
0.3–3.0	614	$1.63/f$	(100)*	6
3.0–30	$1{,}842/f$	$4.89/f$	$(900/f^2)$*	6
30–300	61.4	0.163	1.0	6
300–1,500	—	—	$f/300$	6
1,500–100,000	—	—	5	6

* Plane-wave equivalent power density.

not be aware of or able to control their exposure. There are corresponding SAR limits in [40], but here we are concerned with radiation issues and will confine our study to RF exposure.

Table 9.5
1996 FCC Limits for General Population/Uncontrolled Environments [40]

Frequency Range, *f* (MHz)	Electric Field Strength, *E* (V/m)	Magnetic Field Strength, *H* (A/m)	Power Density *S* (mW/cm^2)	Averaging Time (min.)
0.3–1.34	614	1.63/*f*	(100)*	30
01.34–30	824/*f*	2.19/*f*	(180/f^2)*	30
30–300	27.5	0.073	0.2	30
300–1,500	—	—	*f*/1,500	30
1,500–100,000	—	—	1.0	30

* Plane-wave equivalent power density.

9.9.2 Compliance with RF Exposure Standards

Once MPE regulations are set, the remaining task is compliance. The NCRP [34] and ANSI/IEEE [32] define the rationale behind and methodology of compliance with MPE standards. We present the 1996 FCC limits as an example. Both the electric and magnetic field quantities in Tables 9.2 to 9.5 are total magnitude values. That is, electric field E and magnetic field H in the tables are

$$E = \sqrt{\left|E_x^2\right| + \left|E_y^2\right| + \left|E_z^2\right|} \tag{9.23}$$

and

$$H = \sqrt{\left|H_x^2\right| + \left|H_y^2\right| + \left|H_z^2\right|} \tag{9.24}$$

as can be measured by isotropic probes, such as described by U.S. Patent 4,588,933 [42], or by equivalent probes as described in [32]. Figure 9.13 shows graphs of essentially all of the limits defined in Tables 9.4 and 9.5 with the added provision that the quantity S mW/cm^2 be defined and evaluated in both the near and far fields in terms of the electric fields by

$$S_E = 0.1\left[\left|E_x\right|^2 + \left|E_y\right|^2 + \left|E_z\right|^2\right] / 376.73 \tag{9.25}$$

and in terms of the magnetic fields by

$$S_H = 0.1\left[\left|H_x\right|^2 + \left|H_y\right|^2 + \left|H_z\right|^2\right] 376.73 \tag{9.26}$$

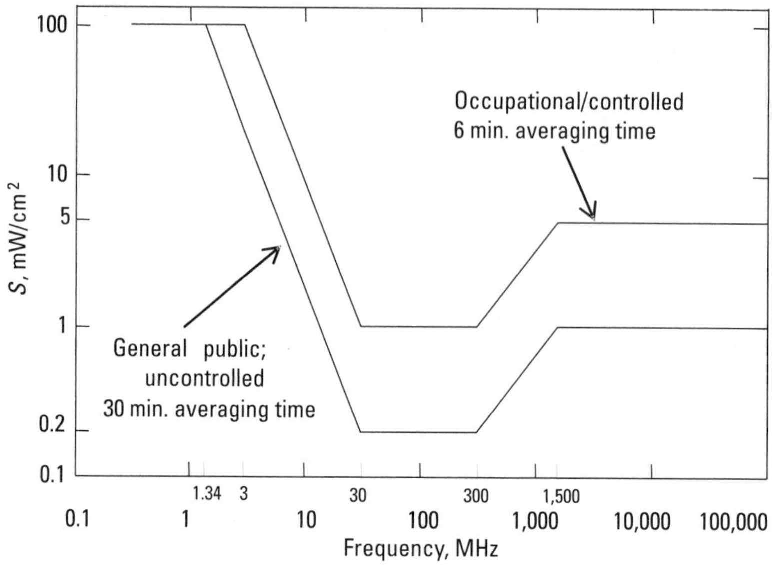

Figure 9.13 A representation of 1996 FCC RF MPE limits.

with E V/m and H A/m. Finally,

$$S = greater\ of \begin{cases} S_E \\ S_H \end{cases} \tag{9.27}$$

In the far field in unbounded media, $S_E = S_H$, and the quantity S is the power density. The 1996 FCC limits show a single trace for S only because those limits do not make a distinction in "far-field equivalent" power density between electric and magnetic fields at the lower frequencies, as do the ANSI/IEEE limits shown in Tables 9.2 and 9.3. The ANSI/IEEE standard must apply S_E to the electric field limit and S_H to the magnetic field limit separately.

The MPE limits are applied to mixed frequency fields by weighting their individual far-field equivalent power densities, as found by (9.27), in accordance with the exposure limit at each frequency. That is, the combined power densities are conditioned on

$$\sum_{i=1}^{n} \frac{S_i}{L_i} = \frac{S_1}{L_1} + \frac{S_2}{L_2} + \frac{S_3}{L_2} + \cdots \frac{S_3}{L_2} = 1 \tag{9.28}$$

where L_i is the exposure limits at the respective frequencies.

[9-8a.mcd] EM fields at some distance from four antennas are found to be 33 V/m at 18.1 MHz, 20 V/m at 47 MHz, 15 V/m at 152 MHz, and 1 V/m at 460 MHz. Evaluate the exposure level with respect to the 1996 FCC general population and occupational MPE limits.

The frequency dependence of the standard is evident in the reduced permissible exposure over the whole-body resonant range from 30 to 300 MHz. This range corresponds to the resonant behavior of human phantoms used in receiver-sensitivity measurements, as shown in Chapter 10.

The presence of boundaries, such as Earth ground, alter the wave impedance so that electric and magnetic fields must be considered separately, even in the far field of the source. This is illustrated by considering a horizontal dipole 15m above the Earth operating at 30 MHz with 1,500W of supplied power. The electric and magnetic fields both obey the boundary conditions at the air-Earth interface, and the magnetic field is enhanced, while the electric field is diminished. When normalized to the MPE of the 1996 FCC standard, the total magnetic field in decibels relative to the standard is shown in Figure 9.14, while the total electric field contours similarly normalized are picture in Figure 9.15. Ignoring the exposure averaging time in the standards, permissible general population exposure levels are the regions outside the "0-dB" contours. Significantly, the magnetic field contours of Figure 9.14 are substantially different from the electric field contours shown in Figure 9.15. Magnetic fields peak at ground level, while electric fields peak a quarter-wavelength above ground. This is a consequence of the ground reflection and has nothing to do with whether the fields are near or far with respect to the dipole. The wave impedance

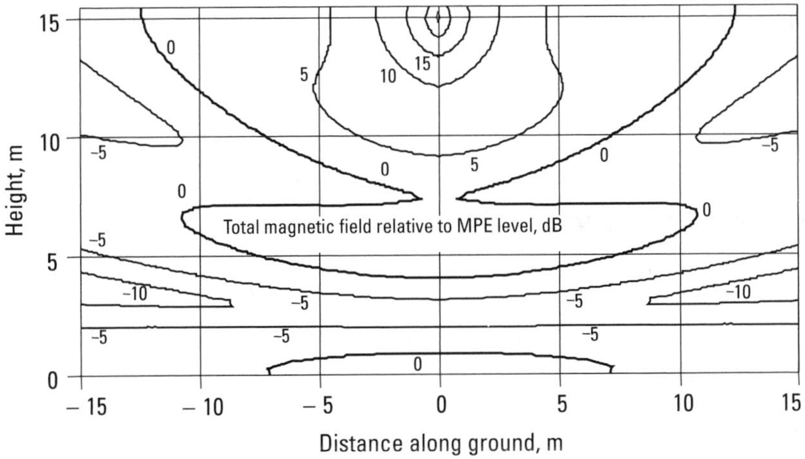

Figure 9.14 Magnetic fields relative to MPE limits. (*Source:* [37].)

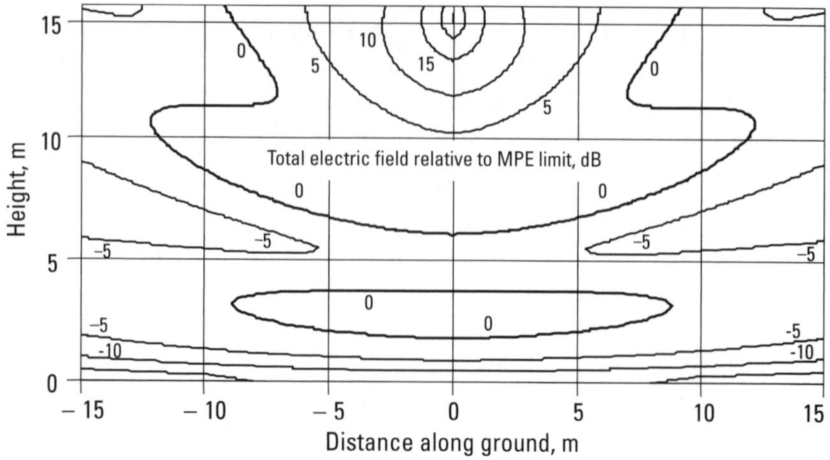

Figure 9.15 Electric fields relative to MPE limits. (*Source:* [37].)

evaluated on the total fields is simply not equal to the intrinsic impedance associated with the medium.

- [9-8b.mcd] Compute the ground reflection factor ($\varepsilon = 7, \sigma = 0.03$ S/m) at 150 MHz as a function of ground incidence angle.

- [9-8c.mcd] Compute the fields from a 30-MHz dipole parallel with and 10m, 11m, 12m, and 13m above the ground ($\varepsilon = 7, \sigma = 0.03$ S/m) in a plane 3m from the dipole. Compare these with free space fields for a dipole 10m above ground.

The standard is written around the maximum of either the electric or magnetic field limit, as calculated by (9.27). That quantity is pictured in Figure 9.16. The "0-dB" contours represent the limits within which either the electric or magnetic fields exceed the MPE level of the standard. If the power transmitted by the dipole were reduced by 5 dB, then the MPE limit contour would be represented by the "5-dB" contour in Figure 9.16. The figure shows that the determination of field levels relative to MPE levels is complex, even for the case of a simple antenna in the presence of a single boundary.

- [9-8d.mcd] Compute the fields in ratio to the FCC 1997 compliance standard from a 30-MHz dipole parallel with and 10m above the ground ($\varepsilon = 20, \sigma = 0.03$ S/m) and radiating 750W. Find the E and H compliance contours.

We saw in Figure 2.6 that the field strength near the ground, though not considering the ground, is a function of the antenna pattern. One way of accounting conservatively for the ground reflection for RF MPE calculations is

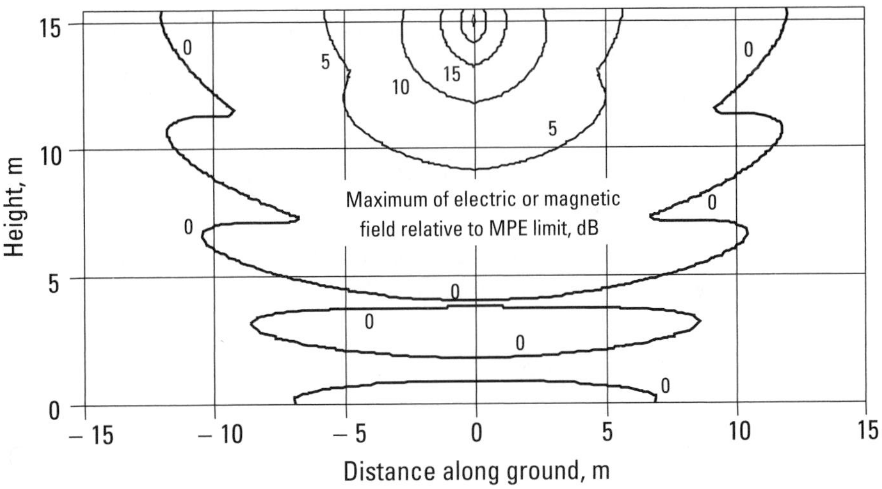

Figure 9.16 Greater of magnetic or electric fields relative to MPE limits. (*Source:* [37].)

to increase the free space field by 6 dB to account for the maximum possible reflection contribution from the ground, then to evaluate the result against an MPE limit.

9.10 Influence of Ground on Yagi Antenna Patterns

The ground contribution (ignoring mutual coupling to ground) can be considered from image theory, and this provides a good method for studying both vertically and horizontally polarized antennas over ground. The relevant geometry is in Figure 6.2, and we use the reflection coefficient defined in (6.6) to (6.8) as the amplitude modifier for the image current. The ground is in the *xy*-plane. We will consider dipoles and Yagi arrays, so, for convenience, the electric radiation pattern of a sinusoidally excited dipole element is simplified to

$$E_{dip} = F(\theta)\frac{e^{-jkr}}{2kr} \tag{9.29}$$

and

$$H_{dip} = \frac{E_{dip}}{\eta_0} \tag{9.30}$$

where $F(\theta)$ is from (2.1), and θ is measured from the axis of the dipole. In the plane of incidence, that is the *xz*-plane, the electric field is *y*-directed for horizontal polarization:

$$\mathbf{E} = \mathbf{y}\left[E_{dip}(\theta_{dir}) + E_{dip}(\theta_{ref})\Gamma_H\right] \tag{9.31}$$

where angles θ_{dir} and θ_{ref} define the direct and reflected rays relative to the *x*-axis. The magnetic field has *x*- and *z*-components

$$\begin{aligned} \mathbf{H} = &H_{dip}(\theta_{dir})\left[-\mathbf{x}\cos(\theta_{dir}) + \mathbf{z}\sin(\theta_{dir})\right] \\ &+H_{dip}(\theta_{ref})\left[-\mathbf{x}\cos(\theta_{ref}) + \mathbf{z}\sin(\theta_{ref})\right]\Gamma_H \end{aligned} \tag{9.32}$$

For vertical polarization, the electric field has *x*- and *z*-components:

$$\begin{aligned} \mathbf{E} = &E_{dip}(\theta_{dir})\left[-\mathbf{x}\cos(\theta_{dir}) + \mathbf{z}\sin(\theta_{dir})\right] \\ &+E_{dip}(\theta_{ref})\left[-\mathbf{x}\cos(\theta_{ref}) + \mathbf{z}\sin(\theta_{ref})\right]\Gamma_V \end{aligned} \tag{9.33}$$

The magnetic field in the plane of incidence is *y*-directed:

$$\mathbf{H} = -\mathbf{y}\left[H_{dip}(\theta_{dip}) + H_{dip}(\theta_{ref})\Gamma_V\right] \tag{9.34}$$

The total field is defined as the square root of the sum of the magnitude squared of the individual field components, and we define the total field wave impedance Z_W as the ratio of the total electric to the total magnetic field:

$$Z_W = \sqrt{\frac{|E_x|^2 + |E_y|^2 + |E_z|^2}{|H_x|^2 + |H_y|^2 + |H_z|^2}} \tag{9.35}$$

The array response is found by using (9.31) to (9.34) for each element of an array of dipoles to represent a Yagi antenna as described in Section 2.9.6. Figure 9.17 shows the wave impedance defined by (9.35) for a three-element, horizontally polarized Yagi antenna three wavelengths above Earth ground, having parameters $\varepsilon = 7$ and $\sigma = 0.005$ S/m. Figure 9.18 shows the wave impedance for the vertically polarized three-element yagi with the same Earth ground parameters. Distances are shown in wavelengths. The dashed lines in the figures represent impedances smaller than η_0, the thin lines are contours equal to η_0,

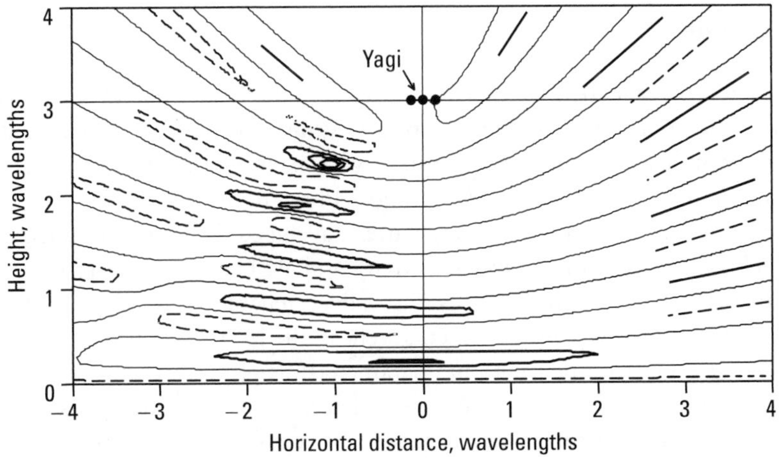

Figure 9.17 Wave impedances of a horizontally polarized yagi above ground.

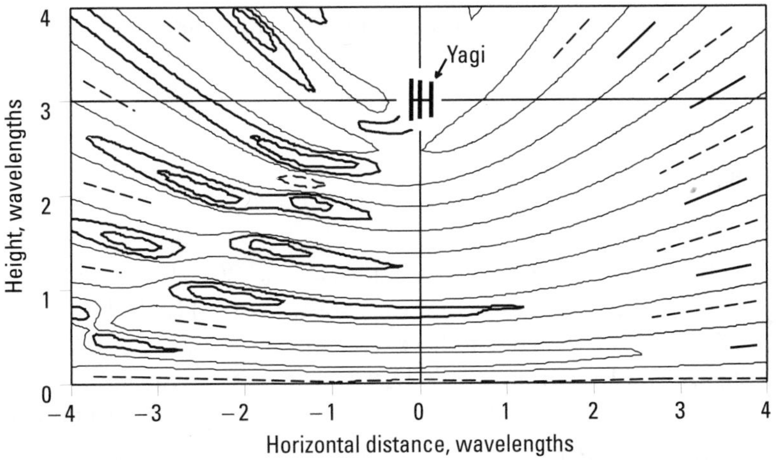

Figure 9.18 Wave impedances of a vertically polarized yagi above ground.

and the thick lines represent impedances greater than η_0. The wave impedance is low near the ground, and a vertical standing wave pattern is evident above the ground.

- [9-9.mcd] Using (9.31) to (9.34), compute the wave impedances as defined by (9.35) at distances of 1m, 3m, 10m, 30m, and 1,000m and 1m above ground for both a horizontally and vertically polarized, 170-MHz dipole three wavelengths above ground.

9.11 Summary

We studied the statistical methods of measuring the field-strength sensitivity of selective call receivers. The accuracy of two statistical methods, the 20 call test and a simplified three-of-three method, were examined in detail for accuracy and repeatability. The equations relating field strength to received power were derived as was a method of reporting pattern-averaged field-strength sensitivity.

The calibration of receiver-sensitivity and PCD transmitter test sites was examined, and the various factors that affect reliability and repeatability of field-strength sensitivity measurements were studied. The effects of variations in the calibration factor due to ground parameters, height errors, and the accuracy of the calibration gain standards were determined. An intercomparison of receiver sensitivity at completely independent test sites validated the analytical methods used here to describe test sites. The effects due the presence of the human body were examined from the point of view of fields external to the body and from the body-internal biological aspects. The averaging methods for receiver sensitivity and for total radiated power were examined. Finally, we examined RF exposure guidelines and regulations from the compliance point of view.

Problems

9.1 A PCD receiver-sensitivity test follows the 20 call test method. The recorded attenuator values are 16, 15, 14, 15, 12, 11, 10, 13, 14, and 15. The PCD was designed to have calling characteristics similar to those in Figure 9.2. Determine the PCD sensitivity if an attenuator setting of 15 corresponds to 14 dBμV/m, and comment on the accuracy of the result.

9.2 A 900-MHz PCD receiver sensitivity is measured at eight equally spaced azimuth angles, with the following results: 14.2, 15, 19, 27.3, 33, 28, 18.5, and 15 dBμV/m. Find (a) the gain-averaged sensitivity, and (b) the absolute gain pattern if the receiver sensitivity is –128 dBm.

9.3 Find the average gain of a PCD antenna if the eight-position receiver-sensitivity values are 14.1, 14.3, 15, 15.5, 16, 15.3, 15.1, and 15.1 at 160 MHz, and the receiver sensitivity is –124 dBm.

9.4 A calibrated antenna has an antenna factor of 9.6 dB. Find the field strength if the recorded signal level is –82 dBm.

9.5 Find the antenna factor for a lossless, resonant, half-wave dipole.

9.6 Determine the antenna factor for a yagi antenna with 8.5-dBi gain.

9.7 An antenna range transmitter generates 1 mW, which is supplied through a cable and attenuator having 10-dB loss to a dipole. An 8-dBi gain antenna is used at the receiver location and connects to a calibrated receiver through a cable having 8 dB of loss. If the site attenuation at 930 MHz is 62 dB, find (a) the transmitter calibration factor for this range, and (b) the signal level at the calibration receiver.

9.8 A 10m-long test site with a metal ground plane floor is proposed for testing the performance of circularly polarized PCS receiver devices at 1,620 and 2,510 MHz placed 1.5m above ground. A pair of phased crossed dipoles at 2m height is used for the transmitting antenna. Find the polarization axial ratio at the receiver location.

9.9 Figure 9.P1 shows a transmitter test site configured for a calibrating site using a dipole at C. Cables A and B have 7- and 8-dB attenuation, respectively, and the receiver antenna has 8.5-dBi gain. Find (a) the site attenuation if the ground parameters are ε = 5.5, σ = 0.001 S/m, and (b) the signal strength reading at receiver RX at 170, 300, and 940 MHz if transmitter TX delivers 10 mW.

9.10 In Figure 9.P1, cable A attenuates 7.3 dB, the receiver antenna has 8.4-dBi gain, cable B attenuates 9.5 dB, and antenna C is an ideal dipole. With 0 dBm generated at transmitter TX, receiver RX indicates –71.3 dBm. Find (a) the transmit calibration factor and (b) the site attenuation at 902 MHz.

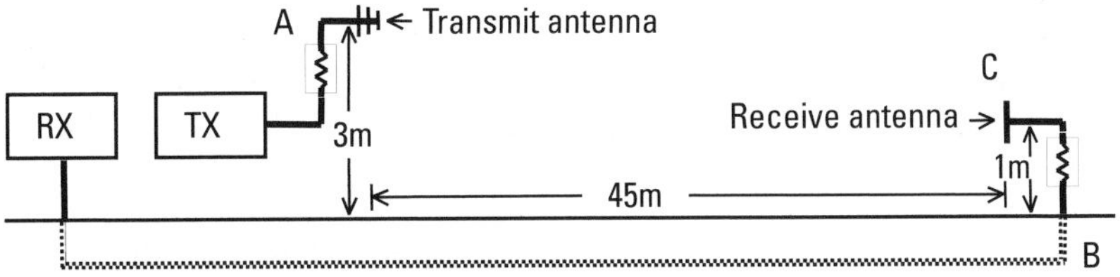

Figure 9.P1 A test-site configuration.

9.11 How much energy is absorbed by 1 kg of water with σ = 1 S/m if 10 V/m is impressed for 600 sec? What is the maximum possible temperature rise [1 cal = 4.18605 J]?

Ans:

$W = \text{SAR } t$; $W = t\sigma E^2 = 600(1)(100) = 60$ J.

$\Delta T = 60{,}000/(4{,}186.05) = 14.3°\text{C}$.

9.12 How much energy is absorbed by 0.25 kg of water at 20°C in a microwave oven cavity that delivers 700W for 200 sec [1 kg-cal ≡ 1/860 kW hr = 4,186.05 J]? What is the maximum temperature rise? Estimate the electric field within the water if $\sigma_{water} = 1.23$ S/m.

Ans:

$W = Pt$; $W = (700)(200) = 140{,}000$ J.

$\Delta T = 140{,}000/(4{,}168.05 \times 0.250)°\text{C} = 134°\text{C}$, but water boils at 100°C, so the maximum rise is 80°C, and the rest of the energy turns water to steam.

$E^2 = P\rho/\sigma$, so $E = \sqrt{700W\ /\ 1.23S\ /m} = 23.9$ V/m (rms)

9.13 A half-kilogram of water initially at 25°C just begins to boil after 240 sec in a microwave oven. If the heat of evaporation of water is 539 cal/g, estimate the microwave's oven power.

Ans:

The temperature must first rise by 75°C, then energy is transferred at 539 cal/g to begin the onset of boil.

$P = (75°\text{C})(0.5)(4{,}186.05)/240 = 654\text{W}$.

9.14 Field strengths near a site with four continuously transmitting antennas are 12 V/m at 21 MHz, 12 V/m at 32 MHz, 9 V/m at 159 MHz, and 4 V/m at 930 MHz. Find the exposure level with respect to the 1996 FCC limits.

9.15 A dipole radiates 175W continuously at 465 MHz. Find the minimum distance at which the exposure level will be in compliance with the 1996 FCC standard.

Ans:

Power density/limit = $1.641P/(4\pi d^2)$/limit, so
$d = \sqrt{1.641 \cdot 175{,}000\ /\ 4\pi\ limit} = 67.6$ and 151.2 cm for the occupational and general population limits. The distance will double if a conservative estimate of a ground reflection must be considered.

9.16 The conducted receiver sensitivity of a GSM mobile phone was measured in a shielded fixture using a three-of-three method at the receiver test port, yielding a receiver sensitivity of –112 dBm. The same mobile phone was tested in a shielded anechoic chamber using a radiated signal to the phone mounted on a phantom human device, and a peak sensitivity of –107 dBm was recorded. In transmit, the antenna gain in the direction of the peak gain was found to be +1 dBi, suggesting that the peak receiver sensitivity should have been –113 dBm. What can account for the 6-dB discrepancy?

Ans:

The antenna gain could vary between the transmitter and receiver bands. Also, receiver measurements are actually "signal-to-noise" determinations, so the result is dependent on the noise level. Since the measurement was made in a shielded environment, external noise can be ruled out, and it is likely that the mobile phone is generating self-interference, which reduces sensitivity by up to 6 dB.

References

[1] "Methods of Measurement for Radio Equipment Used in the Mobile Services," Sec. 8—Reference Sensitivity (selective calling), *IEC 489-6,* 2nd ed., 1987.

[2] "Paging Systems; European Radio Message System (ERMES) Part 5: Receiver Conformance Specification," *ETS 300 133-5,* ETSI, Valbonne, France, July 1992, amended (A1), January 1994.

[3] "Radio Equipment and Systems (RES); European Radio Message System (ERMES) Receiver Requirements," *Final Draft prTBR 7,* ETSI, Valbonne, France, January 1994, amended March 15, 1994.

[4] Siwiak, K., L. Ponce de Leon, and W. M. Elliott III, "Pager Sensitivities, Open Field Antenna Ranges, and Simulated Body Test Devices: Analysis and Measurements," First Annual RF/Microwave Non-Linear Simulation Technical Exchange, Motorola, Inc., Plantation, FL, March 26, 1993.

[5] Siwiak, K., *Radio Wave Propagation and Antennas for Portable Communications,* Workshop Notes, Taipei, Taiwan, Republic of China, October 5–7, 1993.

[6] Babij, T. M., et al., "Accuracy of Near-Field Measurements Involving Phantom Humans on Open Field Test Sites, in *Electricity and Magnetism in Biology and Medicine,* M. Blank, (ed.), San Francisco: San Francisco Press, 1993, pp. 589–592.

[7] Siwiak, K., "UWB Emissions and EMC," EMC2006: 18th International Wroclaw Symposium and Exhibition on Electromagnetic Compatibility, Wroclaw, Poland, June 28–30, 2006.

[8] "Method of Measurement for Radiated RF Power and Receiver Performance," Revision 2.0, CTIA Certification Program Test Plan for Mobile Station over the Air Performance, Cellular Telecommunications and Internet Association, Washington, D.C., March 2003.

[9] Siwiak, K., and W. M. Elliott III, "Use of Simulated Human Bodies in Pager Receiver Sensitivity Measurements," *SouthCon/92 Conference Record*, Orlando, FL, March 11, 1992, pp. 189–192.

[10] FitzGerrell, R. G., "Linear Gain-Standard Antennas below 1,000 MHz," *NBS Technical Note 1098*, May 1986.

[11] "Minimum Standards for Communication Antennas Part I—Base Station Antennas," *EIA/TIA-329-B*, Electronics Industries Association, Washington D.C., September 1989.

[12] Taggart, N. E., and J. F. Shafer, "Testing of Electronics Industries Association Land-Mobile Communication Antenna Gain Standards at the National Bureau of Standards," *IEEE Transactions on Vehicular Technology*, Vol. VT-27, No. 4, November 1978, pp. 259.

[13] Slayton, W. T., "Design and Calibration of Microwave Gain Standards," *NRL Report No. 4433*, November 9, 1954.

[14] "Measurement of Receiver Sensitivity," *ETSI/STC Technical Paper RES04(93)13*, Amsterdam, Netherlands, November 15–17, 1993 (available from the author).

[15] "Test Report of Sensitivity Measurements of Six POCSAG Pagers Working on ERMES Frequencies in an Anechoic Room and on an Open-Area Test Site," *ETSI/STC Technical Paper RES04(94)35*, Dublin, Ireland, September 20–21, 1994 (available from the author).

[16] Stubenrauch, C. F., P. G. Galliano, and T. M. Babij, "International Intercomparison of Electric-Field Strength at 100 MHz, *IEEE Transactions on Instrumentation and Measurement*, Vol. IM-32, No. 1, March 1983, pp. 235–237.

[17] Taylor, Barry N., and Chris E. Kuyatt, NIST Technical Note 1297, 1994 Edition, *Guidelines for Evaluating and Expressing the Uncertainty of NIST Measurement Results*, available at http://physics.nist.gov/Pubs/guidelines.

[18] Hill, C., and T. Kneisel, "Portable Radio Antenna Performance in the 150, 450, 800, and 900 MHz Bands 'Outside' and In-Vehicle," *IEEE Transactions on Vehicular Technology*, Vol. VT-40, No. 4, November 1991, pp. 750–756.

[19] Fujimoto, K., and J. R. James, (eds.), *Mobile Antenna Systems Handbook*, Norwood, MA: Artech House, 1994.

[20] Gandhi, O. P., (ed.), *Biological Effects and Medical Applications of Electromagnetic Energy*, Englewood Cliffs, NJ: Prentice Hall, 1990.

[21] Johnson, C. C., and A. W. Guy, "Nonionizing Electromagnetic Wave Effects in Biological Materials and Systems," *Proceedings of the IEEE*, Vol. 60, No. 6, June 1972, pp. 692–718.

[22] Durney, C. H., H. Massoudi, and M. F. Iskander, *Radio Frequency Radiation Dosimetry Handbook*, 4th ed., USAFSAM-TR-85-73, USAF School of Aerospace, Brooks AFB, TX, 78235, October 1986.

[23] Osepchuk, J. M., (ed.), *Biological Effects of Electromagnetic Radiation*, New York: IEEE Press, 1983.

[24] Balzano, Q., O. Garay, and K. Siwiak, "The Near Field of Dipole Antennas, Part I: Theory," *IEEE Transactions on Vehicular Technology*, Vol. VT-30, No. 4, November, 1981, pp. 161–174.

[25] Balzano, Q., O. Garay, and K. Siwiak, "The Near Field of Dipole Antennas, Part II: Experimental Results," *IEEE Transactions on Vehicular Technology*, Vol. VT-30, No. 4, November 1981, pp. 175–181.

[26] Balzano, Q., O. Garay, and K. Siwiak, "The Near Field of Omnidirectional Helices," *IEEE Transactions on Vehicular Technology*, Vol. VT-31, No. 4, November 1982.

[27] Balzano, Q., and K. Siwiak, "Radiation of Annular Antennas," *Correlations* (Motorola Engineering Bulletin, Schaumburg, IL), Vol. 6, No. 2, Winter 1987.

[28] Balzano, Q., and K. Siwiak, "The Near Field of Annular Antennas," *IEEE Transactions on Vehicular Technology*, Vol. VT-36, No. 4, November 1987, pp. 173–183.

[29] Chuang, H.-R., "Human Operator Coupling Effects on Radiation Characteristics of a Portable Communication Dipole Antenna," *IEEE Transactions on Antennas and Propagation*, Vol. 42, No. 4, April 1994, pp. 556–560.

[30] *IEEE Standard for Safety Levels with Respect to Human Exposure to Radio Frequency Electromagnetic Fields, 3 kHz to 300 GHz*, IEEE C95.1-1991 (revision of ANSI C95.1-1982), New York: Institute of Electrical and Electronics Engineers, April 27, 1992 [also issued as ANSI/IEEE C95.1-1992].

[31] "Evaluating Compliance with FCC Guidelines for Human Exposure to Radio Frequency Electromagnetic Fields," *OET Bulletin 65, Edition 97-01*, Federal Communications Commission, Office of Engineering and Technology, Washington, D.C., August 1997.

[32] *IEEE Recommended Practice for the Measurement of Potentially Hazardous Electromagnetic Fields—RF and Microwave*, IEEE C95.3-1991 (revision of ANSI C95.3-1973 and ANSI C95.3-1981), New York: Institute of Electrical and Electronics Engineers, April 27, 1992.

[33] Chuang, H.-R., "Computer Simulation of the Human-Body Effects on a Circular-Loop-Wire Antenna for Radio-Pager Communications at 152, 280, and 400 MHz," *IEEE Transactions on Vehicular Technology*, Vol. 46, No. 3, August 1997, pp. 544–559.

[34] "Biological Effects and Exposure Criteria for Radio Frequency Electromagnetic Fields," *NCRP Report No. 86*, 1986.

[35] "EMF in Your Environment," *Environmental Protection Agency document 402-R-92-008*, December 1992.

[36] Hare, E., *RF Exposure and You*, Newington CT: The American Radio Relay League, 1998.

[37] Straw, R. D., (ed.), *The ARRL Antenna Book*, 18th ed., Newington CT: The American Radio Relay League, 2003.

[38] Michaelson, S. M., and J. C. Lin, *Biological Effects and Health Implications of Radio Frequency Radiation*, New York: Plenum Publishing, 1987.

[39] Lin, J. C., (ed.), *Electromagnetic Interaction with Biological Systems*, New York: Plenum Publishing, 1989.

[40] Report and Order, *Guidelines for Evaluating the Environmental Effects of Radio Frequency Radiation*, FCC 96-326, ET Docket No. 93-62 Washington, D.C., 20554, August 1, 1996.

[41] "Survey on Data Concerning Biological Effects on the Human Body (Including Implants) of Electromagnetic Waves, in the Frequency Range of 80 MHz–6 GHz.," CENELEC Report, Bruxelles, Belguim, 1995.

[42] Babij, T. M., and H. Bassen, "Broadband Isotropic Probe System for Simultaneous Measurement of Complex E- and H-fields," U.S. Patent 4,588,993, issued May 13, 1986.

10

Simulated Human Body Devices

10.1 Introduction

The simulated human body devices used in receiver sensitivity and mobile phone total radiated power testing will be analyzed here in detail. These devices are often used with body-worn receiver measurements because of the lengthy testing time involved and because people tend not to remain motion free for the duration of the test. We will study the results of an experimental program involving an anthropometrically diverse group of people, which yielded the unexpected result that adult humans are remarkably similar with respect to belt-level body-mounted receiver-sensitivity performance. The standing human body behaves essentially like a lossy wire antenna at frequencies below about 150 MHz. The erect body is resonant to vertically polarized incident fields in the range of about 40 to 80 MHz, depending on the presence and type of ground. That is, the body on a perfectly conducting ground looks like a quarter-wave element with a ground image, so its resonant length is about 3.4m, while in free space the resonant length is 1.7m.

Above the whole-body resonance, humans are well represented analytically by infinitely long cylinders containing saline water. Infinitely long cylinders are also a good representation of the body at all frequencies for horizontal incident polarization. The saline-water cylinders used to represent people in receiver measurements will be analyzed. The implications of body resonance, the impact on body-worn sensitivity measurements, and the consequences for the development of standard measuring methods will be explored. Also, we will study the effects of temperature on the accuracy of field-strength measurements using saline-water-filled simulated body devices.

In mobile phone testing, the total radiated power and the total isotropic sensitivity of the receiver are measured using an anthropomorphically correct human head phantom, which is described in this chapter.

10.2 Field-Strength Sensitivities of Body-Worn Receivers

We will analyze the effects of the human body on the performance of body-mounted radios and paging receivers here. Pagers, or selective call receivers, can be used as miniature wireless magnetic field probes to measure the fields close to humans and phantom human simulated body devices. The statistical techniques for determining receiver sensitivity, described in Chapter 9, are applied here for determining the levels of fields surrounding the phantoms and human test subjects. The performance of two phantom human simulated body devices, SALTY and SALTY-LITE [1, 2] (the physical designs of both devices are shown in Figure 10.1), are analyzed here. The computer simulations use a moment-method computer code, numerical electromagnetic code (NEC) [3], for the low-frequency region, including the whole-body resonance. Analytical infinitely long cylinder representations are implemented at frequencies above the resonance. The computer-generated simulations and the results obtained from the open-field experiment are in satisfactory agreement.

Both SALTY and the recently introduced lightweight SALTY-LITE (62 kg versus 125 kg for SALTY) [4] phantom humans are internationally recognized simulated body device standards [5–7]. SALTY-LITE was first proposed

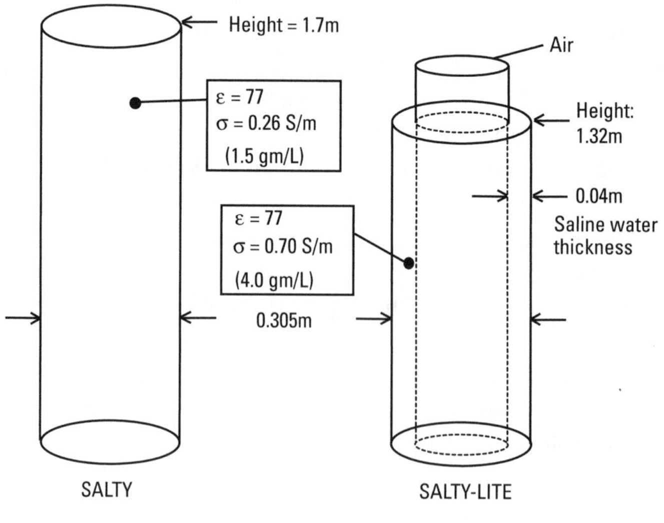

Figure 10.1 Standard human phantoms SALTY and SALTY-LITE. (After [1].)

and used in 1989 by Karlheinz Kraft (private communication, August 13, 1991) for measurements inside a small anechoic chamber at frequencies near 460 MHz. The size and weight of the newly introduced SALTY-LITE simulated body device make it appropriate for use inside small anechoic chambers at frequencies above 100 MHz. A measurement program comparing the performance of belt-level body-mounted selective call receivers characterized the performance using a group of people and the performance on SALTY and SALTY-LITE [1]. Measurements involving people were made at 148, 153, 169, 466, and 929 MHz. The receiver field-strength sensitivities were recorded at eight azimuth angles spaced 45° and up to 12 people per frequency for comparison with SALTY and SALTY-LITE. Results of the trials indicate that paging receiver sensitivities on the new SALTY-LITE are nearly the same as on SALTY at all of the tested frequencies.

10.2.1 Population Sample for Measurements

Eight males and four females were selected to match the weight and height standard deviations, as well as the means of the U.S. and European populations for the age group 19 to 65 years, as shown in Figure 10.2 [8–10]. Because the group of people was varied (46 to 89 kg in weight and 152 to 184 cm in height), we are able to comment on the effect of body dimensions versus pager sensitivities. The measurements were made on a level, 45m-long, open-field antenna range described from an analytical perspective in Chapter 6 and from a repeatability and reliability perspective in Chapter 9.

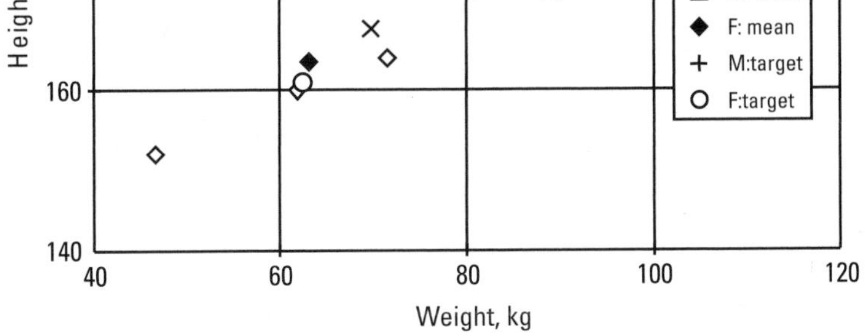

Figure 10.2 Anthropometry of the receiver-sensitivity measurement group. (*Source:* [11].)

The field-strength sensitivities were measured using Motorola BRAVO EXPRESS™ numeric POCSAG receivers operating at 1,200 baud as very sensitive (in the submicroampere per meter range) wireless magnetic field sensors. The front, or "best," position was measured using the IEC (International Electrotechnical Commission) 20-call method [2, 6, 7], described in Chapter 9. To reduce testing time, the first occurrence of three-of-three pages (three correctly received pages in a row) was used to estimate the remaining seven position sensitivities. During each 20-call test, the sensitivity reported by the 20-call method was compared to the level of the first three-of-three occurrence. On the basis of 102 such comparisons, the three-of-three method reports an average of 0.33-dB poorer sensitivity than does the 20-call method. The standard deviation of this difference was 0.73 dB, although differences of as much as 2 dB were observed. The 20 call method and the three-of-three method are described in Chapter 9.

10.2.2 Design of the Measurement Experiment

Both of the statistical methods, 20-call and three-of-three, used to determine sensitivity are biased toward reporting poor receiver sensitivity when the test subject, in this case a human subject, moves during a testing sequence. In this respect, the three-of-three method is influenced to a greater degree than the 20-call method. A typical eight-position test requires 22 minutes of testing time, taxing the ability of human test subjects to remain motionless. Motion of the body during a testing sequence can lead to significant measurement errors. In the standard position adopted, the pager was worn approximately 45° to the right of body center, and the arms of the subject were folded loosely in front. The front position was defined with the pager initially facing the transmitter. Tests performed with subjects holding their arms rigidly motionless at, but not touching, their sides resulted in approximately 1-dB poorer sensitivity when compared with the folded-arms case. That position was extremely difficult to maintain.

To reduce human error, all paging receiver sensitivities in the testing program were recorded automatically using computer-driven test equipment and a rotating platform. The pager alert signal, indicating message reception, was detected automatically through a plastic acoustic tube fixed near the pager. Measurements on SALTY and SALTY-LITE were in a standard configuration with the pager centered 1.0m above the ground and attached to SALTY or SALTY-LITE with no intervening gap [6, 7]. Automatic testing is the preferred and recommended procedure because of the long testing times, the monotony of the test procedure, and the tendency for humans to introduce errors.

10.2.3 Receiver-Sensitivity Measurement Results

Figure 10.3 shows measured sensitivities on 11 people superimposed on top of the measurement on SALTY and SALTY-LITE at 153 MHz. Measurements at 148 and 169 MHz show the same close correspondence between individuals and phantoms and are omitted here for brevity. Eight-point antenna patterns at VHF on all of the test subjects were within ±1.6 dB at the front (0°) position. The back position (180°) was typically 9 dB below the front position and exhibited a total variation of only ±2.7 dB at VHF over all tests with people. The depth of the null can be expected to vary significantly since the patterns points were measured 45° apart. On a gain-average basis, all sensitivities involving people were within a ±0.8 dB range at VHF.

Figure 10.4 extends this comparison to 466 MHz, where both SALTY and SALTY-LITE match the patterns of people well for all azimuth angles. Patterns at UHF on all of the test subjects were within ±1.1 dB at the front (0°) position. The back position (180°) was typically 20 dB below the front position and exhibited a typical variation of about ±6 dB over all tested people. There is a slight asymmetry in the patterns of people at −135° compared with +135°, which occurs because the human body is not symmetric with respect to the paging receiver placement. On a gain-average basis, all sensitivities involving people were within a ±0.6 dB range at UHF.

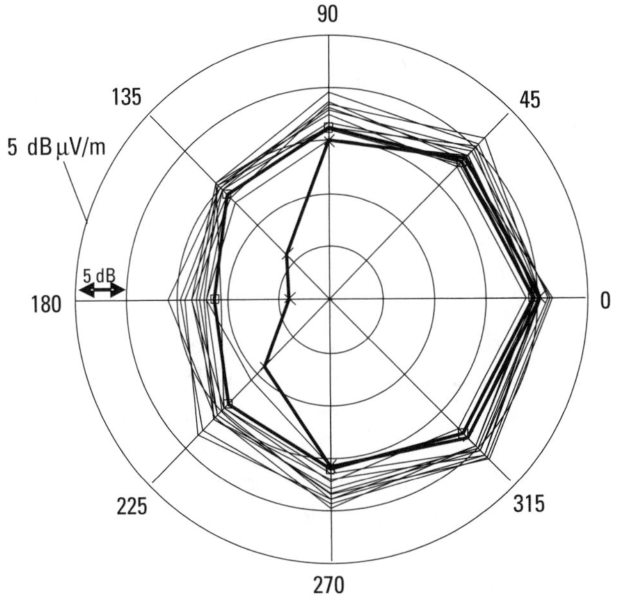

Figure 10.3 Receiver sensitivities on people (–), SALTY (✧), and SALTY-LITE (×) at VHF. (After [4].)

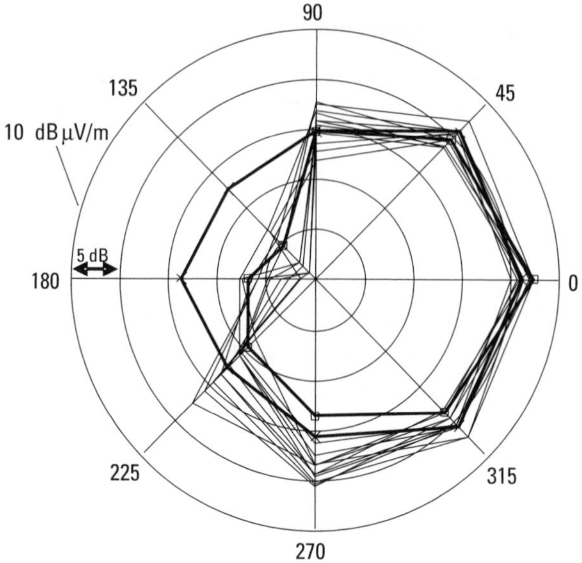

Figure 10.4 Receiver sensitivities on people (–), SALTY (✧), and SALTY-LITE (×) at UHF. (*Source:* [11].)

When compared with antenna patterns on people, the patterns on SALTY and on SALTY-LITE were acceptably close at both VHF and UHF, which means that both of these phantom humans properly simulate people. Eight-position measurements of pagers on SALTY and on SALTY-LITE at 929 MHz also showed excellent correlation comparable to the 466-MHz results presented here. The phantoms and people differ most in the rear (180°) region, but receiver field-strength performance depends on the gain average, which does not weight the depth of pattern nulls excessively.

Attempts were made to correlate the average sensitivities measured on individuals with gender, body weight, and height. Figure 10.5 shows a scatter plot of individual eight-position pager sensitivities averaged using the IEC method [5], given earlier as the gain-average formula, versus (a) individual heights and (b) versus individual weights. A VHF grouping is seen below 15 dBμV/m, and a UHF grouping is seen above 15 dBμV/m. No clear trends could be found at either VHF or UHF in spite of the wide range of body dimensions encountered in the measurement program. It was noticed, however, that the poorer sensitivity performance tended to be associated with the individuals who could not remain still for the lengthy measurements.

Performance on people was measured to design and validate the lightweight SALTY-LITE simulated human (phantom human) test device. The statistical means and standard deviations of the anthropometrically diverse group of people had weights and heights similar to the target population of adult

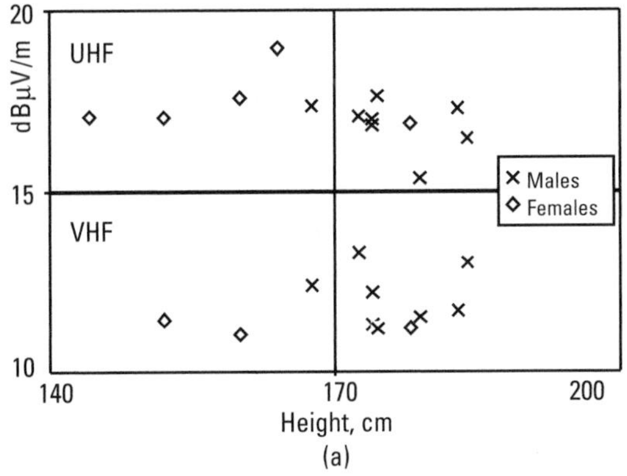

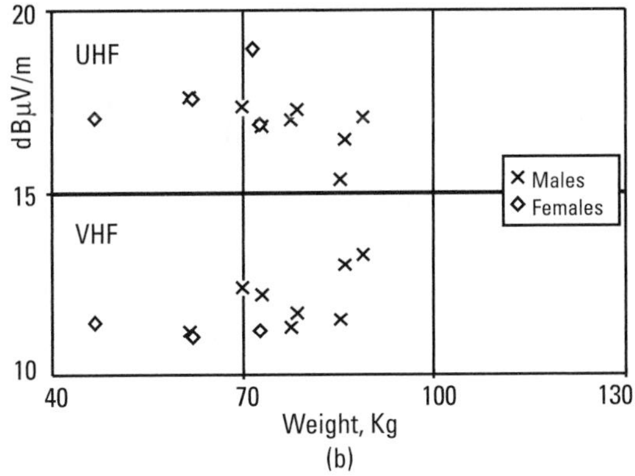

Figure 10.5 Receiver sensitivities versus individuals' (a) heights and (b) weights. (After [4, 11])

humans. The individual on-body tests were averaged over the eight azimuth angles using the gain-average formula (9.14) over eight equally spaced azimuth angles. The gain-averages were then averaged in decibels over the individuals, as in [2], and are compared in Figure 10.6 over frequency with the same eight-position averages measured for the phantom humans according to

$$D = 20\log\left[\frac{H_{avg}(phantom)}{H_{avg}(avg\ of\ humans)}\right] \tag{10.1}$$

where $H_{avg} = E_{avg}/\eta_0$. From Figure 10.6, we see that the gain-averaged magnetic fields around humans and around SALTY and SALTY-LITE are within 1.7 dB above 85 MHz. The SALTY response is 3 dB higher than the human response near the whole-body resonance, which occurs near 60 MHz over Earth ground but matches the human response at 43.5 MHz. SALTY-LITE is shorter in

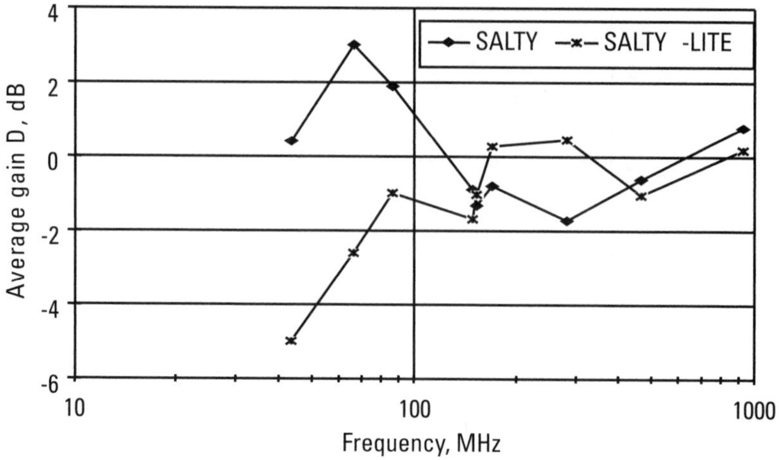

Figure 10.6 The gain-averaged performance of simulated human devices compared with the average of results on people. (After [1].)

stature than adult humans; hence, the response relative to humans seen in Figure 10.6 steadily decreases as frequency decreases below the SALTY-LITE resonance at 90 MHz. *Both SALTY and SALTY-LITE are remarkably similar to humans with respect to external magnetic fields close to and around the body in spite of the radically different internal compositions of the phantoms and of humans.*

10.3 Analysis of Phantom Simulated Human Body Devices

The simulated body devices are analyzed using two distinct methods. The analyses were performed with the phantoms situated in free space, over an Earth ground with relative permittivity $\varepsilon = 12$ and conductivity $\sigma = 0.002$ S/m, and over a perfectly electrically conducting (PEC) ground. The total magnetic fields relative to the incident fields for SALTY were calculated for the three ground configurations. A whole-body resonance, found using a lossy wire antenna model, using a technique suggested by A. W. Guy [12], implemented using moment-method code, was found to be near 70 MHz for the free space case and is consistent with similar calculations reported in [13]. The resonance shifts down to about 60 MHz over Earth ground and to about 40 MHz over a perfectly conducting ground.

At frequencies above the whole-body resonance, the SALTY saline-water column was modeled as a homogeneous lossy dielectric using the methods given in [14], while the SALTY-LITE hollow water column analysis was derived from [15]. Both phantom models were validated using a generalized, multilayered cylinder analysis of [16], which includes the effects of the plastic cylinder walls.

10.3.1 Saline Water

Saline water is used to represent the bulk properties of the human body with respect to the fields external to the body. Water is surprisingly complex as a chemical substance, considering the apparent simplicity of its chemical composition. The electrical properties of water and of water in combination with dissolved salts are treated exhaustively in [17]. Pure water comprises molecules with a dipole moment that exhibits a broad resonance, one of the broadest molecular resonances in nature, in the vicinity of 21 GHz. The exact resonance is temperature dependent. This resonance accounts for the absorptive behavior of water vapor in the atmosphere. The addition of salts, particularly sodium chloride (NaCl) here, results in increased dielectric losses at the frequencies primarily below a few gigahertz. Figure 4.1 showed the effect of salt in the attenuation of seawater as a function of frequency compared with the attenuation of pure water.

The dielectric properties of water are modeled here because of the importance of saline water as a component of the standard human body device. Saline water may be represented over frequency by a Debye formula with corrections for temperature and for saline content [18] and written here as

$$\varepsilon_w = \varepsilon_{\lim} + \frac{\varepsilon_{DC} A - \varepsilon_{\lim}}{1 + j2\pi\tau Bf} - j\frac{\sigma_{NaCl}}{2\pi f \varepsilon_0} \tag{10.2}$$

where $\varepsilon_{lim} = 4.9$ is the limiting dielectric constant at high (optical) frequencies, ε_o is the permittivity of free space, and f is frequency in hertz. The remaining quantities depend on the saline concentration, S parts per thousand, and temperature T°C. The relaxation time constant for the water molecule is

$$\begin{aligned}\tau = \big[&1.1109 \times 10^{-10} - 3.824 \times 10^{-12} T \\ &+ 6.938 \times 10^{-14} T^2 - 5.096 \times 10^{-16} T^3\big] / 2\pi\end{aligned} \tag{10.3}$$

The dc value for the dielectric constant of water as a function of temperature T°C from [19] is

$$\varepsilon_{DC} = 87.740 - 0.40008T + 9.398 \times 10^{-4} T^2 - 1.410 \times 10^{-6} T^3 \tag{10.4}$$

The conductivity due to sodium chloride normality N as a function of temperature is

$$\sigma_{NaCl} = \sigma_n \left[\begin{array}{l} 1 - 0.01962(25-T) + 8.08 \times 10^{-5}(25-T)^2 \cdots \\ +(T-25)N \times 10^{-5} \left[\begin{array}{l} (3.02 + 3.922(25-T)) \cdots \\ +N(1.721 - 0.6584(25-T)) \end{array} \right] \end{array} \right] \quad (10.5)$$

where the value at 25°C is

$$\begin{aligned} \sigma_n &= 10.394N - 2.3776N^2 + 0.68258N^3 \\ &\quad - 0.13538N^4 + 0.010086N^5 \end{aligned} \quad (10.6)$$

The solution normality correction factors A and B are

$$A = 1 - 0.2551N + 0.05151N^2 - 0.006889N^3 \quad (10.7)$$

$$B = 1 + 0.001463NT - 0.04896N - 0.02967N^2 + 0.005644N^3 \quad (10.8)$$

where the normality of the solution N in terms of the salinity S in parts per thousand is

$$N = 0.01707S + 1.205 \times 10^{-5} S^2 + 4.058 \times 10^{-9} S^3 \quad (10.9)$$

Expressions (10.2) to (10.9) provide a convenient set of curve-fitting equations that accurately model saline water over a temperature range of –10° to 40°C and over saline concentrations of up to 160 parts per thousand (grams per liter). Figure 10.7 shows the dielectric constant of saline water at 27°C with a saline concentration of 4g of NaCl per liter of water presented as a Cole-Cole diagram (that is, a plot of the imaginary versus the real parts of the complex dielectric constant). Changing the saline concentration primarily effects the tail of the curve at frequencies below 10 GHz.

A more interesting view of the electrical properties of saline water is portrayed in Figure 10.8, where the real and imaginary components of the relative permittivity are plotted versus frequency. The contribution of the salt to the imaginary part of the saline-water dielectric constant is readily evident at frequencies below a few gigahertz, as is the water molecule relaxation resonance near 21 GHz. Increasing the salt concentration lowers the real part of the dielectric constant very slightly.

Phantom human muscle tissue can be simulated by mixing water with salt to control the conductivity and with other materials to control the dielectric constant. Significant reductions of the real part of the dielectric constant at high frequencies (above 100 MHz) are realized by adding sucrose in large quantities [20] or by adding polyethylene powder [21]. At frequencies below 100 MHz,

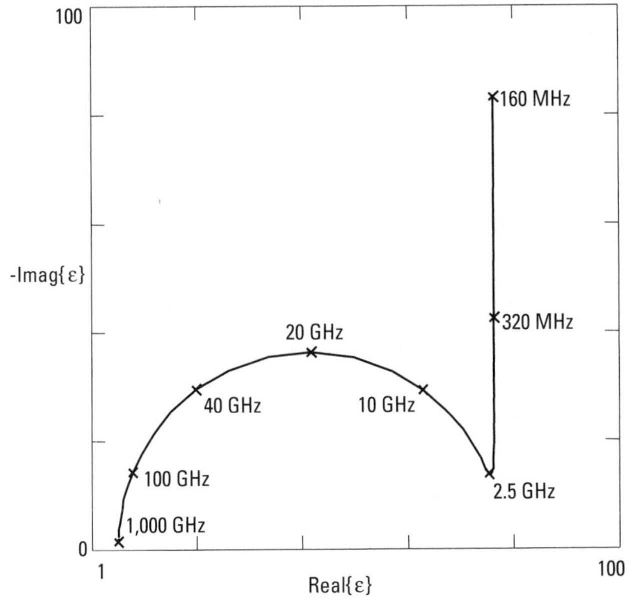

Figure 10.7 Cole-Cole diagram of saline water.

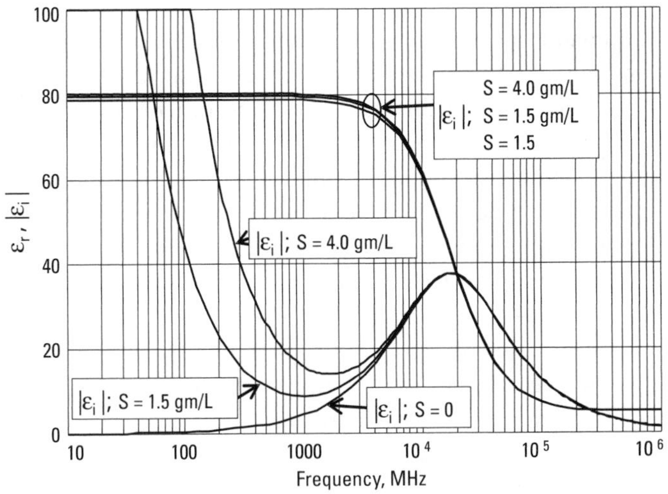

Figure 10.8 Complex relative permittivity of saline water.

aluminum powder is added to increase the dielectric constant. These detailed mixtures are required for in vivo simulations where accurate representations of human biological materials, whose electromagnetic properties are described in [22], are necessary. For fields external to the body, at frequencies above about 30

MHz, saline water in a concentration of 4.0 g/L (SALTY-LITE) is adequate. Historically, a concentration of 1.5 g/L has been used in the SALTY simulated human test device.

- [10-3a.mcd] Using (10.2) to (10.9) to model saline water, calculate the dielectric constant and conductivity of saline water at 0, 1.5, and 4.0 g/L salt concentrations at 25°C. Use frequencies of 30, 160, 280, 460, and 930 MHz.
- [10-3b.mcd] Using (10.2) to (10.9) to model saline water, calculate the dielectric constant and conductivity of saline water at 1.5 and 4.0 g/L salt concentrations over a temperature range 0°C to 40°C. Use a frequency of 170 MHz.

10.3.2 SALTY and SALTY-LITE Human Body Devices

Saline water is used in concentrations of 1.5g and 4.0g of sodium chloride (NaCl) to 1L of water in the two simulated body devices, SALTY and SALTY-LITE, respectively. The 1.5 g/L figure for SALTY is historical and cast in standards [5–7], based loosely on simplified considerations of skin-depth penetration of electromagnetic waves. The saline concentration value of 4.0 g/L for SALTY-LITE was based on matching the electromagnetic skin depth with that of bulk human muscle tissue. Table 10.1 shows the complex dielectric constant and conductivity of bulk tissue at belt level measured using a network analyzer with a dielectric probe; this compares with other findings [20–22]. The table also compares the values for saline water at a salt concentration of 4 g/L.

Using the dielectric parameters from Table 10.1 and using (9.20) for skin depth, we can compare the penetration depth for saline water and muscle tissue. The skin depths of muscle tissue and saline water at concentrations of 1.5 and 4.0 g/L are plotted in Figure 10.9. Clearly, the better match occurs at the higher saline concentration, but the close correspondence between measured field-strength sensitivities on people and on both SALTY and SALTY-LITE simulated body devices shows the relative insensitivity to the body dielectric parameters over a wide range of conductivities.

10.3.3 Lossy Wire Antenna Model of Simulated Body Devices

Both phantom humans were modeled at low frequencies as lossy wire antennas using moment-method code with a technique suggested by Guy [23] and pictured in Figure 10.10. The approximate analysis captures the behavior at frequencies at and below the whole-body resonance for vertical polarization incidence and is in general agreement with published measurements of the body resonance [13].

Table 10.1
Dielectric Parameters of Muscle Tissue and Saline Water at 25°C and a Concentration of 4 g/L

Frequency, MHz	ε_r, **Muscle**	σ_r, **S/m Muscle**	ε_r, **Saline Water**	σ_r, **S/m Saline Water**
100	64	0.45	77	0.70
200	54.5	0.51	77	0.71
400	48.4	0.58	76.9	0.73
800	45.3	0.73	76.8	0.83
1,600	43	1.07	76.5	1.22
3,000	41.6	1.93	75.3	2.50

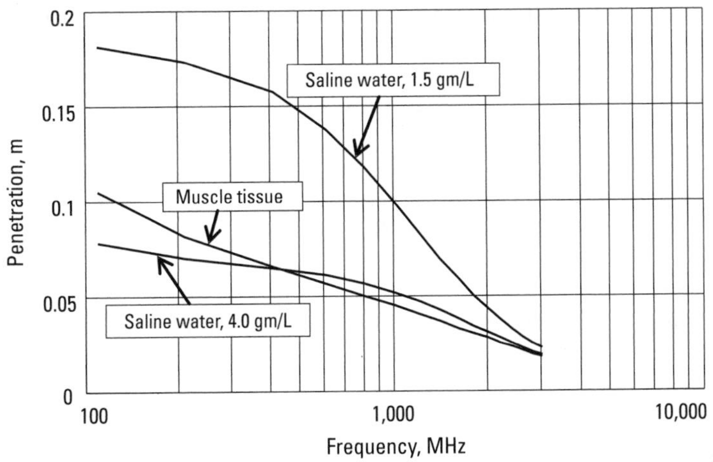

Figure 10.9 Depth of penetration (skin depth) in muscle tissue and in saline-water solutions.

SALTY is a saline-water-filled plastic cylinder modeled as a wire antenna h = 1.7m long using N = 39 segments with a parallel 2.29Ω resistor and a parallel 1,147-pF capacitor at each segment to represent a saline-water column having a dc conductivity of σ = 0.26 S/m. SALTY-LITE is a pair of concentric plastic tubes h = 1.32m high with the intertube layer filled with saline water; it is modeled by N = 30 wire segments, each loaded by a parallel 4.63Ω resistance and a parallel 567-pF capacitance. The diameters of both SALTY and SALTY-LITE are 0.305m, as pictured in Figure 10.1. The capacitance and resistance were chosen by calculating the series capacitance of each segment of the saline-water cylinder using the complex dielectric constant of saline water. The capacitance C and parallel conductance G of each of the N segments is

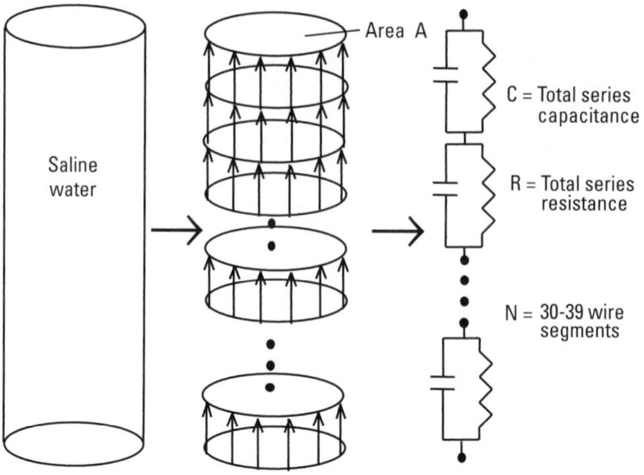

Figure 10.10 Lossy wire antenna model used to analyze the saline-water cylinders. (*Source:* [11].)

$$C = N\frac{\varepsilon_r \varepsilon_0 A}{h} \tag{10.10}$$

$$G = N\frac{\omega \varepsilon_i \varepsilon_0 A}{h} \tag{10.11}$$

and the complex relative dielectric constant of saline water is

$$\varepsilon_w = \varepsilon_r - j\varepsilon_i \tag{10.12}$$

using (10.2) for the electrical properties of saline water.

The whole-body resonance calculated for the SALTY device is shown in Figure 10.11, where the magnetic field 2 cm in front of the water layer is shown in ratio to the incident field. The resonant peaks are evident and migrate from 70 MHz in the free space case through about 60 MHz for Earth ground to 40 MHz in the presence of a perfectly conducting ground. This shift in the resonance of SALTY (and of SALTY-LITE) makes the choice of ground parameters important when selecting a standard testing method. Clearly, the enhancement of the magnetic field is strongly dependent on the resonant behavior of the simulated body devices, particularly at frequencies below about 200 MHz. The preferred choice for open-air receiver-sensitivity testing is to place the simulated body devices on Earth ground, as this most closely represents the case of people in their most natural environment. The resulting field that illuminates the simulated device, as was analyzed in Chapter 6 and detailed in Chapter 9, is preferred and appropriate for testing.

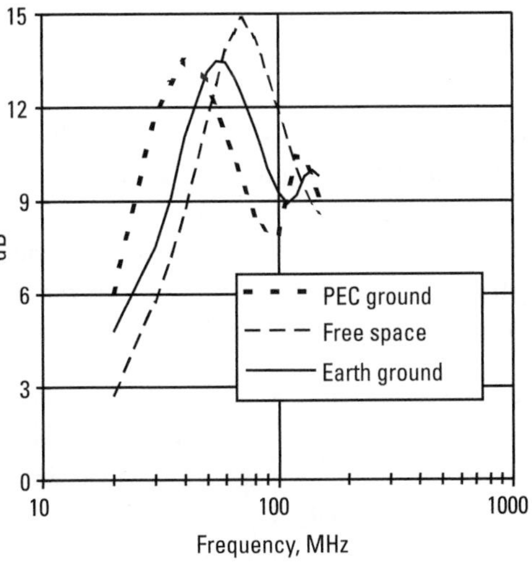

Figure 10.11 Magnetic fields in ratio to the incident wave near SALTY. (After [1].)

10.3.4 Infinite Cylinder Model of Simulated Body Devices for Vertical Polarization

At frequencies above the whole-body resonant frequency, the simulated body devices are modeled using analytical cylindrical expansions for infinitely long cylinders. The infinite cylinder analysis for the ϕ-component of the total magnetic field near the infinite cylinder model of SALTY and SALTY-LITE is carried out according to [14], and the magnetic field in ratio to the incident wave ($E_0 = \eta_0 H_0$) fields can be expressed as

$$H_\phi = -jH_0 \sum_{n=-\infty}^{\infty} j^n \left[J_n'(kr) + A_n H_n^{(2)\prime}(kr) \right] e^{jk\phi} \qquad (10.13)$$

The other field components for normal plane wave incidence are

$$E_z = E_0 \sum_{n=-\infty}^{\infty} j^n \left[J_n(kr) + A_n H_n^{(2)}(kr) \right] e^{jn\phi} \qquad (10.14)$$

$$H_\rho = \frac{jH_0}{kr} \sum_{n=-\infty}^{\infty} nj^{n+1} \left[J_n(kr) + A_n H_n^{(2)}(kr) \right] e^{jn\phi} \qquad (10.15)$$

[10-3c.mcd] Using (10.13) with $A_n = b_n$ and b_n given by (2.27) to model a PEC cylinder, compute the total magnetic fields in ratio to the incident fields around the surface of a cylinder. What is the front-to-back ratio?

What is the gain average? Carry out the computations at 100, 200, 400, and 800 MHz.

The cylinder is centered on the axis of the cylindrical coordinate system. The coefficients A_n are modal reflection coefficients from the cylinder surface, which, for a perfectly conducting cylinder, are given by $A_n = b_n$ of (2.27). The distance from the cylinder center is r, and the wave number is $k = 2\pi/\lambda$. The general expression for A_n in the case of two concentric cylinders [15], where the outer cylinder has a radius a_2 and the inner cylinder has a radius of a_1, is

$$A_n = \frac{-J_n(ka_2)}{H_n^{(2)}(ka_2)}\left[\frac{B_n - \dfrac{J_n'(ka_2)}{J_n(ka_2)}}{B_n - \dfrac{H_n^{(2)\prime}(ka_2)}{H_n^{(2)}(ka_2)}}\right] \tag{10.16}$$

The term in brackets is unity for a perfectly conducting cylinder. Otherwise, the coefficients B_n are

$$B_n = \sqrt{\varepsilon_2}\,\frac{H_n^{(2)\prime}(k_{\varepsilon 2}a_2) - C_n H_n^{(1)\prime}(k_{\varepsilon 2}a_2)}{H_n^{(2)}(k_{\varepsilon 2}a_2) - C_n H_n^{(1)}(k_{\varepsilon 2}a_2)} \tag{10.17}$$

The wave number $k_{\varepsilon 2}$ is in the medium between radii a_1 and a_2. Functions $H_n^{(x)}$ are the Hankel functions of the first and second kind [24, 25], and the prime denotes the derivative of the function with respect to the argument. The coefficients C_n are

$$C_n = \frac{H_n^{(2)}(k_{\varepsilon 2}a_1)}{H_n^{(1)}(k_{\varepsilon 2}a_1)}\left[\frac{\dfrac{J_n'(k_{\varepsilon 1}a_1)}{J_n(k_{\varepsilon 1}a_1)} - \sqrt{\dfrac{\varepsilon_2}{\varepsilon_1}}\,\dfrac{H_n^{(2)\prime}(k_{\varepsilon 2}a_1)}{H_n^{(2)}(k_{\varepsilon 2}a_1)}}{\dfrac{J_n'(k_{\varepsilon 1}a_1)}{J_n(k_{\varepsilon 1}a_1)} - \sqrt{\dfrac{\varepsilon_2}{\varepsilon_1}}\,\dfrac{H_n^{(1)\prime}(k_{\varepsilon 2}a_1)}{H_n^{(1)}(k_{\varepsilon 2}a_1)}}\right] \tag{10.18}$$

written in terms of Hankel functions and the Bessel functions J_n and their derivatives. The wave number $k_{\varepsilon 1}$ is in the medium bounded by radius a_1. The wave numbers $k_{\varepsilon 1}$ and $k_{\varepsilon 2}$ can be written in terms of frequency f in megahertz, the real parts of the dielectric constants ε_1 and ε_2, and the conductivities σ_1 and σ_2:

$$k_{\varepsilon 1} = k\sqrt{\varepsilon_1 - \frac{17{,}975\,j\sigma_1}{f}} \tag{10.19}$$

and

$$k_{\varepsilon 2} = k\sqrt{\varepsilon_2 - \frac{17{,}975 j\sigma_2}{f}} \tag{10.20}$$

Expressions (10.13) to (10.20) give the electric and magnetic fields for $r \geq a_2$, or the region external to the cylinders for a vertically polarized plane wave incident normally on the cylinders. A more general analysis solves the problem in all regions for arbitrary incidence angle and for an arbitrary number of cylinder layers [16].

Figure 10.12 shows the magnetic fields in ratio to the incident wave at 2 cm in front of an infinitely long SALTY-LITE as a function of frequency and with salt (NaCl) concentration as a parameter for vertical incident polarization. Wildly varying fields are evident for the pure-water case (0 g/L salt concentration). These resonances are associated with the transverse dimensions of the several cylinder layers. The resonances are still evident at 1.5 g/L concentration (σ = 0.26 S/m) but are generally damped out at the recommended concentration of 4 g/L (σ = 0.69 S/m). The fields for a PEC cylinder are shown in Figure 10.12 for comparison. As could be shown for SALTY, the calculated magnetic fields external to, but very near, SALTY-LITE differ by less than 3 dB above 100 MHz for the wide range of cylinder conductivities from σ = 0.26 S/m to a PEC.

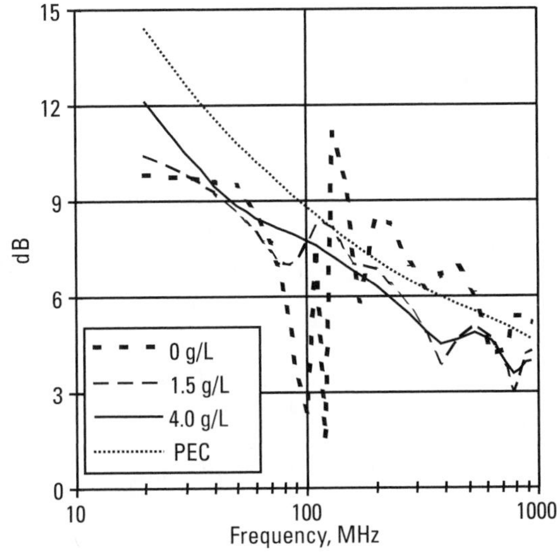

Figure 10.12 Magnetic fields for vertical polarization in ratio to the incident wave near SALTY-LITE. (*Source:* [11].)

[10-3d.mcd] Using (10.16) with A_n for a homogenous lossy dielectric cylinder with radius a = 0.1525m filled with saline water with 1.5 g/L saline solution (ε = 77.23, σ = 0.16), compute the total magnetic fields in ratio to the incident fields around the cylinder at r = 2 cm distance. What is the gain-averaged value? What is the front-to-back ratio? Carry out the computations at 100, 200, 400, and 800 MHz.

10.3.5 Infinite Cylinder Model of Simulated Body Devices for Horizontal Polarization

For horizontal polarization incidence, there is no whole-body resonant frequency in the sense that can be simulated by a lossy wire antenna. The simulated body devices for horizontal polarization are modeled using analytical cylindrical expansions for infinitely long cylinders. The infinite cylinder analysis for the θ-component of the total magnetic field near the infinite cylinder model of SALTY and SALTY-LITE is carried out according to [14]. The total fields in terms of the normally incident (θ = 90°) fields are

$$H_z = H_0 \sum_{n=-\infty}^{\infty} j^n \left[J_n(kr) + A_n^H H_n^{(2)}(kr) \right] e^{jn\phi} \tag{10.21}$$

$$E_\phi = jE_0 \sum_{n=-\infty}^{\infty} j^n \left[J_n'(kr) + A_n^H H_n^{(2)\prime}(kr) \right] e^{jn\phi} \tag{10.22}$$

$$E_\rho = \frac{-jE_0}{kr} \sum_{n=-\infty}^{\infty} nj^{n+1} \left[J_n(kr) + A_n^H H_n^{(2)}(kr) \right] e^{jn\phi} \tag{10.23}$$

The cylinder is on the axis of the cylindrical coordinate system. The distance from the cylinder center is r, and the wave number is $k = 2\pi/\lambda$. The general expression for A_n^H in the case of two concentric cylinders is in [15]. For normal incidence on the cylinder, outer cylinder radius a_2, and inner cylinder radius a_1, we get

$$A_n^H = -\frac{J_n'(ka_2)}{H_n^{(2)\prime}(ka_2)} \left[\frac{B_n^H \dfrac{J_n(ka_2)}{J_n'(ka_2)} - 1}{B_n^H \dfrac{H_n^{(2)}(ka_2)}{H_n^{(2)\prime}(ka_2)} - 1} \right] \tag{10.24}$$

The term in brackets is unity for a perfectly conducting cylinder. Otherwise, the coefficients B_n^H are

$$B_n^H = \frac{1}{\sqrt{\varepsilon_2}} \frac{H_n^{(2)\prime}(k_{\varepsilon 2}a_2) - C_n^H H_n^{(1)\prime}(k_{\varepsilon 2}a_2)}{H_n^{(2)}(k_{\varepsilon 2}a_2) - C_n^H H_n^{(1)}(k_{\varepsilon 2}a_2)} \tag{10.25}$$

The wave number $k_{\varepsilon 2}$ is in the medium between radii a_1 and a_2. Functions $H_n^{(x)}$ are the Hankel functions of the first and second kind [24, 25], and the prime denotes the derivative of the function with respect to the argument. The coefficients C_n^H are

$$C_n^H = \frac{H_n^{(2)}(k_{\varepsilon 2}a_1)}{H_n^{(1)}(k_{\varepsilon 2}a_1)} \left[\frac{\dfrac{J_n'(k_{\varepsilon 1}a_1)}{J_n(k_{\varepsilon 1}a_1)} - \sqrt{\dfrac{\varepsilon_1}{\varepsilon_2}} \dfrac{H_n^{(2)\prime}(k_{\varepsilon 2}a_1)}{H_n^{(2)}(k_{\varepsilon 2}a_1)}}{\dfrac{J_n'(k_{\varepsilon 1}a_1)}{J_n(k_{\varepsilon 1}a_1)} - \sqrt{\dfrac{\varepsilon_1}{\varepsilon_2}} \dfrac{H_n^{(1)\prime}(k_{\varepsilon 2}a_1)}{H_n^{(1)}(k_{\varepsilon 2}a_1)}} \right] \tag{10.26}$$

written in terms of Hankel functions and the Bessel functions J_n and their derivatives. The wave number $k_{\varepsilon 1}$ is in the medium bounded by radius a_1. The wave numbers $k_{\varepsilon 1}$ and $k_{\varepsilon 2}$ are given by (10.19) and (10.20). Expressions (10.21) to (10.26) give the fields in the external region $r \geq a_2$ for a horizontally polarized plane wave incident normally on the cylinders.

10.4 Magnetic Fields Around Simulated Body Devices

The analysis of the previous section can be used to predict the near-field behavior of the SALTY and SALTY-LITE simulated body devices. The very simple low-frequency analysis is useful primarily in studying the whole resonance phenomenon, while the cylindrical wave expansions are useful to predict detailed near-field behavior at frequencies above the resonance (greater than 100 MHz). The analysis of the previous sections was used to study the field-strength sensitivity patterns of selective call receivers as a function of the simulated body device temperature. The calculated field-strength performance of the simulated bodies was also validated by measurements across the frequency range of 30 to 930 MHz.

10.4.1 Temperature Dependence of Simulated Body Devices

This analysis was used to study the measured receiver-sensitivity pattern differences between tests performed in South Florida and those performed in Switzerland using the same set of selective call receivers (see Section 9.5.5). The Florida-based (site "A") measurements showed receiver-sensitivity patterns that had nulls shallower by 4 dB than similar measurements made in Switzerland (site "B"). Gain-averaged field-strength sensitivities agreed to within expected measurement errors. Ambient temperature was suspected to be the culprit. The dielectric constant and conductivity of saline water at a 1.5 g/L salt concentration is shown in Table 10.2.

Also shown in Table 10.2 are the peak-to-null ratios and the eight-position average relative to the incident wave calculated using the analysis of the previous section. Even though the conductivity varied over nearly an octave, the gain-averaged result remained within 0.1 dB. This smoothing behavior of the eight-position gain average validates (9.14) as a stable average for reporting the field-strength sensitivity of a receiver. The measurements in Switzerland were at an ambient temperature of between 0°C and 5°C, while the Florida measurements were at a temperature of 35°C, and in the Netherlands the ambient temperature was 18°C. People, of course, maintain a constant body temperature and would be expected to appear nearly invariant with variations in ambient temperature from the electromagnetic point of view.

10.4.2 Measured and Computed Fields near the Simulated Body Devices

A comparison of analysis to measurements for magnetic fields 2 cm in front of the water layer are shown in Figure 10.13 for SALTY-LITE in the vertically polarized case. The magnetic field is shown in ratio to the incident field and is often called the *body-enhancement value* for body-worn receivers. The same comparison is shown in Figure 10.14 for the SALTY simulated body measurement device. In both Figures 10.13 and 10.14, the measured ratio of total magnetic fields to incident magnetic fields shows agreement with a moment-method

Table 10.2
Dielectric Parameters of 1.5-g/L Saline Water and Resulting Pattern Behavior

Temperature (°C)	ε_r	σ (S/m)	Peak to Null (dB)	Average/ Incident (dB)
0	87.1	0.163	17.0	2.61
5	85.2	0.181	16.3	2.64
10	83.3	0.201	15.5	2.67
15	81.4	0.223	14.7	2.68
20	79.6	0.246	14.0	2.68
25	77.8	0.270	13.5	2.68
30	76.0	0.296	12.9	2.67

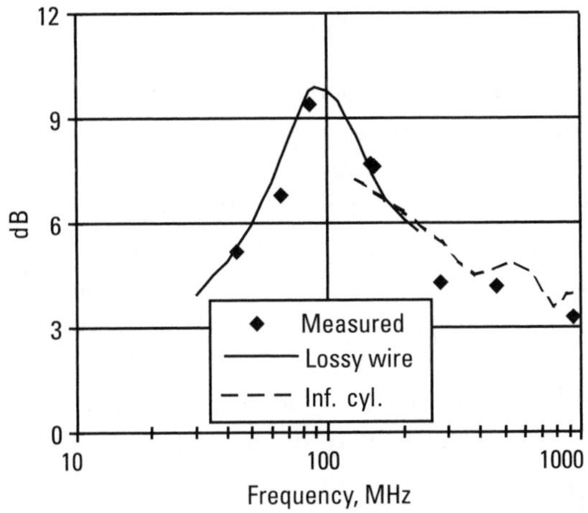

Figure 10.13 Magnetic field enhancement in front of SALTY-LITE. (After [1].)

[3] lossy wire antenna model over Earth ground (solid trace) at low frequencies including the whole-body resonance. An infinite layered cylinder calculation (dashed trace) matches the measure points at frequencies above the whole-body resonance. The region of greatest discrepancy occurs in the transition between the low-frequency and high-frequency models.

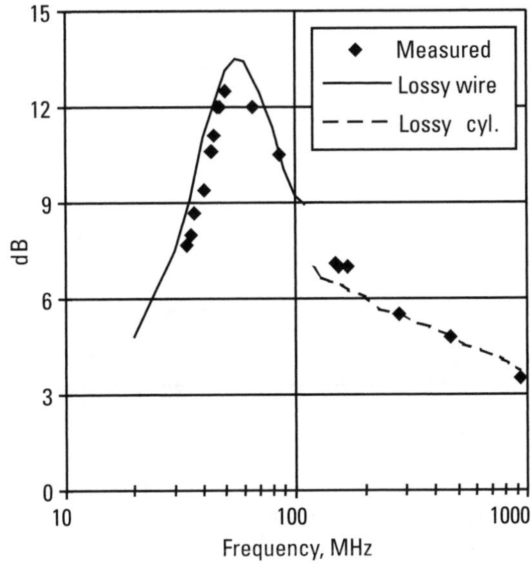

Figure 10.14 Magnetic field enhancement in front of SALTY. (After [1].)

10.4.3 Body Enhancement in Body-Worn Receivers

Figures 10.11 to 10.14 show the body enhancement in the front position for body-worn receivers. From the radio system performance point of view, the more interesting value is the gain-averaged body-enhancement value because this is the value that impacts system designs. It is easy to show, using the analysis developed in Sections 10.3.3 and 10.3.4, that the far-field radiation pattern of a body-worn receiver is nearly omnidirectional at very low frequency. As frequency is increased, the pattern behind the body develops a shadow manifested as a deepening null with increasing frequency. In the high-frequency limit, there is only a forward lobe, with the back half-space essentially completely blocked by the body. For the SALTY body dielectric parameters, the gain-averaged body-enhancement value for the magnetic field is shown in Figure 10.15. Below about 250 MHz, it is advantageous to use vertical polarization for body-mounted devices. Above about 300 MHz, the body has a stronger enhancement in the horizontal than the vertical polarizations. Above about 600 MHz, there is actually a net loss due to the body for the vertical polarization. These findings are consistent with the results in [13, 26]. The relationship between the values for vertical polarization in Figure 10.15 and those on people can be found by comparing them directly with Figure 10.6, where the gain-averaged performance of both SALTY and SALTY-LITE is compared with the gain-averaged result of the average over people. The gain-averaged values shown in Figure 10.15 are calculated using (10.21).

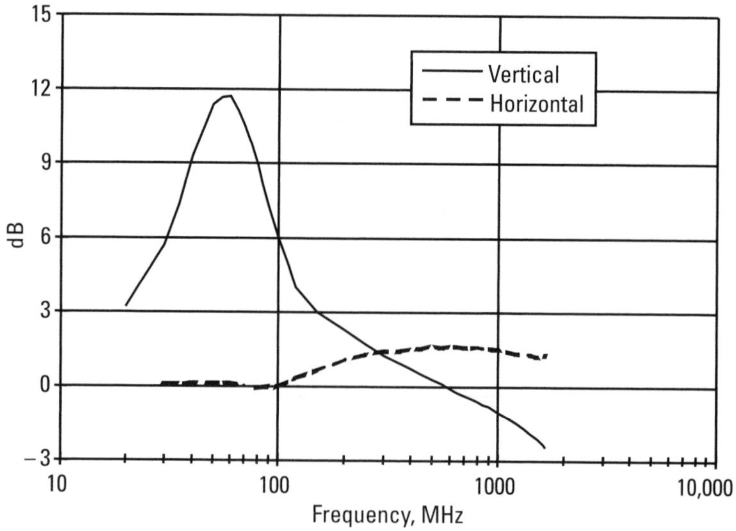

Figure 10.15 Gain-averaged, body-enhanced magnetic field around SALTY, 2 cm from the water layer, for vertical and horizontal polarizations incident.

10.5 Anthropomorphic Simulated Head

Mobile phone testing of total radiated power (TRP) and total isotropic sensitivity (TIS), according to the CTIA Test Plan [27], requires the use of an anthropometric simulated head. The IEEE Standards Coordinating Committee 34 has defined a specific anthropomorphic mannequin as the head model for use in specific absorption rate (SAR) testing. The shape of this head is based on the anthropomorphic data of the 90th percentile adult male head dimensions as published by the U.S. Army, except for the distance between the back of the ear and the head. Figure 10.16 shows the shape of the external shell of the head phantom. The head fluid is a relatively simple sugar-and-water-based mixture. The components of the mixture are shown in Table 10.3.

The tissue simulant RF dielectric properties are shown in Table 10.4 and should be maintained within a tolerance of ±10%. Testing is done in an anechoic chamber.

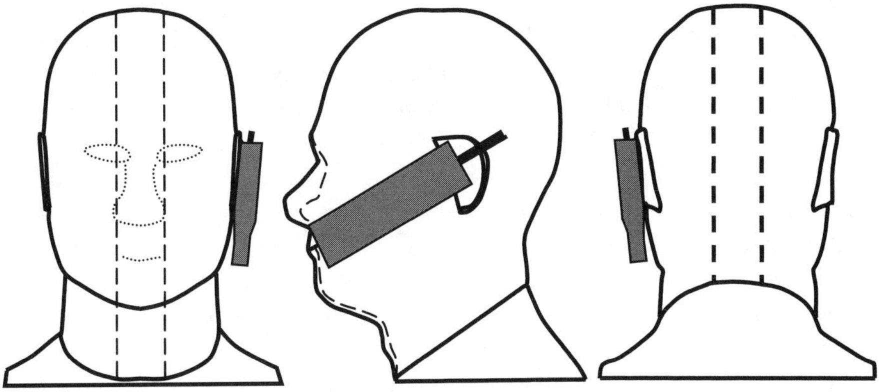

Figure 10.16 Anthropometric simulated head with mobile phone positioned for testing.

Table 10.3
Tissue Simulant Elements and Percentages. *(From* [27].)

Elements	Percent by Weight
Water	45.3
Sugar	54.3
Hydroxyethylcellulosis (HEC)	0.3
Bactericide (Dowicil 75)	0.1

Table 10.4
Tissue Simulant Dielectric Properties. *From* [27].

Frequency Band (MHz)	Relative Dielectric Constant	Conductivity (S/m)
835	48	0.70
1,900	41	1.7

10.6 Summary

A lossy wire antenna model satisfactorily predicts the magnetic fields near a homogenous saline-water-column (SALTY) and a hollow saline-water-column (SALTY-LITE) phantom human simulated body device at frequencies that include the whole-body resonance. Homogeneous and multilayered infinitely long cylinder models yield satisfactory analytical results at frequencies above the resonance. SALTY and SALTY-LITE compare very well with people on an eight-position pattern-average basis. The gain-averaged field sensitivities on SALTY and SALTY-LITE are within about 2 dB of the average of people in a standard pose, and the gain-averaged sensitivities among the anthropometrically diverse group of adults is within a range of ±0.8 dB at VHF and ±0.6 dB at UHF. The findings show no significant correlation between pager sensitivity and either the gender, body weight, or height of the test subjects. *People are remarkably similar with respect to receiver sensitivities at belt level over a wide range of body dimensions, when measured carefully and in a standard position.* The gain-averaging method of reporting the field-strength sensitivity of body-worn radio receivers tends to smooth out effects due to temperature that would otherwise increase the variance of sensitivity measurements between different test sites. The gain-average sensitivity is the appropriate sensitivity value for use in system designs in a multipath environment. At frequencies below about 250 MHz, the vertical polarization is enhanced on the body more than the horizontal polarization because of the whole-body resonance. At frequencies above about 300 MHz, the enhancement for horizontal polarization is greater than that for the vertical polarization.

Mobile phone performance in the 835 and 1,900 MHz bands is measured using an anthropometrically correct simulated head phantom. The head with a mobile phone at the ear is rotated in three dimensions to determine total radiated power and total isotropic sensitivities.

Problems

10.1 Write an expression for, and find the loss tangent of, saline water at 160 MHz. Find the Q at the molecular resonance frequency for saline concentrations of 0, 1.5, and 4.0 g/L at 25°C.

10.2 Referring to Figures 10.13 and 10.14, estimate the Q of SALTY and SALTY-LITE at their respective resonant frequencies.

10.3 Suggest a radio circuit that might exhibit a response similar to that of a SALTY device.

Ans:

A series LCR circuit resonant at 75 MHz and having the same $Q = 1.1$ as SALTY.

10.4 Explain how a selective call receiver with a loop antenna can be used as a magnetic field probe.

10.5 Find the Debye resonant frequency and the Q at resonance for 25°C water.

Ans:

At 25°C, $1/(2\pi\tau) = 19.65$ GHz. $Q = 1/\tan\delta = 1.1$.

10.6 A 1.2m-tall, 0.1m-diameter, thin-wall plastic column is filled with saline water having a salt concentration of 6 g/L. Calculate the parameters for a low-frequency (100 MHz) lossy wire model of this fixture and estimate the self-resonant frequency and Q of the device at resonance.

Ans:

At 100 MHz, $\varepsilon_r = 76.29$, $\sigma = 1.05$ S/m; $C = \varepsilon_r\varepsilon_0 A/h = 4.42$ pF, $G = \omega\varepsilon_r\varepsilon_0 A/h = 0.00685$ S/m, which can be segmented into N segments each of NC pF capacitors parallel with 146/NΩ resistors. Approximately $N = 30$ segments are sufficient.

$F = 0.95 \times 299.79/(2 \times 1.2) = 120$ MHz $Q = 1/\tan\delta = 0.49$ for 6 g/L.

10.7 Find the skin depth for saline water with saline concentrations of 1.5 and 4.0 g/L and compare with the result for bulk muscle tissue (Table 10.1) at 200, 400, 800, and 1,600 MHz at room temperature.

10.8 Find the skin depth for 6-g/L saline-water solution at 25°C at 100, 300, and 1,000 MHz.

10.9 Find the skin depth for 1.5- and 4.0-g/L saline-water solutions at 170 MHz at 0°C, 15°C, and 30°C.

10.10 Find the time needed to bring 250 ml of water, initially at 20°C, to the onset of boiling in a 700W microwave oven.

Ans:

P,[watts] = W, [joules] × t, [seconds]; 1 J = 4.18 cal. $t = W/P = 4.18 \times 250 \times (100 - 20°C) = 83{,}600$ J. So, $t = 83{,}600/700 = 112$ sec. Additional energy must be supplied to heat the water container and further "heat of evaporation" is needed to change the 100°C water to steam.

10.11 Assuming black-body radiation characteristics, approximate the net power radiated by a human adult in an ambient temperature of 23°C.

Ans:

Assume a body surface area $A = 2\ \text{m}^2$, and $T_{body} = 37°C + 273$. $P_{radiated} = A\sigma T^4_{body}$, where $\sigma = 0.567 \times 10^{-7}\ \text{W/km}^2$ is Stefan's constant. $P_{absorbed} = A\sigma T^4_{ambient}$, so $P_{net} = (2\text{m}^2)(\sigma)(T^4_{body} - T^4_{ambient}) = 177\text{W}$. The actual value averages about 100W.

10.12 Assuming a human dissipates an average of 100W continually, find the required fuel equivalent.

Ans:

$P = Wt$, where $W = (100\text{W})(86{,}400\ \text{sec}) = 8{,}640{,}000$ J per day; with 4,180 J/kcal, $W_{kcal} = 8{,}640{,}000/4{,}180 = 2{,}067$ kcal (or "food calories").

10.13 How many food calories are consumed when a human athlete expends 0.75 hp (horsepower) for one hour?

Ans:

$W = (0.75\ \text{hp})(746\ \text{W/hp})(3{,}600\ \text{sec}) = 2{,}014{,}200$ J, so $W_{kcal} = W/4{,}180 = 484.4$ kcal (one bottle of beer or almost four bowls of cereal).

References

[1] Siwiak, K., et al., "Computer Simulations and Measurements of Electromagnetic Fields Close to Human Phantoms,, *Electricity and Magnetism in Biology and Medicine*, M. Blank, (ed.), San Francisco: San Francisco Press, 1993, pp. 589–592.

[2] Siwiak, K., L. Ponce de Leon, and W. M. Elliott III, "Pager Sensitivities, Open Field Antenna Ranges, and Simulated Body Test Devices: Analysis and Measurements," *First Annual RF/Microwave Non-Linear Simulation Technical Exchange*, Motorola, Inc., Plantation, FL, March 26, 1993, (available from the author).

[3] Burke, G. J., and A. J. Poggio, "Numerical Electromagnetics Code (NEC)—Method of Moments," *NOSC TD 116*, Lawrence Livermore Laboratory, Livermore, CA, January 1981.

[4] Siwiak, K., and W. M. Elliott III, "Use of Simulated Human Bodies in Pager Receiver Sensitivity Measurements," *SouthCon/92 Conference Record*, Orlando, FL, March 11, 1992, pp. 189–192.

[5] "Test Site Annex X," *IEC 12F/WG7 Draft, Appendix to Document 489*, January 1993.

[6] "Paging Systems; European Radio Message System (ERMES) Part 5: Receiver Conformance specification," *ETS 300 133-5*, ETSI, Valbonne, France, July 1992, amended (A1), January 1994.

[7] "Methods of Measurement for Radio Equipment Used in the Mobile Services," *Sec. 8—Reference Sensitivity (Selective Calling), IEC 489-6*, 2nd ed., 1987.

[8] Pheasant, S., *Bodyspace*, London: Taylor and Francis, 1988.

[9] *Anthropometry of U. S. Military Personnel*, DOD-HDBK-743 METRIC, October 3, 1980.

[10] Panero, J., and M. Zelnik, *Human Dimensions and Interior Space*, London: The Architectural Press Ltd., 1979.

[11] Siwiak, K., *Radio Wave Propagation and Antennas for Portable Communications*, Workshop Notes, Taipei, Taiwan, Republic of China, October 5–7, 1993.

[12] Guy, A. W., "Analysis of EMP Induced Currents in Human Body by NEC Method of Moments," *Proc. of the Twelfth Annual International Conference of the IEEE Engineering in Medicine and Biology Society*, Philadelphia, PA, November 1–4, 1990, pp. 1547–1548.

[13] Durney, C. H., H. Massoudi, and M. F. Iskander, *Radio Frequency Radiation Dosimetry Handbook*, 4th ed., USAFSAM-TR-85-73, USAF School of Aerospace, Brooks AFB, TX, 78235, October 1986.

[14] Harrington, R. F., *Time Harmonic Electromagnetic Fields*, New York: McGraw-Hill, 1961.

[15] Misra, D. K., "Scattering of Electromagnetic Waves by Human Body and Its Applications," Ph.D. Dissertation, Michigan State University, 1984.

[16] Ponce de Leon, L., "Modeling and Measurement of the Response of Small Antennas near Multilayered Two or Three Dimensional Dielectric bodies," Ph.D. Dissertation, Florida Atlantic University, 1992.

[17] Franks, F., (ed.), *Water, a Comprehensive Treatise*, Vol. 1, 2, and 3, New York: Plenum Press, 1972.

[18] Stogryn, A., "Equations for Calculating the Dielectric Constant of Saline Water," *IEEE Trans. on MTT*, August 1971 (corrected), pp. 733–737.

[19] Malmberg, C. G., and A. A. Maryott, "Dielectric Constant of Water from 0° to 100°C," *J. Res. Nat. Bur. Stand.*, Vol. 20, 1956, pp. 1–8.

[20] Hartsgrove, G., A. Kraszewski, and A. Surowiec, "Simulated Biological Materials for Electromagnetic Radiation Absorption Studies," *Bioelectromagnetics*, Vol. 8, 1987, pp. 29–36.

[21] Chou, CK, et al., "Formulas for Preparing Phantom Muscle Tissue at Various Radiofrequencies," *Bioelectromagnetics*, Vol. 5, 1984, pp. 435–441.

[22] Johnson, C. C., and A. W. Guy, "Nonionizing Electromagnetic Wave Effects in Biological Materials and Systems," *Proceedings of the IEEE*, Vol. 60, No. 6, June 1972, pp. 692–718.

[23] Guy, A. W., "Analysis of EMP Induced Currents in Human Body by NEC Method of Moments," *Proc. of the Twelfth Annual International Conference of the IEEE Engineering in Medicine and Biology Society*, Philadelphia, PA, November 1–4, 1990, pp. 1547–1548.

[24] Abramowitz, M., and I. Stegun, *Handbook of Mathematical Functions*, New York: Dover Publications, 1972.

[25] Jahnke, E., and F. Emde, *Tables of Functions*, New York: Dover Publications, 1945.

[26] Chuang, H.-R., "Computer Simulation of the Human-Body Effects on a Circular-Loop-Wire Antenna for Radio-Pager Communications at 152, 280, and 400 MHz," *IEEE Transactions on Vehicular Technology*, Vol. 46, No. 3, August 1997, pp. 544–559.

[27] "Method of Measurement for Radiated RF Power and Receiver Performance," Revision 2.0, CTIA Certification Program Test Plan for Mobile Station over the Air Performance, Cellular Telecommunications and Internet Association, Washington, D.C., March 2003.

11

Loops, Dipoles, and Patch Antennas

11.1 Introduction

We begin our investigation into small antennas used with personal communications devices by exploring the nature of quality factor Q. This leads to a primer on small antenna fundamental limitations, which acquaints us with the bandwidth and performance limitations fundamental to miniature antenna structures. We learn that there is a minimum possible Q for an antenna and that this leads to a fundamental limitation on fractional bandwidth. We learn further that although the fundamental limit is calculable, it does not leave us with a design recipe for antenna structures that achieve the minimum Q. The radiation and Q properties of small loops and small dipoles are then presented. Next, examples of practical communication antenna applications are shown in view of the fundamental limitations. We explore in detail the performance of electrically small loops, dipoles, and microstrip patch antennas. We also investigate the radiation of ultrawideband (UWB), that is, nonharmonic, short-duration signals. Next, the performance of the helix-radio dipole antenna is investigated. Finally, coupling between a dipole antenna and a radio case is presented to illustrate the performance of a practical application of the dipole antenna to a communications device.

11.2 A Look at Quality Factor Q

The factor Q was introduced in the second decade of the twentieth century to represent the ratio of reactance to resistance as a "figure of merit" for inductors [1]. It has since come to represent the "quality factor" associated with resonant structures. The concept of Q can be applied to the complete realm from atoms to planets in both mechanical and electrical systems. Here, we explore this versatile parameter because an understanding of Q leads to an understanding of fundamental limitations in miniature antenna design.

11.2.1 Definition of *Q*

Q is the reciprocal of the loss tangent or dissipation factor, D_ε, in dielectric materials as defined in Section 3.2.1. At any two terminals of an electrical circuit, Q is the tangent of the phase angle of the impedance. The related quantity "power factor" is the cosine of the phase angle of that same impedance. In dielectric materials, $Q = \omega\varepsilon_0\varepsilon_r/\sigma$. When applied to a resonant network, Q is uniquely related to the resonant frequency as a measure of the sharpness of resonance. Once so identified, Q is seen to relate closely to an oscillatory wave train of continuously decreasing amplitude. The defining relationship is

$$Q = 2\pi \frac{\textit{Total stored energy}}{\textit{Energy dissipated in one cycle}} \tag{11.1}$$

In a circuit, Q is defined in terms of a damped voltage or current envelope,

$$I(t) = I_0 e^{-t\pi/Q} \tag{11.2}$$

which is superimposed on a sinusoid,

$$I_s(t) = I(t)\cos(\pi t) \tag{11.3}$$

as shown in Figure 11.1. From that current or voltage waveform, Q is defined as π divided by the *logarithmic decrement*, which is the natural logarithm of the ratio of two adjacent current peaks. In terms of the energy envelope,

$$Q(t) = \frac{2t\pi}{2\ln\left|\frac{I_0}{I(t)}\right|} \tag{11.4}$$

When applied to resonant circuits, on a plot of the magnitude of impedance versus frequency of a simple resonant circuit, Q is

$$Q = \frac{f_0}{f^+ - f^-} \tag{11.5}$$

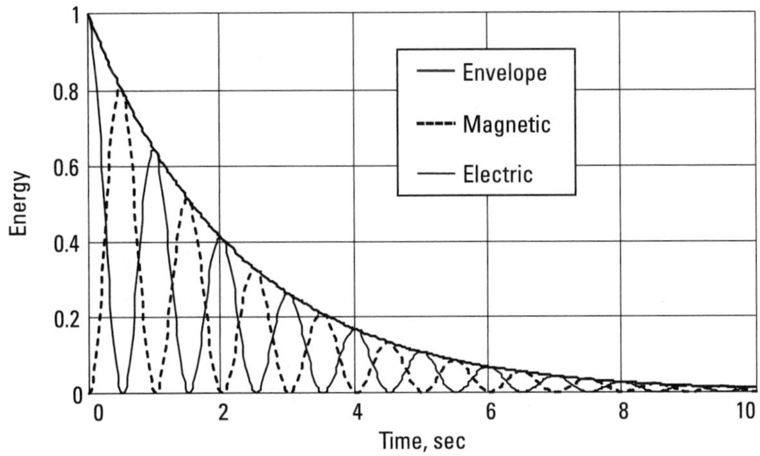

Figure 11.1 Electric and magnetic energies in a damped sinusoid. (*Source:* [2].)

where f_0 is the resonant frequency for which the impedance of a series resonant circuit, or, equivalently, the admittance of a parallel resonant circuit, is at a minimum, and f^+ and f^- are, respectively, the upper and lower frequencies, where the magnitude of the impedance, or admittance, is $\sqrt{2}$ relative to the value at f_0. The frequencies f^+ and f^- are the "half-power" points, and (11.5) is the 3-dB bandwidth Q. The frequencies f^+ and f^- are also where the magnitude of the reactance (susceptance) equals the resistance (conductance) in a series (parallel) resonant circuit. It is easy to show, in the case of the simple series resistor-capacitor-inductor (RCL) resonant circuit, that, using (11.1), with $Q = \omega L/R$ evaluated at $\omega^2 = 1/LC$, is exactly the same as Q given by (11.5). Expression (11.1) defines Q at all frequencies, whereas (11.5) is specific to the resonant frequency. When the resistance includes the source resistance of the circuit driving the RCL circuit, the determined Q is called Q_L, the loaded Q. Since driving circuits are generally impedance matched to the load circuit, the net series resistance doubles, so $Q_L = Q/2$.

[11-2.mcd] Calculate the impedance of a series RCL circuit and plot the magnitude of the impedance versus frequency. What is the Q based on the 3-dB bandwidth; that is, what is the Q calculated from resistance equal to the magnitude of the reactance as in (11.5)? What is the Q calculated from (11.1) at the resonant frequency?

Useful physical pictures emerge from the application of (11.1) to physical materials and electrical circuits. For example, in a simple resonant circuit, the ratio of the maximum energy stored in either the capacitor or the inductor to the energy dissipated per cycle is $Q/2\pi$. We will see in the analysis of the simple loop antenna that this leads to a simple expression for the peak voltage across the capacitor in terms of the capacitive reactance, the Q, and the power delivered to the loop. Thus, for a transmitting loop, the "voltage rating" of the resonating capacitor may be the power-limiting design parameter.

11.2.2 Values of *Q*

In our use of Q, we imply that Q is large compared with unity. We further imply that the largest values are the "best." Whereas the definitions do relate stored, or reactive, energy to dissipated energy, we should be careful to identify the "loss" mechanism and to distinguish between dissipative losses and energy transferred to a useful load, such as radiation in free space. Nevertheless, Table 11.1 presents the range of Q's that may be encountered.

The concept of Q can be used to measure the energy lost in successive bounces of a golf ball from a hard surface to illustrate the behavior of stored and dissipated energy in mechanical systems. Piano strings, quartz resonators, and the planet Earth are likewise resonant mechanical systems that have frictional losses. The Earth's rotational rate has decreased about 1 ms in the last 60 years. As a matter of academic interest, we can compute a $Q = 6 \times 10^{12}$ of this rotational body from (11.4). The rotational decrease is, of course, not monotonic or constant, but the exercise points out that, over the long term, the Earth is a "hi-Q circuit," hence a good time reference. Today's industrial and commercial time standards based on atomic transitions are achieving Q's of between 10^{12} and 10^{13}, while the best primary standards have Q's in the range of 10^{14} to 10^{15} with limiting values around 10^{18}. By comparison, the Q of water at its dipole moment resonance is about 1.1 and is reflected in the broad resonance near 20 GHz, shown in Figure 10.8.

Table 11.1
The Range of Q's Encountered in Nature

Material	Range of Q
Water (at resonance)	1.1
Golf ball	22
Small antennas	10–200
Piano strings	1,200
Quartz resonator	$10^5 - 3 \times 10^6$
Planet Earth	6×10^{12}
Spectral lines	$10^{12} - 10^{18}$

11.3 Primer on Fundamental Limitations in Small Antenna

We begin the investigation of electrically small antennas by studying a theoretical problem that, irrespective of any physical design, expresses the limitations of small antenna behavior. This analysis, based on the work of L. J. Chu [3], reveals fundamental limitations in the relationship between the maximum physical dimensions of an antenna, the Q, and directive gain. H. A. Wheeler viewed the same problem in terms of *radiation power factor*, which is numerically equal to the reciprocal of Q [4, 5]. Recently, J. S. McLean [6] proposed an alternate approach to define a limit on antenna performance. The concept of Q related to UWB antennas was examined by H. Schantz [7]. Our problem is that small portable communication devices restrict the physical antenna size for contained antennas to dimensions smaller, often significantly smaller, than a wavelength. As a consequence, we often encounter high-Q structures; hence, useful bandwidths are limited, and losses for practical implementations tend to be high.

11.3.1 Fields of Radiating Structures

The antennas considered here are electrically (and physically) small vertical dipole and horizontal loop structures. Figure 11.2 shows the radiation field patterns for the TM_{01}, TM_{02}, and TM_{03} modes of free space. The odd-numbered modes are symmetric about the horizontal plane (0° and 180°), while the even-numbered modes are antisymmetric and have zero field contribution in the horizontal plane. The patterns of practical antennas, even small ones, can be represented as a summation of these modes, each multiplied by a modal amplitude in much the same way that periodic voltage waveforms can be represented by a summation of sinusoidal harmonics having complex harmonic amplitudes.

Intuitively, we can see that any far-field radiation pattern can be synthesized from a properly weighted summation of the orthogonal TM_{0n} modes. In fact, the modes may be summed up to yield an arbitrarily large directive gain on the horizon independent of the radius enclosing the antenna. This is the case of "supergain" antennas. We emphasize that the analysis does not lead to any detail about the construction of such an antenna, but simply describes its fundamental behavior. In antennas in portable and personal communications devices, often the object of design is not so much pattern synthesis as finding the widest bandwidth structures without incurring unnecessary dissipative losses. It is instructive, then, to study Chu's analysis as a primer on small antenna fundamentals.

11.3.2 Modal Impedances of Free Space Modes

Following Chu's method, we seek an expression for the modal impedances of the radiating modes of an antenna that is entirely enclosed in a sphere of

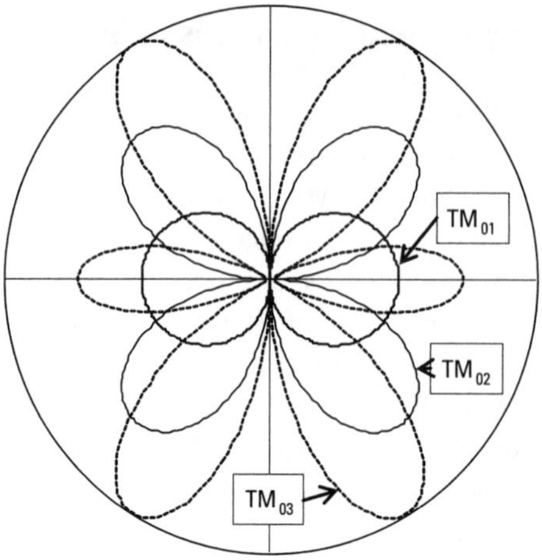

Figure 11.2 Modal radiation patterns for the TM_{01}, TM_{02}, and TM_{03} modes of free space.

diameter $2r$. The fields outside the sphere can be expressed in terms of a complete set of spherical vector waves. The fields from the antennas under consideration require only the n transverse magnetic waves (TM_{0n}) to describe the azimuthally omnidirectional field having the specified polarization. Chu showed that the behavior of the resulting TM_{0n} modes can be likened to a set of equivalent cascade series capacitances and shunt inductances terminated with a unit resistance [3].

As a foundation for our investigation of the behavior of electrically small antennas, we follow Chu's analysis, also detailed in [8, 9], with emphasis on deriving the Q of radiating modes. The modal wave impedances Z_n^{TM} for the TM_{0n} modes are given by the ratio of the radiated electric to magnetic fields, which are here normalized to the free space intrinsic impedance. So,

$$Z_n^{TM} = \frac{j[\rho h_n]'}{\rho h_n} \tag{11.6}$$

where $\rho = kr$, and h_n are the spherical Bessel function of argument kr. The $[\cdots]'$ operation denotes differentiation with respect to ρ.

By duality, the transverse electric (TE_{0n}) modal impedances are

$$Z_n^{TE} = \frac{\rho h_n}{j[\rho h_n]'} \tag{11.7}$$

Using the recurrence formulas for Bessel functions, the TM impedance for outward traveling waves can be written as a continued fraction. For the first mode, letting $\rho = kr$,

$$Z_1^{TM} = \frac{1}{jkr} + \frac{1}{\left[\frac{1}{jkr} + 1\right]} \quad (11.8)$$

Expression (11.8) is comparable to the impedance of a unit load resistance with a shunt inductance and with a series capacitance. By duality, the TE case replaces the series capacitors and shunt inductors of the TM case by equal-valued shunt inductors and series capacitors, respectively. The load resistance in the TE case is in series with the last TE capacitance. Q^{TM} is now defined as the ratio of reactance to resistance in (11.8). Extending the duality, Q^{TE} = Q^{TM}. The expressions for the higher-order modal impedances, given in [8, 9], become tediously long for all but the first mode, TM_{01}, because additional shunt inductors and series capacitors are added in the equivalent circuit representations.

11.3.3 Quality Factors Q_n of Free Space Modes

For all but the first mode (a simple dipole mode for TM and a simple loop mode for TE), the calculation of stored electric energy is exceptionally tedious, so Chu developed a method for arriving at the Q_n for the higher-order modes based on a simple equivalent series RCL representation. Again, the details are in [3]; here, it is important to note only the behavior of the Q_n. Figure 11.3 shows Q_n calculated for the first few modes of free space as a function of the radius in wave numbers, kr, of the sphere enclosing the radiating structure. The Q_n increases dramatically as the radius of the sphere decreases. Evidently, an antenna that excited only the lowest n = 1 mode, if we knew how to construct such an antenna, would have the lowest possible Q for its size. The Q here is defined as the ratio of 2π times the larger of the mean electric or mean magnetic stored energy to the power dissipated in radiation, as given by (11.1). When $kr << 1$, the Q of the first mode can be approximated by

$$Q_1 = \frac{1}{[kr]^3} \quad (11.9)$$

which agrees in first order with McLean's alterative approach in [6].

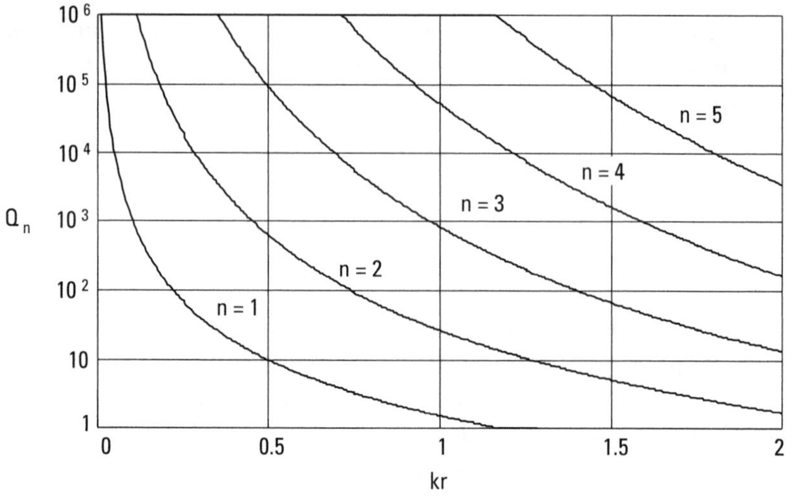

Figure 11.3 Q_n factors for the first five modes of free space.

The Q increases dramatically as the radius kr of the sphere containing the antenna gets small. As also seen in Table 11.2, for a fixed value of kr, Q_n likewise increases dramatically for increasing values of n.

The case $kr = 0.39$ corresponds to a sphere about one-eighth of a wavelength in diameter, while $kr = 1$ corresponds to a sphere nearly one-third of a wavelength across. The behavior shown in Table 11.2 is important to note because, as pointed out earlier, we have not, and from this discussion cannot, provided any guidance on how to construct an antenna that radiates the modes we want in the proportions we desire. Furthermore, the enormous stored energies swapping between electric and magnetic fields every half-cycle associated with the high Q_n extend around the antenna structure beyond a radius of $r = \lambda/2\pi$, even though the physical antenna structure is contained within a sig-

Table 11.2
Q_n for the First Five Modes and for Two Values of kr

n	Q_n for $kr = 0.39$	Q_n for $kr = 1$
1	19	1.5
2	2,103	26
3	5.1×10^5	836
4	2.2×10^8	5.1×10^4
5	1.4×10^{11}	5.0×10^6

nificantly smaller sphere of radius r. We will revisit this concept later in the analysis of small loops and dipoles.

In practical cases, antennas, even physically and electrically small ones, will inevitably produce multiple spherical modes. The combined Q of multiple modes is given by

$$Q = \frac{\sum_n A_n^2 \frac{n(n+1)}{2n+1} Q_n}{\sum_n A_n^2 \frac{n(n+1)}{2n+1}} \tag{11.10}$$

which is a summing up of all the mean electric (or magnetic, if it is greater) energy stored in all the n simplified RCL equivalent circuits divided by the mean dissipation. The modal coefficients A_n are specified by the desired radiation characteristics or, as in the case for most personal communications devices, are determined from the boundary conditions on the surface of the antenna structure. For example, in the radiation pattern synthesis problem, a set of modal excitation coefficients can be found, which results in an arbitrarily high directive gain if such an arbitrarily high Q can be achieved and is acceptable. In our case, we want the lowest possible Q. Clearly, given (11.10), the lowest Q occurs when only the $n = 1$ mode exists. However, how to construct such an antenna is not defined physically, and we are left with practical antennas that (in the lossless case) exhibit a Q greater than the minimum Q.

Practical antennas are subject to the physical design constraints associated with small radio housings and will excite at least several modes to some extent. From (11.10) and from Table 11.2, if the TM_{03} mode is excited with as little as 0.1% of the energy of the TM_{01} mode, the resulting total Q for a structure within a sphere of diameter $2r = 4$ cm at 930 MHz ($kr = 0.39$) is over 1,300 compared with the minimum value of 19 given for $n = 1$ in Table 11.2. A practical antenna will exhibit a radiation Q greater than the minimum value precisely because some inevitable small fraction of a higher-order mode is excited. For such high values of Q, the fractional bandwidth of a driven antenna is given by twice the reciprocal of Q.

11.3.4 Small Antenna Bandwidth Limitations

The bandwidth, Q, and losses are closely interrelated in small antennas. Most often, in the case of antennas contained within the housing of a small personal communications device, the antennas are confined within a sphere that is significantly smaller in radius than $r = \lambda/2\pi = 1/k$. A fundamental relationship between the maximum dimension $2r$ of an electrically small antenna, its maximum fractional bandwidth BW, and radiating efficiency η can be written as

$$BW = \frac{2(kr)^3}{\eta} \tag{11.11}$$

where k is the wave number. Expression (11.11) implies that instantaneous bandwidth may be gained at the price of radiation efficiency once the size of the antenna is constrained. The reactive fields in the close vicinity of the antenna are still present and still extend out to a distance greater than $r = 1/k$, despite the confined physical size of the antenna. Objects within this radius will interact strongly with the antenna and may result in additional losses in performance.

11.3.5 Superdirectivity in Small Antennas

The normal directivity of an antenna of radius a is defined as [8]

$$D_{normal} = 2ka + (ka)^2 \tag{11.12}$$

which is obtained using only the spherical modes that contribute to radiation. For a large aperture, that is, $ka >> 1$, this equals

$$D = \frac{4\pi A_e}{\lambda^2} \tag{11.13}$$

and the directivity, D, is the same as for a large-radius, uniformly illuminated, circular aperture of radius a, hence, area $A_e = \pi a^2$. Recall from (1.43) that A_e is the effective aperture, and D is the gain when efficiency is 100%.

Applying (1.41) and recognizing that the field-strength pattern of the TM_{01} mode is $\sin(\theta)$, the directivity of small loops and dipoles limits to $D = 1.5$, which for small antennas far exceeds the normal directivity given in (11.12). Antennas exhibiting directivities in excess of the normal directivity given by (11.12) are called "supergain" antennas [3, 8]. They radiate with a set of modal coefficients for which the directivity is enhanced. By (11.10) and from [8], the directive gain of an antenna of arbitrary size is unbounded as long as an arbitrarily high Q is possible and acceptable. Electrically small loops and dipoles are supergain antennas in the sense of (11.12) and, indeed, exhibit all of the narrow bandwidth and high-Q behavior associated with supergain antennas.

11.4 Antennas for Personal Communications

Antennas for small radio devices are, by their nature, often electrically small antennas. These antennas may be magnetic loops, electric dipoles, or transmission line antennas, such as low-profile wires and microstrip patch antennas. The special issues relating to UWB antennas, that is, antennas for use with short-duration pulses will be considered later. The loops and transmission line antennas are often internal to the radio case, while electric dipoles are most often appendages to the radio case or use the radio housing as one of the dipole elements. We saw in Chapter 1 that the close near fields of simple loops and dipoles are duals of each other. The fundamental behavior of loops and dipoles has been analyzed thoroughly and early in the literature [10, 11]. Later treatments detailed specific behavior for self-resonant structures and for electrically small devices [3, 8, 9]. More recent studies investigated the very close near fields of loops and dipoles and gave results that are valid to within a fraction of a wire diameter of the antenna surface [12–19]. An explanation of the FORTRAN code for evaluating the close near fields is included as Appendix A for the case of dipoles and helical dipoles (based on [16–18]) and as Appendix B for fat wire loops (based on [12–14]). The FORTRAN code, along with executable files for the Windows operating system, is included on the CD-ROM included with this book.

Here, we will review the fundamental behavior of elementary loops and dipoles and their relationship to the fundamental limitations developed in Section 11.3. The dipoles used with portable communications devices very often comprise a helically wound element in addition to the radio housing, particularly at frequencies lower than 800 MHz. Thus, the radio case and the antenna element form a dipole pair that is small compared to a wavelength. We will also explore resonant half-wavelength dipoles that are used with portable radios at frequencies above about 800 MHz and study the effects of coupling to the radio case. The analogous behavior of loops and dipoles will be pointed out.

11.4.1 Loops and Their Characteristics

The small loop antenna is characterized by a radiation resistance proportional to the fourth power of the loop radius. The reactance is inductive, hence, is proportional to the antenna radius. It follows that the Q is expected to be inversely proportional to the third power of the loop radius. For the geometry shown in Figure 11.4, using the analysis of [20], the electrically small loop, having a diameter $2b$ and a wire diameter $2a$, exhibits a feed-point impedance given by

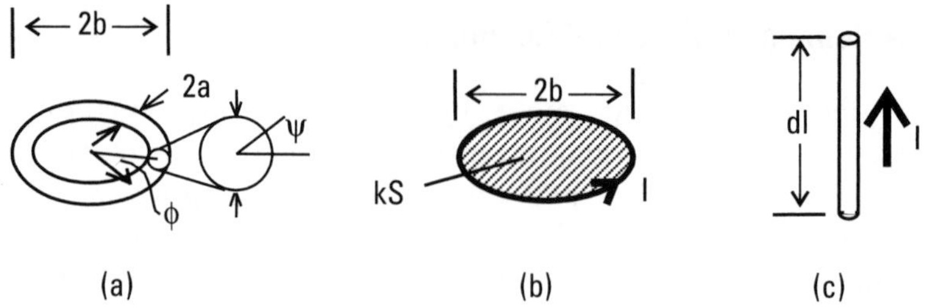

Figure 11.4 Small-antenna geometry showing (a) detail of the wire loop diameter, (b) the parameters *kSI* making up the loop moment, and (c) the elementary dipole moment *Idl*.

$$
\begin{aligned}
Z_{loop} &= \eta_0 \frac{\pi}{6}(kb)^4 \left[1 + 8(kb)^2\right]\left[1 - \frac{a^2}{b^2}\right]\cdots \\
&+ j\eta_0 kb\left[\ln\left[\frac{8b}{a}\right] - 2 + \frac{2}{3}(kb)^2\right]\left[1 + 2(kb)^2\right]
\end{aligned}
\tag{11.14}
$$

including dipole mode terms valid for $kb < 0.4$.

The corresponding unloaded Q of the loop antenna, ignoring the dipole mode terms, is

$$
Q_{loop} = \frac{\frac{6}{\pi}\left[\ln\left[\frac{8b}{a}\right] - 2\right]}{(kb)^3}
\tag{11.15}
$$

which for $b/a = 10$ becomes

$$
Q_{loop} = \frac{4.5}{(kb)^3}
\tag{11.16}
$$

which has the proper limiting behavior for a small loop radius. The Q of the small loop given by (11.15) is indeed larger than the minimum possible Q predicted by (11.9) for a structure of its size. It must be emphasized that the actual Q of such an antenna will be much smaller than given by (11.16) due to unavoidable dissipative losses not represented in (11.14) to (11.16). We can approach the minimum Q but never get smaller, except by introducing dissipative losses.

The simple loop may be resonated with a series capacitor having a magnitude of reactance equal to the loop reactance. The reactive part of the loop impedance is inductive, and the inductance is given by $\text{Im}\{Z_L\} = \omega L$. So,

$$L = \frac{\eta_0 kb\left[\ln\left[\frac{8b}{a}\right] - 2\right]}{\omega} \tag{11.17}$$

which, with the substitution $\eta_0 k/\omega = \mu_0$, reduces to

$$L = \mu_0 b\left[\ln\left[\frac{8b}{a}\right] - 2\right] \tag{11.18}$$

The capacitance required to resonate this small loop at frequency f is $C = 1/(2\pi f)^2 L$. The loop may be coupled to a radio circuit in many different ways, including methods given in [21, 22].

When used in transmitter applications, the small loop antenna is capable of impressing a substantial voltage across the resonating capacitor. For a power P delivered to a small loop with loaded Q of Q_L and with X_C resonating the reactance given by (11.14), it is easy to show that the peak voltage across the resonating capacitor is,

$$V_p = \sqrt{2X_C Q_L P} \tag{11.19}$$

by recognizing that

$$V_p = \sqrt{2} I_{rms} X_C$$

where I_{rms} is the total RMS loop current,

$$I_{rms} = \sqrt{\frac{P}{\text{Re}\{Z_{loop}\}}}$$

along with (11.1) expressed as the ratio of reactance to resistance at the resonant frequency in (11.15).

Transmitter power levels as low as 1W delivered to a moderately efficient, small-diameter ($\lambda/100$) loop can result in peak values of several hundred volts across the resonating capacitor. This is not intuitively expected: the small loop is often viewed as a high-current circuit, which is often described as a short-circuited ring. However, because it is usually implemented as a resonant circuit with a resonating capacitor, it can also be an extremely high-voltage circuit, as shown by (11.19). Care must be exercised in selecting the voltage rating of the resonating capacitor, even for modest transmitting power levels, just as care must be taken to keep resistive losses low in the loop structure.

[11-4.mcd] Calculate the Q and bandwidth of a loop antenna resonated by a series capacitor and operating at 30 MHz. The loop is $2b = 10$ cm in diameter and constructed of copper tubing $2a = 1$ cm in diameter with conductivity $\sigma_{cu} = 5.7 \times 10^7$ S/m [see (3.9)]. What is the voltage across the resonating capacitor if 1W is supplied to the loop? What value is the resonator capacitance? Find the loop current. Calculate the efficiency of this transmitting loop.

11.4.2 Gap-Fed Loop

A detailed analysis of the fat, gap-fed wire loop [12–14] reveals that the current density around the circumference of the wire, angle ψ in Figure 11.4(a), is not constant. An approximation to the current density along the wire circumference for a small diameter loop is

$$J_\phi = \frac{I_\phi}{2\pi a}\left[1 - 2\cos(\phi)(kb)^2\right]\left[1 + M\cos(\psi)\right] \tag{11.20}$$

where I_ϕ is the loop current, which has cosine variation along the loop circumference, and where the variation around the wire circumference is shown as a function of the angle ψ. M is the ratio of the first to the zero order mode in ϕ and is not a simple function of loop dimensions a and b, but can be found from analysis [18] and also from the FORTRAN programs described in Appendix B. For the small loop, M is negative and of order a/b, so (11.20) predicts that there is current bunching along the inner contour ($\psi = 180°$) of the wire loop. This increased current density results in a corresponding increase in dissipative losses in the small loop. We can infer that the shape of the conductor formed into a loop antenna will impact the loss performance in a small loop.

The small loop fed with a voltage gap has a charge accumulation at the gap and will exhibit a close near electric field. For a small loop of radius b in the xy-plane, the fields at $(x, y) = (0,0)$ are derived in [12] and given here as

$$E_\phi = -j\frac{\eta_0 kI}{2} \tag{11.21}$$

where I is the loop current and

$$H_z = \frac{I}{2b} \tag{11.22}$$

Expression (11.22) is recognized as the classic expression for the static magnetic field within a single-turn solenoid. Note that the electric field given by (11.21) does not depend on any loop dimensions but was derived for an electrically small loop. The wave impedance, Z_w, at the origin is the ratio of E_ϕ to H_z from (11.21) and (11.22):

$$Z_w = -j\eta_0 kb \tag{11.23}$$

In addition to providing insight into the behavior of loop probes, (11.21) to (11.23) are useful in testing the results of numerical codes like the Numerical Electromagnetic Code (NEC) often used in the numerical analysis of wire antenna structures (see [23]).

[11-4b.mcd] Starting with (11.20) with $M = 1$ and applying the method of Section 1.3, find the near E and H fields of a small loop and demonstrate that (11.21) to (11.23) are true.

When the small loop is used as an untuned field probe, the current induced in the loop will have a component due to the magnetic field normal to the loop plane, as well as a component due to the electric field in the plane of the loop. A measure of E-field-to-H-field sensitivity is apparent from (11.23). The electric-field-to-magnetic-field sensitivity of a simple small loop probe is proportional to the loop diameter. The small gap-fed loop, then, has a dipole moment, which complicates its use as a purely magnetic field probe.

11.4.3 Near Fields of an Elementary Loop

The fields of an elementary current loop of radius b can be written in terms of the loop enclosed area, $S = \pi b^2$, and a constant excitation current, I (when I is RMS, then the fields are also RMS quantities). The fields are "near" in the sense that the distance parameter r is far smaller than the wave number k, but far larger than the loop dimension $2b$. Hence, this is not the close near-field region. For $r >> b$, the near fields were given earlier by (1.29) to (1.31). The term kIS in those expressions is often called the loop moment and is analogous to the similar term Idl associated with the dipole moment. Expressions (1.29) to (1.31) are not inconsistent with (11.21) and (11.22). They are simply valid for different ranges of the distance variable r. The exact analysis giving (11.21) to (11.23) and (1.29) to (1.31) in the appropriate limits is found in [12] and [14] and can be calculated using the FORTRAN programs in Appendix B.

For the loop antenna, only the θ-component of the magnetic field and the ϕ-component of the electric field survive into the far field. The functional form of the radiation field of the infinitesimal loop is the same as for the TM_{01} radiating mode of free space. The field components varying with the square and the cube of distance are associated with the enormous reactive fields of the small, high-Q antenna. These fields exceed the radiating field components to a distance beyond $r = 1/k$.

Expressions (11.21) and (11.22) serve to emphasize that the near fields given by (1.29) to (1.31) are not valid in the close near-field region, where r approaches the surface of the wire. The magnetic and the electric fields in the close near-field region do not vary with the inverse second or third power of distance, as is pointed out in [12, 13]. In fact, the magnetic field amplitude at the loop wire surface is finite and can be found using (1.10) to be exactly equal to the current density on the wire surface.

11.4.4 Dipoles and Their Characteristics

The dipole antenna has been analyzed for harmonic wave excitations in detail in [10, 11] and also in work that revealed the near-field behavior to within a wire diameter of the antenna surface [15, 16]. Those results have been experimentally verified [17] and extended to include a detailed look at helical dipoles [18, 19]. Here, we are concerned with the small dipole antenna impedance behavior, which is characterized by a radiation resistance proportional to the square of the dipole height $dl = 2h$ seen in Figure 11.4(c). The reactance is capacitive and, hence, is inversely proportional to the antenna height, with an additional dependence on the height-to-diameter ratio, $2h/2a$. It follows that the Q is inversely proportional to the third power of the antenna height, as is expected given the limitation expressed by (11.9). The impedance at the midpoint of a short dipole having a triangular current across the length $2h$ is

$$Z_{dipole} = \frac{\eta_0}{6\pi}(kh)^2 - j\frac{\frac{\eta_0}{\pi}\left[\ln\left[\frac{h}{a}\right] - 1\right]}{kh} \tag{11.24}$$

A practical short dipole is center fed, and the current profile is triangular, becoming zero at the tips. The field strength at every point from such a dipole is reduced to one-half of the field strength from the short current element. The corresponding radiation resistance is one-quarter that of a uniform current. The corresponding unloaded Q of the dipole antenna is

$$Q_{dipole} = \frac{6\left[\ln\left[\frac{h}{a}\right] - 1\right]}{(kh)^3} \tag{11.25}$$

For $h/a = 10$, (11.25) has the expected inverse third power with size behavior predicted by (11.9) for small antennas:

$$Q_{dipole} = \frac{7.9}{(kh)^3} \tag{11.26}$$

Comparing the Q for a small dipole (11.26) with the Q of a small loop (11.16), we see that the loop Q is smaller, by nearly half, even though the same ratio of antenna dimension to wire diameter was used. If the same wire length to diameter were used, the loop Q would be somewhat smaller. We conclude that the small loop utilizes the smallest sphere that encloses it more efficiently than does the small dipole. Indeed, the thin dipole is essentially a one-dimensional structure, while the small loop is essentially a two-dimensional structure.

11.4.5 Near Fields of Dipoles

When a uniform current distribution exists across the antenna length dl, the close near fields of an elementary dipole can be written in terms of the dipole length, dl, and a constant excitation current, I. Those fields were expressed by (1.26) to (1.28) for the region $r >> dl$ Again, as in the case of the infinitesimal loop, this is not the close near-field region since $r >> dl$. The fields given by (1.26) to (1.28) are for a uniform current across the length of the dipole as compared with the triangular-shaped current used in the determination of the dipole impedance and Q in (11.24) to (11.26). For the dipole antenna, only the θ-component of the electric field and the ϕ-component of the magnetic field survive into the far field.

As in the case of the infinitesimal loop, the infinitesimal dipole near fields given by (1.26) to (1.28) are not valid in the close near-field region, where distance r approaches the surface of the wire. The magnetic and electric fields in the close near-field region do not vary with the inverse second or third power of distance, as is pointed out in the detailed analysis of [16–18]. In fact, the magnetic field amplitude at the dipole wire surface is finite and can be found using (1.10) to be exactly equal to the current density on the wire surface.

There is a strong resemblance between the elementary dipole fields and the fields of the elementary loop. The magnetic loop moment kIS corresponds to the electric dipole moment Idl, and the forms of the electric and magnetic fields have corresponding forms for the three geometric components. This is an expression of the duality of the two elementary antenna types. In fact, as we saw in Chapter 1, if an elementary loop and elementary dipole were collocated and excited so that $kIS = Idl$, the radiation field everywhere in space would be circularly polarized. The fields of the loop and the dipole are orthogonal and perfect duals.

The fields of a sinusoidally excited dipole of length $2H$ can be expressed in closed form for both the near and far regions as detailed in [24, 25] for the specific case where the wire is infinitesimally thin. Using the approach of [25], the current on the wire is

$$I = \sqrt{2} I_{rms} \sin(k[H - |h|]); \quad |h| \leq H \tag{11.27}$$

where I_{rms} is the RMS feeding-point current of the dipole, which relates to power by the radiation resistance R_{rad} given by the real part of (2.36) evaluated for the self-impedance. For a resonant half-wave dipole the power P delivered to the $2H = \lambda/2$ dipole with $R_{rad} = 73.08\Omega$, radiation resistance is $P_{rad} = 73.08\ [I_{rms}]^2$. With the current I_{rms} specified, we can apply (1.25), integrating over the current source length $2H$, to obtain the vector potential and use (1.15) to solve for the magnetic field. A closed-form analytical solution is available when the current I is exactly sinusoidal. The electric field is then found from (1.1). Finally, with the algebraic details contained in [25], the RMS fields for a center-fed, sinusoidally excited, half-wave dipole aligned on the z-axis and centered at the origin, in cylindrical coordinates, are

$$E_z = \frac{-jI_{rms}\eta_0}{4\pi}\left[\frac{e^{-jkR_1}}{R_1} + \frac{e^{-jkR_2}}{R_2} - 2\cos(kH)\frac{e^{-jkr}}{r}\right] \tag{11.28}$$

$$E_\rho = \frac{jI_{rms}\eta_0}{4\pi}\left[\frac{z-H}{\rho}\frac{e^{-jkR_1}}{R_1} + \frac{z+H}{\rho}\frac{e^{-jkR_2}}{R_2} - \frac{2z}{\rho}\cos(kH)\frac{e^{-jkr}}{r}\right] \tag{11.29}$$

and

$$H_\phi = \frac{jI_{rms}}{4\pi\rho}\left[e^{-jkR_1} + e^{-jkR_2} - 2\cos(kH)e^{-jkr}\right] \tag{11.30}$$

where

$$\rho = \sqrt{x^2 + y^2} = r\sin(\theta)$$

and

$$z = r\cos(\theta)$$

are the usual cylindrical coordinates in terms of the Cartesian components x and y, and spherical components r and θ. The distances from the dipole ends are

$$R_1 = \sqrt{\rho^2 + (z - H)^2} \tag{11.31}$$

and

$$R_2 = \sqrt{\rho^2 + (z + H)^2} \tag{11.32}$$

Expressions (11.28) to (11.30) give the electric and magnetic fields both near to and far from the dipole carrying a sinusoidal current distribution. When (11.30) is specialized to a half-wavelength and evaluated in the far-field region and compared to the magnetic field of an isotropic source radiating the same total power (equal to $P_{rad} = R_{rad} I_{rms}^2$) as the half-wave dipole, (11.30), along with $\rho = r\sin(\theta)$, reduces exactly to the half-wave dipole far-field pattern given by (2.1) evaluated for a half-wavelength dipole. Expressions (11.28) and (11.29) for the electric field similarly reduce to (2.1) in the far field when restated in spherical coordinates and after considerable algebraic manipulation.

Interestingly, the fields of the sinusoidally excited dipole have solutions in terms of spherical sources located at the two ends of the dipole and at the center. The source at the center vanishes when $2H = \lambda/2$. The character of the field is vastly different from that of the infinitesimal dipole given earlier by (1.26) to (1.28). Specifically, the inverse third power with distance terms are absent. The magnetic field (11.30) appears to become unbounded as the radial distance decreases. This is indeed the case for the infinitesimally thin wire postulated for this case, and this permits the relatively simple form of the solution. For a finite wire of radius a, the current becomes a current density $J = I_{rms}/(2\pi a)$, and the magnetic field remains bounded. We postulated the case of a sinusoidally excited dipole to arrive at a simple fields solution. A dipole, whether a half-wave long or very short, with vanishingly thin wire would also exhibit a vanishingly small bandwidth, as seen in (11.24) for the short dipole. We note, however, that when the wire assumes a finite thickness, the solution for the fields is no longer trivial (see [16–18]). FORTRAN programs implementing a more general dipole solution are described in Appendix A.

11.4.6 A Ferrite-Loaded Loop Antenna

Let us examine a small ferrite-loaded loop antenna with dimensions $2h = 0.04$m and $2a = 0.005$m, at a wavelength of about $\lambda = 8.6$m, as pictured in Figure 11.5. When the permeability of the ferrite is sufficiently high, this antenna behaves like an ideal magnetic dipole. The magnetic fields are strongly confined to the

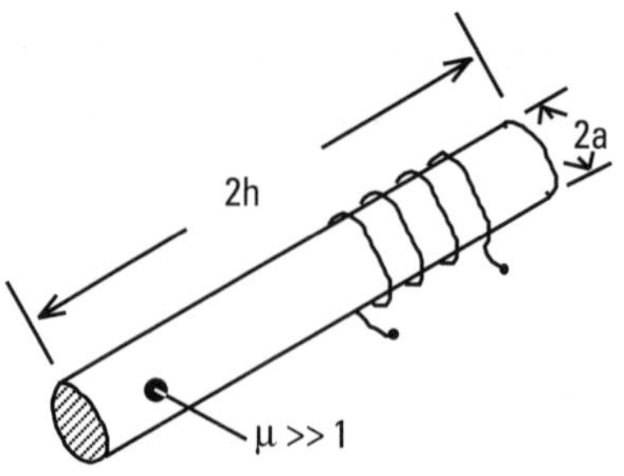

Figure 11.5 A ferrite-loaded loop antenna.

magnetic medium and behave as the dual of the electric dipole excited by a nearly triangular current distribution. We can therefore analyze its behavior using (11.24) and (11.25) developed for the elementary dipole. The minimum ideal loaded Q of this antenna is, using (11.25), 2×10^6. The corresponding bandwidth of such an antenna having no dissipative losses would be $2/Q = 35$ MHz/10^6 = 35 Hz!

A practical ferrite antenna at this frequency has an actual loaded Q of nearer to 150, so an estimate of the actual antenna efficiency is

$$Q_{actual}/Q_{loaded} = 0.000075 = -41.2 \text{ dB}$$

and the actual resultant bandwidth is 200 kHz. Such an antenna is typical of the type that would be used in a body-mounted paging receiver application. The resulting field strength is calculated using (9.5) with the above efficiency, a receiver having a 50Ω input sensitivity of 0.08 μV (−129 dBm), and a body-enhancement directivity of 8 (corresponding to 9 dB, as seen in Figure 10.14). The calculated field-strength sensitivity is $E = 3.7$ μV/m. This is typical of a front-position body-mounted paging or personal communications receiver sensitivity in this frequency range (see also Table 11.4 later in this chapter).

11.5 Transmission Line Antennas

We turn our attention now to some of the low-profile structures that find application in miniature personal communications devices. These structures include transmission line antennas like the microstrip patches, inverted "L" and "F" wire antennas, and the open ring radiator. They are described in texts devoted to the subject of small antennas, such as [26–30], and in more general treatments, such as [24, 31]. These antennas are most suitable when the maximum dimension of the antenna is a quarter- or half-wavelength, with dielectric loading of not much more than $\varepsilon_r = 40$. For practical small radios, this usually restricts the operating frequencies to bands above 800 MHz.

The desire for miniaturization and the use of increasingly higher frequencies have made the consideration of microstrip patch antennas important in communications devices. Here, we will include useful design equations for the rectangular and circular patches, which can be implemented on small computers using mathematical "document" processors such as Mathcad by MathSoft, Inc., and MATLAB from The MathWorks, Inc. Electromagnetic design software like HFSS from Ansoft Corporation and FDTD from Remcom provide solutions by direct numerical methods. Patch antennas may take on a variety of shapes but are generally resonant lengths of conductor suspended by a dielectric material above a conducting ground plane. Because the patch-to-ground-plane distance is usually very small, the Q is generally large (25–200), and bandwidth is limited. The two most popular shapes, the rectangular and the circular patch, can be analyzed in a straightforward manner. The presented equations are a good starting point for design, however, in small radio devices. The ground plane behind the patch elements is usually not much greater in size than the patch. Practical results, particularly the gain, will differ significantly from predictions when using small ground planes. Additionally, there are surface wave losses [32], which are usually ignored for the range of parameters normally encountered.

11.5.1 Rectangular Microstrip Patch Antennas

Various techniques have been explored in analyzing microstrip patch antennas, including transmission line models, cavity models, full modal expansions, and various numerical techniques. Most recently, finite difference time domain (FDTD) analysis has emerged as a numerical technique of choice, as for example in [33]. The cavity model will be shown here because it leads to useful results for both circular and rectangular patch antennas. For the rectangular patch, we follow essentially the method outlined by R. E. Collin [31].

The geometry of the microstrip patch having length a along the x-axis, width b along the y-axis, and a feed point displaced p along x is shown in Figure 11.6; the relevant parameters are:

ε_r	relative dielectric constant;
$\tan\delta$	dielectric loss tangent;
σ	patch conductance (S/m);
δ_t	effective total loss tangent;
h	patch dielectric thickness (m);
a	patch length along x (m);
b	patch width along y (m);

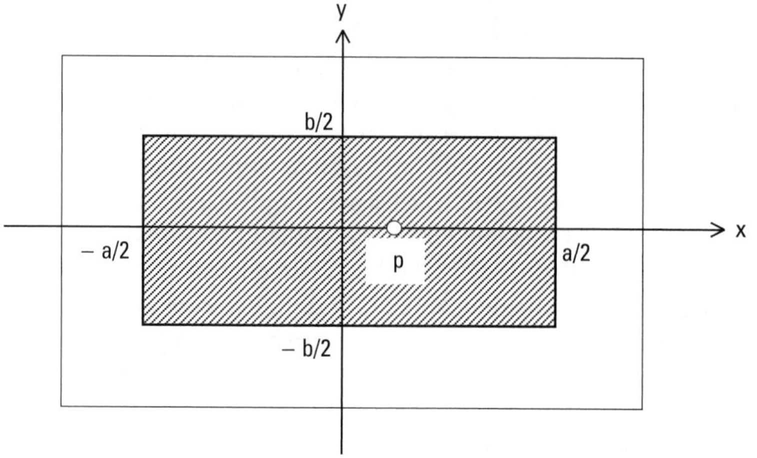

Figure 11.6 Geometry of the rectangular microstrip patch antenna.

p origin to feed point (along x) (m);

r_o feed-post radius (m).

The rectangular microstrip is analyzed by assuming that predominately transverse magnetic modes exist in a cavity bounded by the rectangular patch. Suitable values for the z-directed electric field take the form of rectangular cavity mode fields, and the "ideal" cavity is now driven by a current probe located at $x = p$, which must satisfy the wave equation in the cavity. The total energy stored in the cavity and the cavity losses can now be calculated. For the predominant TM_{10} mode, the total stored energy, W, is derived by integrating over the square of the TM_{10} mode electric field in the cavity [31], giving

$$W = \frac{\varepsilon_0 \varepsilon_b hab}{16} \tag{11.33}$$

where ε_0 is the permittivity of free space. The effective relative dielectric constant of the insulating material between the patch and the ground plane is ε_b and is given by the commonly used microstrip approximation

$$\varepsilon_b = \frac{\varepsilon_r + 1}{2} + \frac{\varepsilon_r - 1}{2}\left[1 + \frac{10h}{b}\right]^{-1/2} \tag{11.34}$$

The cavity radiates from the edges at $x = \pm a/2$ as if there were radiating slots there. The radiation resistance of the pair of slots is given approximately by

$$R_r = \left[90 \left[\frac{\lambda}{b} \right]^2 + 120 \frac{\lambda}{b+\lambda} \right] \frac{1}{2} \tag{11.35}$$

The radiation losses can now be introduced in terms of the loss factors δ_r:

$$\delta_r = \frac{P_r}{\omega W} \tag{11.36}$$

where ω is radian frequency, and P_r is the radiated power by the two slots at each end, $x = \pm a/2$, of the microstrip patch:

$$P_r = \frac{h^2}{4R_r} \tag{11.37}$$

Dielectric losses are proportional to the dielectric loss tangent:

$$\delta_\varepsilon = \tan \delta \tag{11.38}$$

Patch conductor losses are found from the power dissipated in the patch conductor and the ground plane and are written as a conductor loss tangent:

$$\delta_c = \frac{P_c}{\omega W} \tag{11.39}$$

where the dissipated power is found by integrating over the square of the tangential magnetic field for the TM_{10} mode. The power loss P_c is

$$P_c = \frac{\frac{b}{a} \pi^2}{8\omega\mu_0 \sqrt{2\sigma\omega\mu_0}} \tag{11.40}$$

Generally, there are other losses, such as from the feed post; these are usually not significant. The total loss tangent of the rectangular microstrip patch is given by the sum of the individual loss tangents,

$$\delta_t = \delta_c + \delta_\varepsilon + \delta_r \tag{11.41}$$

These loss tangents are recognized as the reciprocals of the respective Q factors.

Finally, the feed-point impedance for the TM_{10} mode, including the inductance of the feed post of radius r_o, is

$$Z_{10} = j\omega\mu_0 h \frac{\frac{2}{ab}\sin\left[\frac{\pi p}{a}\right]^2}{\left[\frac{\pi}{a}\right]^2 - \varepsilon_b\left[1 - j\delta_t\right]k^2} + j\omega L \tag{11.42}$$

where the feed-post inductance is given by

$$L = -\frac{\mu_0 h}{2\pi}\ln(kr_o) \tag{11.43}$$

Patch length a and feed-point location p are designed based on resonance. The selection of width b is often constrained by availability of space in a miniature telecommunications device. A practical choice of width b is suggested in [30] as

$$b = \frac{\lambda}{2}\sqrt{\frac{2}{\varepsilon_r + 1}} \tag{11.44}$$

based on the observation that wider patches exhibit increased efficiency, but higher-order modes that might distort the antenna pattern may result from a width that is too great. Equations (11.33) to (11.44) are a convenient set of expressions to use for a preliminary design of a rectangular half-wave patch antenna.

The directivity of a half-wave rectangular microstrip patch is given by

$$D = \frac{2}{1 + g_{12}}\frac{4}{I_1}\left[\frac{b\pi}{\lambda}\right]^2 \tag{11.45}$$

where the normalized mutual conductance g_{12} between the two radiating slots is

$$g_{12} = \frac{2R_r}{\eta_o\pi}\int_0^{\pi}\left[\sin\left[\frac{b\pi}{\lambda}\cos(u)\right]\tan(u)\right]^2 J_0\left[\frac{2\pi a}{\lambda}\sin(u)\right]\sin(u)du \tag{11.46}$$

$J_0(z)$ is the Bessel function of zero order, and

$$I_1 = \int_0^{\pi}\left[\sin\left[\frac{b\pi}{\lambda}\cos(u)\right]\tan(u)\right]^2 \sin(u)du \qquad (11.47)$$

In (11.46), R_r is the radiation resistance of a pair of slots, as given by (11.35).

11.5.2 Circular Microstrip Patch Antennas

The geometry of the microstrip patch having radius a and a feed point displaced p from the center and along x is shown in Figure 11.7; the parameters for the circular patch are:

ε_r	relative dielectric constant;
$\tan\delta$	dielectric loss tangent;
σ	patch conductance (S/m);
δ_t	effective total loss tangent;
h	patch dielectric thickness (m);
a	patch radius (m);
p	feed point from center (m);
r_o	feed-post radius (m).

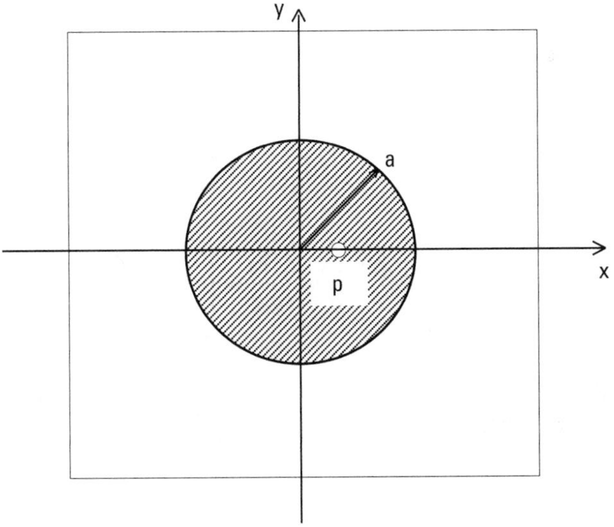

Figure 11.7 Geometry of the circular microstrip patch antenna.

The circular microstrip is analyzed by assuming that predominately transverse magnetic modes exist in a cavity bounded by the circular patch. The analysis takes the same form as for the rectangular patch, except cylinder functions are used to describe the field. The total energy stored in the cavity and the cavity losses are found for the predominant TM_{11} mode. The total stored energy is derived by integrating over the square of the TM_{11} mode electric field in the cavity, giving

$$W = \frac{\varepsilon_0 \varepsilon_r \pi_h}{2} \int_0^a J_1^2\left[k\rho\sqrt{\varepsilon_r}\right]\rho d\rho \tag{11.48}$$

where $J_1(z)$ is the Bessel function of order 1, and ε_0 is the permittivity of free space. The effective radius a_e of the microstrip patch is given by the approximation

$$a_e = a\sqrt{1 + \frac{2h}{\pi a \varepsilon_r}\left[\ln\left[\frac{a\pi}{2h}\right] + 1.7726\right]} \tag{11.49}$$

The circular cavity radiates from the edges at $r = a$ as if there were radiating slots there, and the radiated power is given approximately by

$$P_r = \frac{\left[kahJ_1\left[ka\sqrt{\varepsilon_r}\right]\right]^2}{16\eta_0} I_1(ka) \tag{11.50}$$

where

$$I_1(ka) = 4\int_0^{\pi}\left[\left[J_0(x\sin(ka)) - \frac{J_1(x\sin(ka))}{x\sin(ka)}\right]^2 + \left[\cos(ka)\frac{J_1(x\sin(ka))}{x\sin(ka)}\right]^2\right]\sin(ka)d(ka) \tag{11.51}$$

The radiation losses can now be introduced in terms of the loss factors δ_r and radian frequency ω:

$$\delta_r = \frac{P_r}{\omega W} \tag{11.52}$$

Dielectric losses are proportional to the dielectric loss tangent:

$$\delta_{\varepsilon} = \tan\delta \tag{11.53}$$

Patch conductor losses are found from the power dissipated in the patch conductor and the ground plane:

$$\delta_{c} = \frac{P_c}{\omega W} \tag{11.54}$$

where the dissipated power is found by integrating over the square of the tangential magnetic field for the TM_{11} mode:

$$P_c = C\int_0^a \left[\left[\frac{J_1\left[k\rho\sqrt{\varepsilon_r}\right]}{\rho}\right]^2 + \left[\frac{J_0\left[k\rho\sqrt{\varepsilon_r}\right] - J_2\left[k\rho\sqrt{\varepsilon_r}\right]}{2}k\sqrt{\varepsilon_r}\right]^2\right]\rho\, d\rho \tag{11.55}$$

and

$$C = \sqrt{\frac{\omega\mu_0}{2\sigma}}\frac{\pi}{\left[\omega\mu_0\right]^2} \tag{11.56}$$

The total loss tangent of the circular microstrip patch is given by the sum of the individual loss tangents:

$$\delta_t = \delta_c + \delta_\varepsilon + \delta_r \tag{11.57}$$

As for the rectangular patch earlier, these are recognized as the reciprocals of the respective Q factors.

The feed-point impedance of the TM_{11} mode circular patch is

$$Z_{11} = \left[\frac{-j\left[\dfrac{J_1\left[kp\sqrt{\varepsilon_r}\right]}{a_e}\right]^2 h2.775k\eta_0}{\varepsilon_r - \left[\dfrac{1.84118}{ka_e}\right]^2\left[1 + j\delta_t\right]}\right] + j\omega L \tag{11.58}$$

and the feed-post inductance L was given earlier as (11.43).

The directivity of the TM_{11} mode circular microstrip patch is

$$D = \frac{8}{I_1(ka)} \tag{11.59}$$

with the denominator of (11.59) given by (11.51).

For both the rectangular and circular microstrip patch antennas, the usable fractional bandwidth stated in terms of the maximum acceptable VSWR (Voltage Standing Wave Ratio) S is

$$BW = \frac{S-1}{\sqrt{S}}\delta_t \tag{11.60}$$

Rectangular and circular microstrip patch antennas exhibit essentially the same performance for resonant patches that have approximately the same surface areas and the same electrical parameters. As an illustrative example, the impedances and gains of microstrip patch antennas were computed using (11.33) to (11.59).

The patch parameters and performances are shown in Table 11.3. The impedance performance of the two patches is shown in Figure 11.8 as return-loss plots. Clearly, the two patches perform nearly identically. It is evident from Table 11.3 that neither of these antennas is electrically small. In fact, if we ignore that an infinite ground plane is required and use only the patch dimension, the rectangular patch fits into a sphere or radius $kr = 0.71$, and the circular patch fits in a sphere $kr = 0.57$. Neither is particularly small electrically, yet both are fairly high-Q antennas. This is the price to be paid for shrinking the thickness dimension to obtain a low-profile antenna.

- [11-5a.mcd] Compute the Q and efficiency of a rectangular patch antenna described in Table 11.3.
- [11-5b.mcd] Compute the Q and efficiency of a circular patch antenna described in Table 11.3.
- [11-5c.mcd] Estimate the directivity of a rectangular patch by approximating the radiating patch edges as dipoles over a ground plane.

11.6 Practical Considerations in Small Antennas

Very little practical design information is generally available in the area of small antennas used with personal communications devices, particularly in the frequency range below 800 MHz. For transmitting radios, the antennas are usually dipoles involving two elements: the radio case and an appendage to the radio case, like a whip or a helically wound whip. The combination of the case and

Table 11.3
Microstrip Patch Antenna Parameters

Patch	Rectangular	Circular
Size	$a = 53.1$ mm $b = 50$ mm $t = 2.54$ mm	$2a = 58.55$ $t = 2.54$ mm
Area	Area = 2,656 mm^2	Area = 2,693 mm^2
Relative permittivity	10.2	10.2
tan δ	0.001	0.001
Total Q	166	165
Efficiency (dB)	−1.6	−1.7

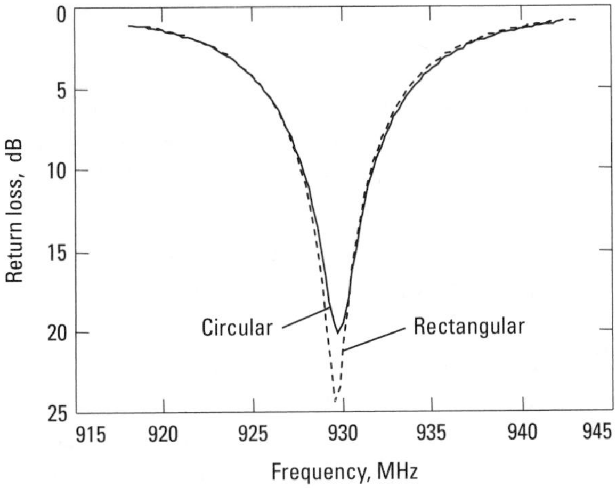

Figure 11.8 Return loss of rectangular and circular microstrip patches.

whip forms an off-center fed dipole. The complexity of the geometry, particularly when the radio case is held in the hand or belt mounted, precludes analysis by simple methods. Complete solutions require tedious numerical computations, as, for example, those by H.-R. Chuang [34], and, oftentimes, good results can be obtained by modeling only portions of the body [35].

11.6.1 Helix-Radio Dipole

Antennas used with portable transceivers at VHF usually take the form of a helix mounted on a radio case, as depicted in Figure 11.9. The helical portion of this

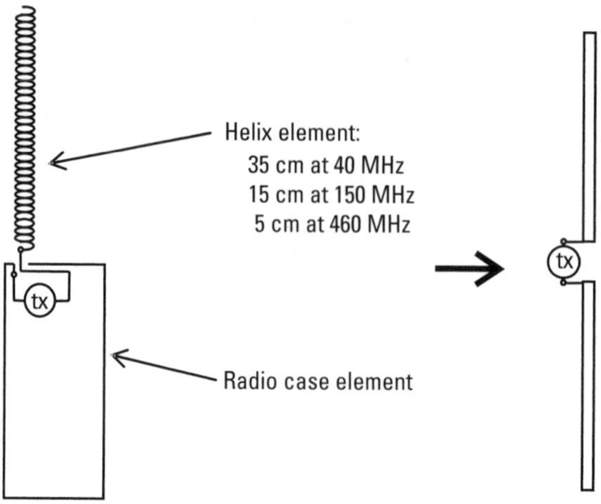

Figure 11.9 The helix-radio dipole.

dipole is approximately 35 cm long for operation in the 40-MHz range, about 12 to 18 cm long in the 150-MHz band, and about 4 to 6 cm long in the 460-MHz range. The radio cases range typically between 10 and 20 cm long. This means that the total dipole length in these bands is less than half a free space wavelength. At the lowest frequencies, the dipole is significantly shorter than a half-wave. Above 800 MHz, the helix-radio combination is generally longer than a half-wave, so the radiation pattern no longer has a peak at the horizon. This results in a significant performance penalty and is the reason why a coaxial dipole is often used at these frequencies.

Loss mechanisms in the helix-radio dipole include resistive losses in the helix winding and resistive losses in the mass of circuit metal and radio case components that act as the second element of the helix-radio dipole or as a counterpoise element for the helix. Finally, there is significant loss introduced when the radio case is held in the hand since the hand is wrapped around one of the dipole elements. Table 11.4 lists the typical helix-radio antenna gains at the head level [36] that might be expected in practice and compares them with typical belt-mounted radio-paging field-strength sensitivities. The paging receiver antennas are loops internal to the radio case, which is between 6 and 9 cm in total length. The helix-radio performance at the belt is 6 to 12 dB worse than at head level. Table 11.4 reports average performances over all azimuth directions. For helix-radio field-strength sensitivities, a –121-dBm receiver sensitivity is assumed, and this roughly corresponds to a 12 dB-SINAD performance. Radio performance above 800 MHz using a coaxial dipole improves to about –3 dBi at head level and –6 to –10 dBi on the belt. Remarkably, the paging receiver field-strength sensitivities are not different from those of a larger helix-radio dipole. The paging receiver field-strength performance, stated here at the 99%

Table 11.4
Helix-Radio Dipole and Paging Receiver Performance Compared

Frequency Band (MHz)	Helix-Radio at Head Level, Average Gain (dBi)	Paging Receiver at Belt, Average Gain (dBi)	Helix-Radio Field-Strength Sensitivity (dBμV/m)	Paging Receiver Field-Strength Sensitivity (dBμV/m)
40	–25 to –35	–32 to –37	13 to 23	12 to 17
85	—	–26	—	13
160	–9 to –15	–19 to –23	9 to 15	10 to 14
300	—	–16	—	10
460	–4 to –8	–12	13 to 18	12
800 to 960	–6 to –14	–9	21 to 29	18 to 28

calling rate, benefits from a receiver sensitivity [27] that is in the range of –120 to –129 dBm, depending on the coding protocol and frequency band.

11.6.2 Mutual Coupling of a Dipole with a Radio Case

A physically small antenna often used with portable radios in the frequency bands higher than 800 MHz is the center-fed coaxial dipole. Here, we are concerned with assessing the performance of such an antenna in the vicinity of a radio case. This is important because the coaxial dipole is not in a free space environment; it is very near a radio housing, which is on the order of a half-wavelength structure. The physical consequence of this proximity, pictured in Figure 11.10, is that the dipole cannot be isolated from the radio case. To obtain an order-of-magnitude feel for the coupling, we can study a similar, but tractable, problem involving two collinear dipoles separated a distance s along the mutual axis.

The mutual impedance between two parallel dipoles $2h$ long is found from an application of the reciprocity theorem and the solution of an integral involving the current $I(z)$ on one dipole along the z-axis and the z-component of the electric field $E_z(z, X, Z)$ on the other dipole:

$$Z_{12}(X,Z) = \frac{1}{I_0^2}\int_{-h}^{h} I(z)E_z(z,X,Z)dz \tag{11.61}$$

where the dipoles are displaced X along the x-axis and Y along the y-axis. The mutual coupling for the sinusoidally excited dipole is given by (2.36) to (2.39) and specialized to the half-wave dipole in (2.40). One resonant half-wave dipole

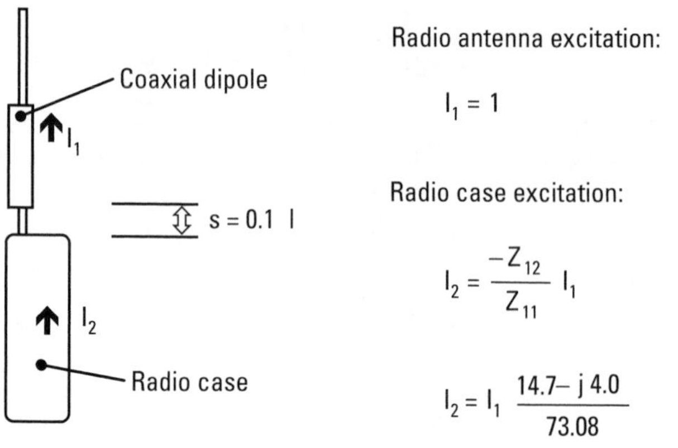

Figure 11.10 A radio case excited by mutual coupling to a dipole. (*Source:* [2].)

is centered at the origin of the coordinate system, so its current is approximately given by

$$I(z) = I_0 \cos(kz) \tag{11.62}$$

where I_o is an arbitrary excitation current, and k is the wave number. The z-component of the electric field on the z-axis due to another half-wave dipole centered at $x = X$ and $z = Z$ is approximately

$$E_z(z, X, Z) = \frac{j\eta_0 I_0}{4\pi}\left[\frac{\exp\left\{-jk\sqrt{(Z+h+z)^2+X^2}\right\}}{\sqrt{(Z+h+z)^2+X^2}} + \frac{\exp\left\{-jk\sqrt{(Z-h+z)^2+X^2}\right\}}{\sqrt{(Z-h+z)^2+X^2}}\right] \tag{11.63}$$

Equation (11.61) is general, while (11.62) and (11.63) are approximations useful in estimating the effect of the mutual coupling of a half-wave dipole to a radio case that is approximately a half-wavelength long. The mutual impedance is calculated from (11.61), using (11.62) and (11.63), and is plotted in Figure 11.11. There are many analytical approaches to computing mutual-coupling impedances [24, 31]. With today's powerful desktop computational capabilities, numerical solutions to (11.61) or method-of-moments solutions [23] are readily accessible. The coupling drops off very rapidly, as may be expected for end-coupled dipoles. For the separation of $s = 0.1$ wavelengths shown in Figure 11.10, the mutual impedance is $Z_{12} = 14.7 - j4.0\Omega$, compared with the nominal 73.08Ω for the resonant half-wave dipole in isolation.

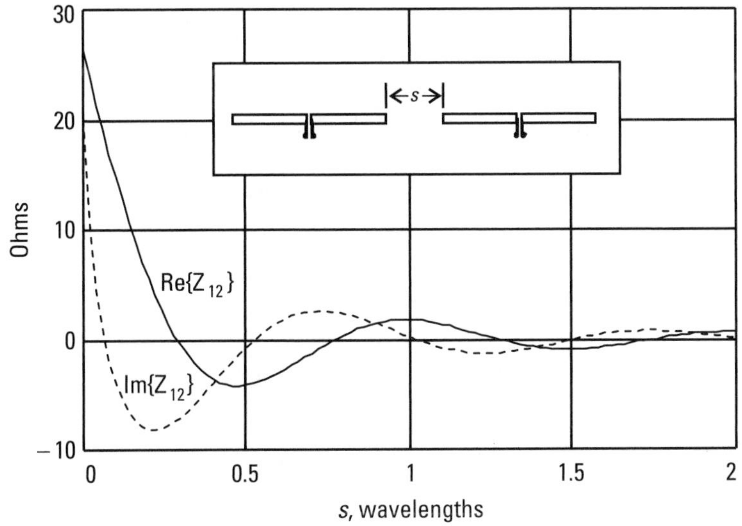

Figure 11.11 Mutual coupling between collinear dipoles. (*Source:* [2].)

The current I_2 of charges flowing in the "dipole" representing the radio case is depicted in Figure 11.6, and its amplitude is found, using reciprocity, from

$$I_2 = \frac{-Z_{12}}{Z_{11}} I_1 \tag{11.64}$$

The dipole excitation current is I_1. The far-field radiation pattern of the dipole in combination with the parasitically excited dipole representing the radio case is shown in Figure 11.12. The undistorted free space dipole pattern is shown for comparison. Pattern distortion shown in Figure 11.8 is typical for a coaxial dipole in the presence of a radio case [27]. A more detailed analysis would reveal that I_2 has a strong frequency dependence because the dipole impedance and the mutual impedance change as they move through the resonance bandwidth of the dipole.

[11-6.mcd] Calculate the mutual coupling between two half-wavelength dipoles as given by (11.61) to (11.63) using direct numerical integration. Find the self-impedance by letting $X = Y = 0$. Compute the mutual impedance for 0.1 wavelength separation along the dipole axis ($X = 0$, $Y = 0.61$). Compute several orientations of parallel dipoles, including an echelon configuration. Compare the with published results.

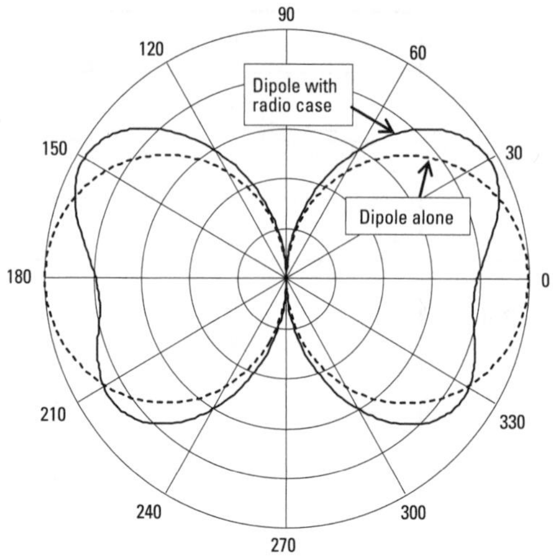

Figure 11.12 Influence of the radio case on the radiation pattern of a dipole. (*Source:* [2].)

Mutual coupling is not the sole mechanism for impressing currents on the housing and circuitry of a radio feeding the coaxial dipole. The coaxial dipole has a sleeve balun that doubles as one of the dipole elements. This balun does not completely suppress currents along the dipole coaxial feed line, which, in addition to the mutual-coupling mechanism, produce the mechanism for distorting the radiation pattern of the radio-antenna combination.

11.7 UWB Antennas

When radiated, short UWB pulse signals manifest some interesting and different behavior when compared with the narrowband signals. UWB radio technology and some details about the radiation and propagation properties are provided in [37], and that approach is followed here. The radiation of signals supplied to an antenna involves an acceleration of charges (except for traveling wave antennas); hence, there is a partial time-delay derivative operation on the current density from the various parts of the antenna involved in the radiating process. Maxwell's equations are still obeyed but must be viewed in the time rather than the frequency domain. We also need to abandon the narrowband harmonic simplifications to Maxwell's equations.

11.7.1 Radiation of Short Pulses

Time-harmonic solutions (sine waves) are not the only solutions to Maxwell's equations (see [38]). They are the most commonly used in narrowband and moderately wideband systems, but they are not always useful for UWB systems. With UWB, we are concerned with solutions that involve wide-bandwidth functions. A heuristic approach is taken here with the goal of understanding the UWB radiation process and how it differs from the time-harmonic radiation solution. Although UWB under FCC regulations may include bandwidths as narrow as 5% (a 500-MHz bandwidth at 10 GHz) [39], we consider here the short-time signal problem: UWB signals with bandwidths greater than about 40%.

Radiation in free space occurs when the current velocity changes, that is, the charges are accelerated. Charge acceleration is expressed mathematically by a partial derivative in time and delay of the electric current. Therefore, we expect the shape of the transmitting antenna current to be different from the radiated signal shape. The difference between wideband and narrowband signals is that the transmitted signals in narrowband systems are sine waves, so their time derivatives are also sine waves. The signal shape for narrowband signals remains sinusoidal throughout the radiation process. The wider the signal bandwidth, the more a signal shape differs from its time derivative in amplitude versus time.

We begin by defining the geometry of an arbitrarily shaped dipole antenna that supports surface currents $\mathbf{J}(\mathbf{r}',\tau)$ at points described by a vector $\mathbf{r}'$ pointing from the coordinate origin to the current density point and by a retarded time variable $\tau = t - R/c$. The solution to the dipole current density on an arbitrarily shaped antenna is well beyond the scope of this work. C. L. Bennett and G. F. Ross [40] provide us with a space-time integral equation that allows us to express the radiated fields due to antenna currents in time-space rather than frequency. Thus, transient, or short-pulse, antenna excitations can be considered rather than the steady state time-harmonic solution appropriate for narrowband techniques. We point out that although we are not employing time-harmonic solutions, the waveforms that we do chose must be solutions of Maxwell's equations. To that end, Gaussian pulses are sometimes employed, and we choose a cosine modulated Gaussian pulse expressed by the equation

$$g_0(t) = \exp\left(-0.5t^2 / u_B^2\right)\cos\left(2\pi f_C t\right) \tag{11.65}$$

as the signal supplied to the terminals of the antennas. The choice is made for mathematical convenience, and there is no loss of generality in the following descriptions for other signal shapes. The surface currents on the dipole to a feed-point current $I_T(t)$ supplied to a feed region of incremental length Δh and radius a, thus

$$I_T(t) = \Delta h 2\pi a J(0,t) \tag{11.66}$$

and $I_T(t)$ is proportional to signal $g_0(t)$ from (11.65), so that

$$I_T(t) = (\Delta h 2\pi a)\exp\left[-(\pi t B)^2 \log(e)/2\right]\cos(2\pi f_C t) \qquad (11.67)$$

and B is characterized here for the signal 10 dB bandwidth, as used in the FCC bandwidth definition for UWB. The current $I_T(t)$ is the signal supplied to the transmitting antenna, and as will be shown next, the radiated field strength is a weighted sum of time-delay derivatives of this current.

11.7.2 Far-Field of an Arbitrary UWB Antenna

C. L. Bennet and G. F. Ross give the general expression for the magnetic far-zone field as [40]

$$\mathbf{H}(r,t) = \frac{1}{4\pi rc}\int_V \frac{\partial}{\partial \tau}\mathbf{J}(\mathbf{r}',\tau)\times \mathbf{a}_r dV' \qquad (11.68)$$

The volume integral reduces to an integration over the surface current density $\mathbf{J}$, the retarded time variable is $\tau = t - R/c$, c is the speed of propagation, and $\mathbf{a}_r$ is a unit vector pointing in the direction of radiation. The retardation time R/c is explicitly the propagation time required for a disturbance to travel the distance R at velocity c. The concept becomes obscured for sine wave solutions: delayed sine waves are still sine waves. The electric far-zone field is

$$\mathbf{E} = -\eta_0 \mathbf{a}_r \times \mathbf{H} \qquad (11.69)$$

and $\eta_0 = 376.73$ is the intrinsic free space impedance. Equation (11.68) includes a partial derivative in τ, which includes both time and delay. Because of this, the radiated field signal shapes will be different from the signal shape supplied to the feed point of the antenna. *This is a key difference from narrowband solutions, where the signal shape is sinusoidal everywhere.* We should add that this time-delay derivative is not always present, as for example in some traveling wave antennas, which we will not consider here.

The fields calculated from currents on an antenna have been studied for over a century (see [10] and [11]). These solutions are exceptionally difficult, even for simple wire antennas. A complete solution will not be attempted here and is not needed for a heuristic understanding of the problem. Referring to Figure 11.13, a current is supplied to the feed point of a symmetrical wideband dipole. Radiation occurs from all parts of this antenna, as is represented by (11.68), but we will approximate this distributed source by a source at the feed point and sources at each dipole end. This three-source model very closely resembles the exact expressions for radiation from a thin wire dipole carrying a sinusoidal current distribution, as was seen in (11.63), and allows us to see the effects of time delay due to the finite size of an antenna. The approximate shape of the magnetic fields radiated from the dipole aligned with the z-axis and having effective length $2h_a$ is

$$H_\varphi(r,t) = \frac{\sin(\theta)}{4\pi rc}\frac{h_a}{2}\frac{\partial}{\partial t}\{I_z(t) + I_z(t - [1-\cos(\theta)]h_a / c) + I_z(t - [1+\cos(\theta)]h_a / c)\} \qquad (11.70)$$

where $I_z(t)$ is the dipole feed-point current. The usual spherical (r, θ, ϕ) coordinates are employed, with $\sin(\theta)$ being the projection of the z-directed current density on the θ direction at the observation point. The magnetic fields in (11.70) appear to emanate from three sources (see Figure 11.13): one at the feed point and one at each dipole end. The total magnetic field comprises time-derivative components that are delayed by t, as well as by $\cos(\theta)h_a/c$. Thus, two processes contribute to the magnetic-field signal shape, a time and distance delay. *Both processes tend to lengthen the signal in time; hence, both processes will contribute to a narrowing of the radiated bandwidth compared to the bandwidth of the signal supplied to the antenna.* The delay term $\tau = t - \cos(\theta)R/c$ in (11.68) and (11.70) means that the radiated signal shape as a function of time will vary, depending on where the observation point is measured by angle θ along the z-axis relative to the antenna.

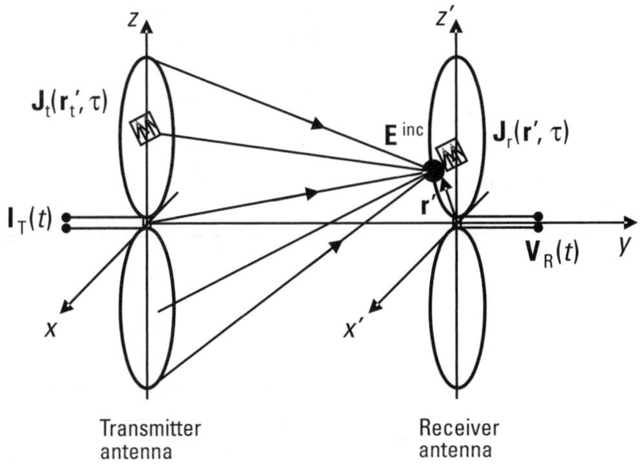

Figure 11.13 Wideband transmitting and receiving antennas.

The far-field radiated signals for an antenna feed-point current of (11.65), using an FCC-defined UWB 10-dB bandwidth of 4.5 GHz and center frequency $f_c = 4.5$ GHz at $\theta = 90°$, 45°, and 30°, are shown compared with the current at the feed point. Note that for this 100%-wide bandwidth signal, the signal shape varies with angle θ. As the observation angle θ moves toward the z-axis, the magnetic field waveform changes shape and its amplitude diminishes.

11.7.3 Receiving UWB Signals

The receiving process is most easily understood when the receiving antenna is an infinitesimal "field probe" of Δh incremental length oriented along the z-axis and $2\pi a$ in circumference. The received voltage can be deduced from the Lorentz force law which relates the force **F** exerted on a charge q by the incident electric (**E**) and magnetic (**H**) fields on an antenna having length Δh:

$$\mathbf{F} = \mathbf{E}q + q\mu_0 \mathbf{v} \times \mathbf{H} \tag{11.71}$$

The antenna velocity v here is zero, so the magnetic field term drops out. The force per charge acting over the incremental antenna length is in units of voltage, so:

$$V_R(t) = -\int_{-\Delta h/2}^{\Delta h/2} \frac{1}{q}\mathbf{F}\cdot d\mathbf{l} = -\int_{-\Delta h/2}^{\Delta h/2} \mathbf{E}^{\text{inc}} \cdot d\mathbf{l} = -E^{inc}(t)\Delta h \tag{11.72}$$

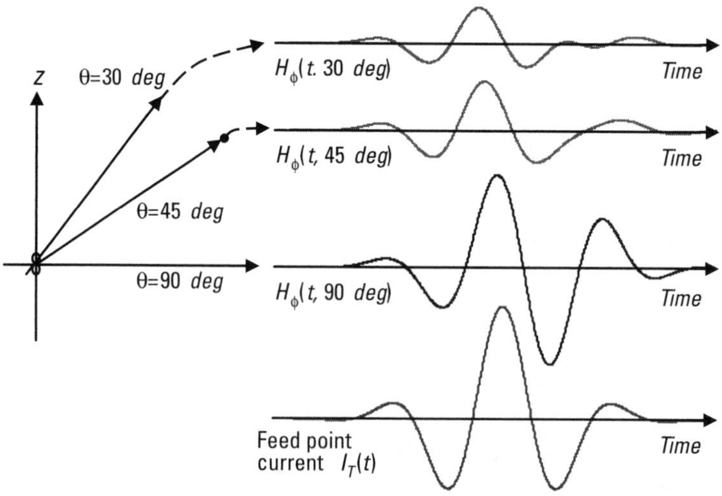

Figure 11.14 Radiated signal shapes vary with elevation angle.[[AQ: cite in text?]]

A more general result invokes the principle of reciprocity in electromagnetic theory. The principle allows us to write an expression for the voltage signal at the receive terminal of the antenna in terms of:

- The current density $\mathbf{J}(\mathbf{r}', t)$ that would be present if the receive antenna were used to transmit a feed-point current I_T;
- The electric field $\mathbf{E}^{\text{inc}}$ due to the transmit antenna that would be incident on the receive antenna location without the receive antenna present.

Figure 11.13 shows the relationships between these various terms. From the reciprocity principle, the open-circuit voltage at the receiving antenna terminals is

$$V_R(t) = \frac{-1}{I_T(t)} \int_V \mathbf{E}^{\text{inc}} \cdot \mathbf{J}(\mathbf{r}', t) dV' \tag{11.73}$$

The integration is performed over the current densities on the surface of volume V. From Figure 11.13, it is clear that the received voltage signal V_R is related to the transmitted current signal I_T through several processes that can alter the shape of the signal. First, I_T gives rise to a current density distribution $\mathbf{J}$ on the transmitting antenna. Depending on the transmitter antenna dimensions, this current density has time delays associated with it. The current densities produce the electric field $\mathbf{E}^{\text{inc}}$ at the field point on the receiving antenna that is proportional to the time-delay derivative of the transmitting antenna current density. Finally, the received voltage V_R is a summation of weighted electric field points, and another set of signal time delays is encountered in this summation due to the physical dimensions of the receiving antenna. A reasonable approximation for the voltage received by a wideband dipole is

$$\begin{aligned} V_r(t) = -h_a\big(E_z(t) + E_z(t - [1 - \cos(\theta)]h_a / c) \\ + E_z(t - [1 + \cos(\theta)]h_a / c)\big) \end{aligned} \tag{11.74}$$

Thus, the received signal V_R comprises weighted sum of time-delayed derivatives of the transmitted current I_T.

Approximations (11.70) and (11.74) are justified by measurements compared with both FDTD and simplified analyses [41]. The measurements were carried out on a 2-GHz center frequency system having a 10-dB bandwidth of about 150%. Measurements and simulations of signals in a UWB link between two diamond-shaped dipoles oriented for maximum signal transfer were performed for the geometry shown in Figure 11.15. A broadband antenna, patented by R. W. Masters, comprises two flat, conducting, radiating elements shaped like triangles and arranged in the same plane with their bases parallel [42]. They are fed symmetrically at the base. Measurements and FDTD calculations are shown in Figures 11.16 and 11.17. Figure 11.16 shows the measured and simulated pulses at the transmitter and receiver for an impulse stimulus. Figure 11.17 extends this comparison for a transmitted bipolar pulse. In both cases, there is a close correspondence between measurements and analysis. Figures 11.18 and 11.19 show a comparison of waveforms calculated using the simplified analysis and those using the FDTD method. The simplified analysis gives results that are nearly identical to the detailed FDTD calculations, and by comparison to the measurements we can conclude that the approximations of (11.70) and (11.74) are justified.

11.8 A Simple UWB Antenna

UWB antennas need not be complicated, as illustrated in the "two-cent" solution pictured in Figure 11.20. The design is essentially a fat dipole and is based on a study of elliptical element antennas [43]. The fortuitous size of the U.S. one-cent piece results in a bandwidth that coincides with the U.S. 3.1–10.6-GHz UWB band. While the practicality of coin antennas might be questioned, simplicity of design has been demonstrated.

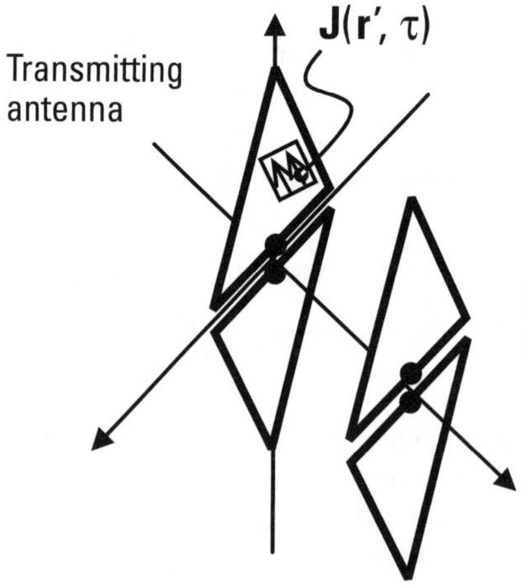

Figure 11.15 Antennas arranged for FDTD analysis. (After [37].)

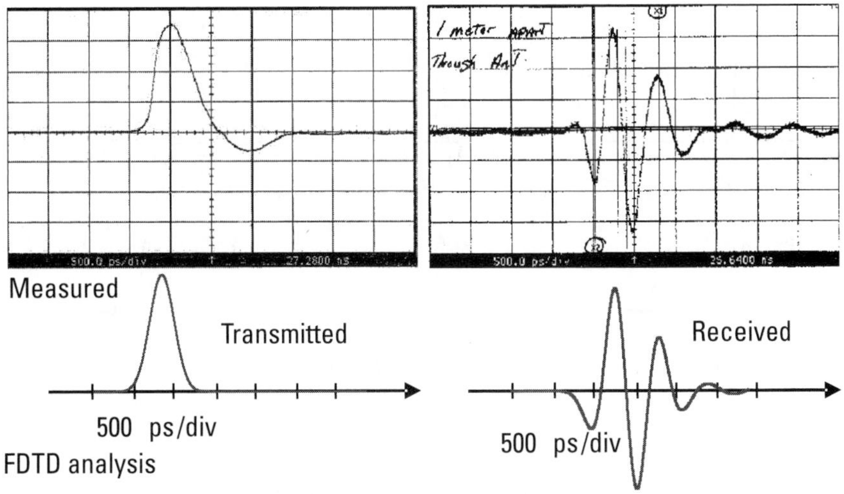

Figure 11.16 Measured and simulated pulses for a unipolar excitation current [41].

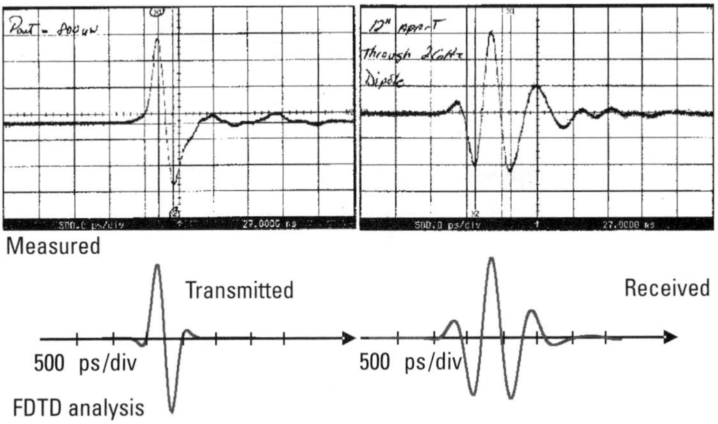

Figure 11.17 Measured and simulated pulses for a bipolar excitation current [41].

The matching is accomplished by adjusting the spacing between the two circular elements. Performance is seen in Figure 11.21. This wideband dipole has a broadside gain the exceeds 0 dBi from 3 to 9.5 GHz and is matched better than 10-dB return loss from 3 GHz to past 12 GHz. The drop-off in gain at the upper end of the band is due to losses in the semirigid coaxial feed line, which also contributes to the lower return loss above 10 GHz.

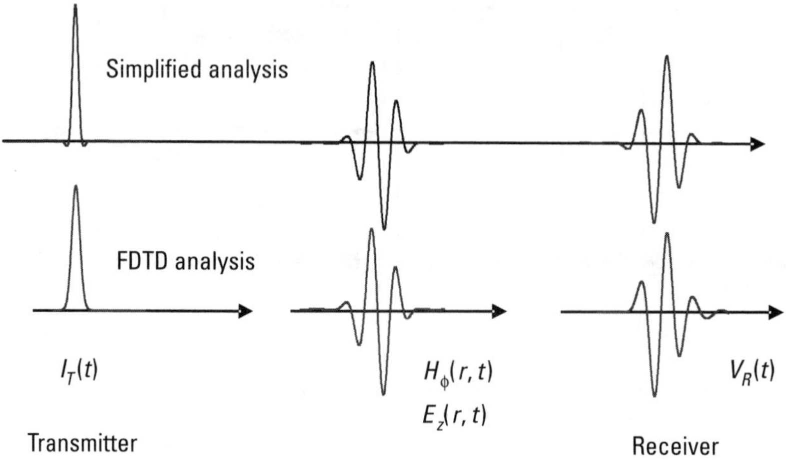

Figure 11.18 Simplified analysis compared to FDTD results for a unipolar excitation current. (After [37].)

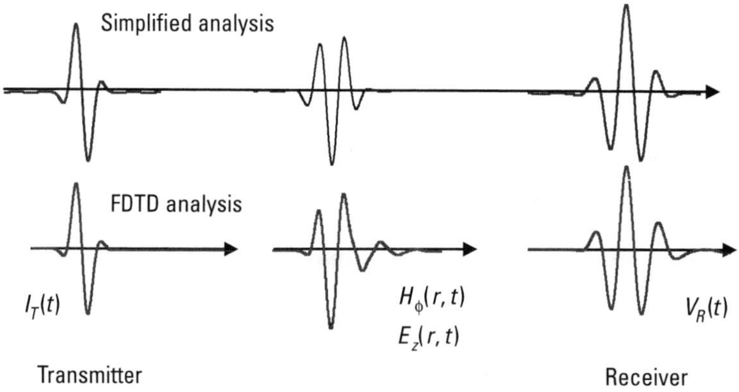

Figure 11.19 Simplified analysis compared to FDTD results for a bipolar excitation current. (After [37].)

11.9 Summary

The concept of the quality factor, Q, was defined and explored in terms of antenna performance. We explored the fundamental limitations of small antennas by relating size, Q, bandwidth, and efficiency. Small loops and dipoles were examined and found to behave as predicted based on the fundamental limitations. The near fields of a sinusoidally excited, half-wave dipole were shown to be substantially different from the near fields of an infinitesimal dipole. Next,

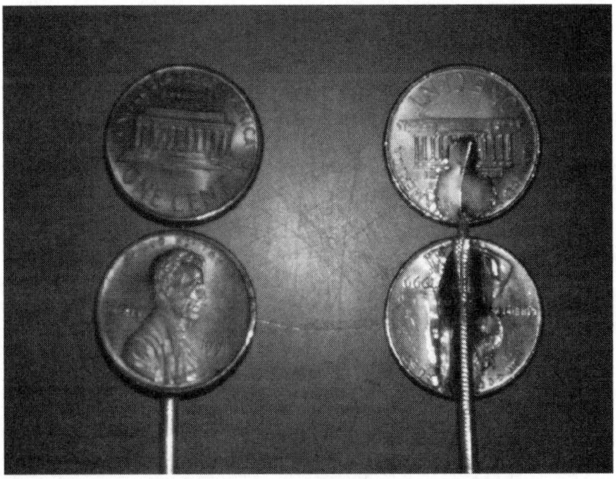

Figure 11.20 UWB dipole.

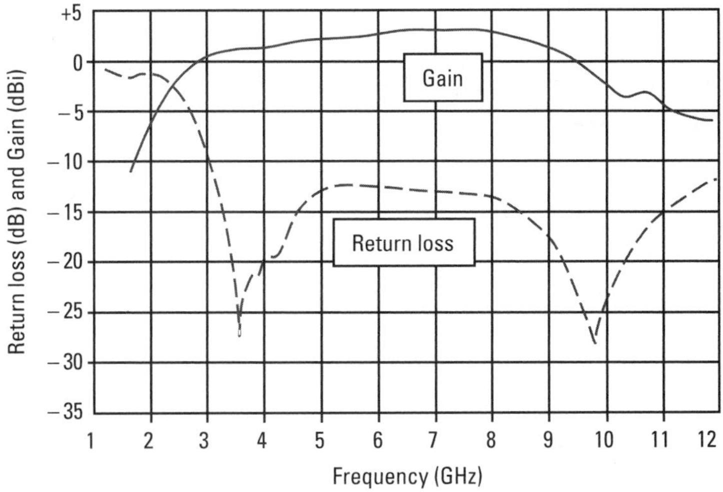

Figure 11.21 Gain and match of UWB dipole.

the ferrite antenna was examined as an electrically small antenna in terms of fundamental limitations and practical implementation. The rectangular and circular microstrip antennas were analyzed, and detailed design equations were presented. The helix-radio dipole's performance was studied. The coaxial dipole in the presence of a radio case was examined as an example of a physically small antenna that, in practice, cannot operate in the free space ideal because of unavoidable mutual coupling to the radio case. Finally, an introduction to

UWB antennas led us to an understanding of the radiation process of antennas fed by short-pulse signals.

Problems

11.1 A golf ball dropped to a solid nondeforming surface rebounds to 72% of its original height h. Energy $W = mgh$ where m is mass, and g is the acceleration due to gravity. Find Q.

Ans:

$Q = 2\pi mgh / mg(h - 0.72h) = 2\pi/(1 - 0.72) = 22.44$

11.2 The Earth has rotational potential energy equal to one-half the square of the angular velocity ω multiplied by the moment of inertia I_o, which equals $0.4\ m_e R_e^2$. Earth's rotational velocity has been slowing at about 0.00164 sec per century due mainly to tidal forces. Find the Q of Earth as a rotating body.

Ans:

$\omega_0 = 1$ rotation/day, $\omega_1 = 1 - 0.00164/(100 \times 365.24 \times 24 \times 3{,}600)$ rotation/day, and

$$Q_{Earth} = \frac{2\pi}{1 - \dfrac{0.5 I_0 \omega_1^2}{0.5 I_0 \omega_0}}$$

So, $Q_{Earth} = 6.045 \times 10^{12}$.

11.3 A dielectric has the properties

$$\varepsilon_r = 4.9 + \frac{73.4}{1 + jf/f_0}$$

Find Q at $f = f_0$.

Ans:

$Q = |\mathrm{Im}(\varepsilon_r)|/\mathrm{Re}(\varepsilon_r) = 0.882$

11.4 A miniature antenna with $kr = 0.39$ excites modes $n = 1$, 3, and 5 with TM modal amplitudes 1, 10^{-3}, and 10^{-6} and TE modal amplitudes 1, 10^{-4}, and 10^{-7}. Find the antenna Q, and estimate the antenna pattern if the electric and magnetic sources are superimposed.

Ans:

Using (11.10) and Table 11.2 for Q_{TE} and for Q_{TM}, $Q = 1/(1/\ Q_{TE} + 1/\ Q_{TM})$.

The antenna pattern is primarily that of the dominant $n = 1$ mode, $\sin(\theta)$. If the electric and magnetic current element sources are superimposed, the pattern is $\sin(\theta)$, and the polarization is circular everywhere.

11.5 A new ferrite material with $\mu = 500$ and loss factor $\tan\delta = 0.005$ is for a ferrite rod loop antenna 1 cm in diameter and 10 cm long at 1.7 MHz. The antenna wire windings and resonating capacitor have $Q = 300$. Estimate the antenna Q, efficiency, and the 3-dB bandwidth.

11.6 Two antennas of Problem 11.5 are used for a cordless telephone link at 1.7 MHz. If the transmitted power is 100 mW and the receiver sensitivity is 0.2 μV referenced to 50Ω (–121 dBm), find the maximum free space range of the telephone. What additional factors would affect the actual range?

11.7 A loop antenna with loop reactance 500Ω and conductor resistance 1.2Ù is series-resonated by a capacitor with a Q of 200. If the radiation resistance is 2Ω, find the loaded Q and the radiation efficiency.

Ans:

$Q_{loop} = X/(R_{rad} + R_{loss})$; $Q_{loaded} = 0.5/(1/\ Q_{cap} + 1/\ Q_{loop}) = 43.86$

Efficiency = $2\,Q_{loaded}/\ Q_{radiation} = 87.72/(500/2) = 0.351$.

11.8 The loop of Problem 11.7 is supplied with 150W. Find the peak capacitor voltage and the dissipated power.

Ans:

From (11.19), $Vp = \sqrt{(2 \times 500 \times 43.86 \times 150)} = 2{,}565\text{V}$.

11.9 Consider the system in Figure 11.P1 of two capacitors, $C_1 = 3$ pF and $C_2 = 1$ pF, connected in a series arrangement with a voltage source of $V_1 = 1{,}000$V as in Figure 11.P1(a). After the system has reached steady state, the circuit is reassembled instantaneously, with the voltage source absent and with the capacitors arranged in parallel as in Figure 11.P1(b), connected to form a loop 2.5 cm in diameter from 4-mm-diameter copper wire.

Find the Q of the loop, the radiated electric field, and the voltage induced in the matched 50Ω load of an ideally oriented, resonant, half-wave dipole 3m away.

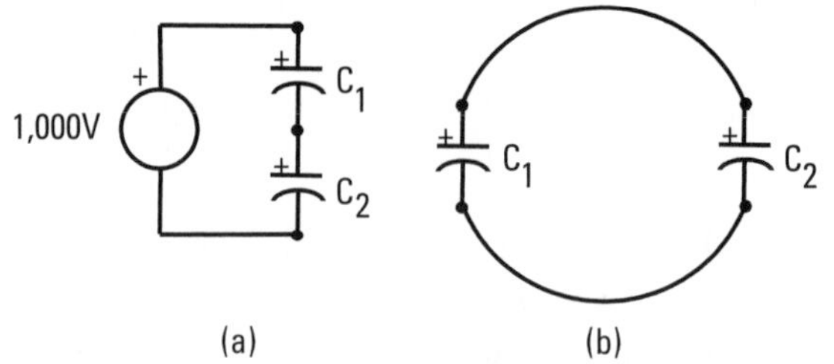

Figure 11.P1 Two capacitors are (a) charged and (b) discharged.

Ans:

P11-9.mcd.

11.10 A loop 0.5m in diameter and constructed from 2-cm-diameter copper wire is gap-resonated with a 75-pF capacitor and subjected to a 10-V/m field at the resonant frequency. Find the resonant frequency and the peak voltage across the capacitor.

Ans:

P11-10.mcd.

11.11 A channel 6 TV transmitter with a video carrier at 83.25 MHz produces a 122-dBμV/m field at a distance of 1.5 km. Find (a) the power delivered to the matched load of a small, 50% efficient, series resonant loop with an unloaded Q of 400, and (b) the peak resonating capacitor voltage when the loop is unloaded.

Ans:

(a) $E = 1.259$ V/m, $P_r = (0.5)1.5(\lambda^2/4\pi)E^2/\eta_0 = 3.2$ mW; (b) $V = 25.2$V.

11.12 A channel 69 TV transmitter with a video carrier at 801.25 MHz produces a 136-dBμV/m field at a distance of 1.5 km. Find (a) the power delivered to the matched load of a small, 50% efficient, series resonant loop with an unloaded Q of 150, and (b) the peak resonating capacitor voltage when the loop is unloaded.

Ans:

(a) $E = 6.31$ V/m, $P_r = (0.5)1.5(\lambda^2/4\pi)E^2/\eta_0 = 0.86$ mW; (b) $V = 10.1$V.

11.13 A quarter-wavelength patch antenna is a half-wave-wide-by-quarter-wave-long conductor over a ground plane. One half-wave edge is grounded; the other forms a radiating slot. Compute the directivity by emulating the structure with a dipole slightly above a conducting half-space.

Ans:

P11-13.mcd. $G = 5.16$ dBi.

11.14 Emulate a half-wave-by-half-wavelength patch using two resonant, parallel, half-wavelength dipoles, each parallel to a conducting half-space. Use mutual-coupling theory to estimate the gain of the model as a function of separation between the dipoles and the half-space.

11.15 A received short-duration pulse signal that was emitted by a UWB dipole is different in shape from the signal fed to the transmitting dipole. Why? What causes the distortion?

Ans:

The emitted fields resemble a weighted sum of time-delay derivatives of the transmitter antenna feed-point signal. The voltage at the receiver dipole terminals is an integral of time-delayed copies of that field-strength signal. Different angular relationships between the transmitting and receiving dipoles will produce different receiver voltage shapes.

11.16 The receiver signal shape is different from the transmitted signal shape in UWB short-pulse systems. Can such UWB links be considered linear?

Ans:

A system is linear if a signal increase at the input (transmit antenna terminals) results in a proportional increase at the output (receiver antenna terminals). The input and output signal shapes may be different.

References

[1] Green, E. I., "The story of *Q*," *American Scientist*, Vol. 43, October 1955, pp. 584–594.

[2] Siwiak, K., *Radio Wave Propagation and Antennas for Portable Communications*, Workshop Notes, Taipei, Taiwan, Republic of China, October 5–7, 1993.

[3] Chu, L. J., "Physical Limitations of Omni-directional Antennas," *J. Appl. Phys.*, Vol. 19, December 1948, pp. 1163–1175.

[4] Wheeler, H. A., "Fundamental Limitations of Small Antennas," *Proceedings of the IRE*, Vol. 35, No. 12, December 1947, pp. 1479–1484.

[5] Wheeler, H. A., "Small Antennas," *IEEE Transactions on Antennas and Propagation*, Vol. AP-23, No. 4, July 1975, pp. 462–469.

[6] McLean, J. S., "A Re-examination of the Fundamental Limits on the Radiation Q of Electrically Small Antennas," *IEEE Transactions on Antennas an Propagation*, Vol. 44, No. 5, May 1996, pp. 672–675.

[7] Schantz, H., *The Art and Science of Ultrawideband Antennas*, Norwood, MA: Artech House, 2005.

[8] Harrington, R. F., "Effect of Antenna Size on Gain, Bandwidth, and Efficiency," *J. Research Nat. Bur. Stand., D. Radio Prop.*, Vol. 64D, No. 1, January-February 1960, pp. 1–12.

[9] Harrington, R. F., *Time Harmonic Electromagnetic Fields*, New York: McGraw-Hill, 1961.

[10] Pocklington, H. C., "Electrical Oscillations in Wires," *Proc. of the Cambridge Phys. Soc.*, Vol. 9, 1897, pp. 324–333.

[11] Hallén, E., "Theoretical Investigation into Transmitting and Receiving Qualities of Antennae," *Nova Acta Regiae Soc. Ser. Upps.*, Vol. 2, November 4, 1938, pp. 1–44.

[12] Balzano, Q., and K. Siwiak, "Radiation of Annular Antennas," *Correlations* (Motorola Engineering Bulletin, Schaumburg, IL), Volume 6, No. 2, Winter 1987.

[13] Siwiak, K., and Q. Balzano, "Radiation from Annular Antennas," *1986 IEEE/VTS Conference*, IEEE Cat. No. 86CH2308-5, 1986, pp. 15–25.

[14] Balzano, Q., and K. Siwiak, "The Near Field of Annular Antennas," *IEEE Transactions on Vehicular Technology*, Vol. VT36, No. 4, November 1987, pp. 173–183.

[15] Siwiak, K., et al., "Evaluation of the Electric Field in Close Proximity to Current Sources," *Third Annual Conference of the Bioelectromagnetics Society*, Washington, D.C., August 9–12, 1981.

[16] Balzano, Q., O. Garay, and K. Siwiak, "The Near Field of Dipole Antennas, Part I: Theory," *IEEE Transactions on Vehicular Technology*, Vol. VT-30, No. 4, November 1981, pp. 161–174.

[17] Balzano, Q., O. Garay, and K. Siwiak, "The Near Field of Dipole Antennas, Part II: Experimental Results," *IEEE Transactions on Vehicular Technology*, Vol. VT-30, No. 4, November 1981, pp. 175–181.

[18] Balzano, Q., O. Garay, and K. Siwiak, "The Near Field of Omnidirectional Helices," *IEEE Transactions on Vehicular Technology*, Vol. VT-31, No. 4, November 1982, pp. 173–185.

[19] Siwiak, K., Q. Balzano, and O. Garay, "The Near Field of Dipole and Helical Antennas," *1981 Meeting of the Vehicular Technology Society*, Washington, D.C., April 1981.

[20] King, R. W. P., and C. W. Harrison Jr., *Antennas and Waves: A Modern Approach*, Cambridge, MA: MIT Press, 1969.

[21] Dunlavy, J. H., Jr., "Wide Range Tunable Transmitting Loop," *U.S. Patent 3,588,905*, June 28, 1971.

[22] Hart, T., "Small, High-Efficiency Loop Antennas," *QST*, June 1986, pp. 33–36.

[23] Burke, G. J., and A. J. Poggio, "Numerical Electromagnetics Code (NEC)—Method of Moments," *Lawrence Livermore Laboratory, NOSC Technical Document 116 (TD 116)*, Vol. 1, Vol. 2, January 1981.

[24] Balanis, C. A., *Advanced Engineering Electromagnetics*, New York: John Wiley and Sons, 1989.

[25] Jordan, E. C., and K. G. Balmain, *Electromagnetic Waves and Radiating Systems*, 2nd ed., Upper Saddle River, NJ: Prentice Hall, 1968.

[26] Hirasawa, K., and M. Haneishi, *Analysis, Design, and Measurement of Small and Low-Profile Antennas*, Norwood, MA: Artech House, 1992.

[27] Fujimoto, K., and J. R. James, (eds.), *Mobile Antenna Systems Handbook*, Norwood, MA: Artech House, 1994.

[28] James, J. R., and P. S. Hall, (eds.), *Handbook of Microstrip Antennas*, Vol. 1, London: Peter Peregrinus, 1989.

[29] James, J. R., and P. S. Hall, (eds.), *Handbook of Microstrip Antennas*, Vol. 2, London: Peter Peregrinus, Ltd., 1989.

[30] Bahl, I. J., and P. Bhartia, *Microstrip Antennas*, Norwood, MA: Artech House, 1980.

[31] Collin, R. E., *Antennas and Radiowave Propagation*, New York: McGraw-Hill, 1985.

[32] Wood, C., "Analysis of Microstrip Circular Patch Antennas," *IEE Proceedings*, 128H, 1981, pp. 69–76.

[33] Kashiwa, T., and I. Fukai, "Analysis of Microstrip Antennas on a Curved Surface Using Conformal Grids FD-TD method," *IEEE Transactions on Antennas and Propagation*, Vol. 42, No. 3, March 1994, pp. 423–427.

[34] Chuang, H.-R., "Human Operator Coupling Effects on Radiation Characteristics of a Portable Communication Dipole Antenna," *IEEE Transactions on Antennas and Propagation*, Vol. 42, No. 4, April 1994, pp. 556–560.

[35] Tofgård, J., S. N. Hornsleth, and J. B. Andersen, "Effects on Portable Antennas of the Presence of a Person," *IEEE Transactions on Antennas and Propagation*, Vol. 41, No. 6, June 1993, pp. 739–746.

[36] Hill, C., and T. Kneisel, "Portable Radio Antenna Performance in the 150, 450, 800, and 900 MHz Bands "Outside" and In-Vehicle," *IEEE Transactions on Vehicular Technology*, Vol. VT-40, No. 4, November 1991, pp. 750–756.

[37] Siwiak, K., and D. McKeown, *Ultra-wideband Radio Technology*, London: John Wiley and Sons, 2004.

[38] Henning, H. F., "A General Concept of Frequency and Some Applications," *IEEE Transaction on Information Theory*, Vol. IT14, No. 3, May 1968, pp. 375–381.

[39] U.S. 47 C.F.R. Part 15(f), Ultra-Wideband Operations, FCC Report and Order, April 22, 2002.

[40] Bennett, C. L., and G. F. Ross, "Time-Domain Electromagnetics and Its Applications," *Proceedings of the IEEE*, Vol. 66, No. 3, March 1978, pp. 299–318.

[41] Siwiak, K., T. M. Babij, and Z. Yang, "FDTD Simulations of Ultra-wideband Impulse Transmissions," *IEEE Radio and Wireless Conference: RAWCON2001,* Boston, MA, August 19–22, 2001, available at http://rawcon.org.

[42] Masters, R. W., "Antenna," *U.S. Patent 2,430,353,* November 4, 1947.

[43] Schantz, H., "Planar Elliptical Element Ultra-wideband Dipole Antennas," *2002 IEEE Antennas and Propagation Society International Symposium,* San Antonio, TX, June 16–21, 2002, Vol. 3, pp. 16–21.

12

Radio Communications System Designs

12.1 Introduction

The design of a personal communications system is essentially the problem of specifying a desired link performance criterion, manipulating the available parameters to ensure that the criterion is satisfied within the boundaries of a specified geographic region, and proving that the design coverage was obtained. For one-way devices, the parameters available are fixed-site effective radiated power referenced to an isotropic source (EIRP), propagation path-loss predictions, statistical descriptions of the signal in the vicinity of the personal communications device (PCD), and the field sensitivity of the personal communications receiver in the presence of the noise and interference at the design point. For two-way devices, the return path can be found from reciprocity, except the noise at the fixed site is not necessarily the same as the noise at the PCD. In Chapter 2, fixed-site antennas were described, including pattern distortions, and in Chapter 3, transmission line losses were analyzed; hence, EIRP can be found at the fixed site. Chapters 5, 6, and 7 treated various scenarios of radio wave propagation at frequencies identified in Chapter 4 as appropriate for PCD systems, so the path attenuation can be specified when the geometry of the problem is given. Chapter 4 additionally introduced signaling protocols because the signaling methods specifically address certain radio channel propagation distortion effects. A statistical description of waves subjected to diffractions and reflections was described in Chapter 5 for orbiting satellite systems and in Chapter 8 for terrestrial systems. A design strategy based on the statistical description of waves was also introduced in Chapter 8, along with the effects of simulcasting and the diversity reception of signals. The coupling of waves to a body-mounted PCD was examined in Chapters 9 and 10 for small antennas, presented in

Chapter 11. In this chapter, we will explore the design of PCD systems to highlight some of the design tools and parameters available for effecting successful and economical solutions.

Since receiver sensitivity is really a specification of the signal-to-noise ratio, the noise in the vicinity of receivers is first examined so that a suitable sensitivity figure can be determined. Next, the statistical description of waves is applied to arrive at a median or mean field strength required for a desired link performance. The system design parameters depend on the nature of the system. For one-way or broadcast services like radio paging, site design and simulcast transmissions (multiple transmitters radiating the same signal on essentially the same frequency and at essentially the same time), if applicable, are then specified to ensure that the required field strength is met or exceeded, with the needed reliability in the coverage area of the PCD. For two-way systems, various forms of antenna diversity may be employed. In two-way systems, the fixed site can use high transmitter power to close the link. However, the PCD is generally a handheld device operating from battery power; hence, its transmitter EIRP is limited, so diversity receiver antennas may be used at the fixed end of the link to help balance the up- and downlinks. Finally, the designs are proven by statistically sampling field-strength measurements within the coverage area.

12.2 Noise

The performance of a personal communications radio system is inevitably tied to the ratio of desired signal to noise. Noise in this context includes thermal noise caused by random vibrations of charged carriers in a lossy conductor, "man-made" and meteorological noise, and unwanted signals. In personal communications applications, thermal and man-made noise play the most important roles in large-scale systems, while unwanted interfering signals play the most critical role in small and microcellular systems. The former are usually called *propagation limited systems*, and the later are often called *interference limited systems.* The distinction can be important, as often receiver sensitivity is a costly parameter that is critical in propagation limited systems (such as in satellite links) but somewhat less important in interference limited systems.

12.2.1 Thermal Noise

A resistive element is a two-terminal device that may be characterized by its resistance *R*. This resistance may contain charged carriers (chiefly electrons) that have random motion, which causes a random voltage to appear across the resistance. The model of a resistor is, then, an equivalent noise voltage source in series with the resistance *R*. From quantum mechanics, it can be shown that the equivalent noise power in watts per Hertz generated in any ideal coherent amplifier of electromagnetic waves is given by [1]

$$N_o = hf\left[\frac{1}{e^{hf/k_bT}-1}+1\right] \tag{12.1}$$

where h in J·sec is the Plank constant, k_b in J/K is the Boltzmann constant, f is frequency in hertz, and T is the absolute temperature (K) of the resistance. The first term in the brackets of (12.1) is black-body radiation in a single propagation mode, while the second term, +1, arises from spontaneous emission in the amplifier. When frequency f is small enough ($<10^{10}$ Hz), the exponential can be replaced by 1 plus its argument, and (12.1) reduces to

$$N_o = k_bT \tag{12.2}$$

It has become customary to define the system noise temperature in terms of an equivalent noise temperature defined as N_o/k_b. Room temperature is 290K, while 2.726K is the isotropic background remnant of the Big Bang.

12.2.2 Noise and Noise Temperature in the Radio Spectrum

Noise levels encountered in the radio spectrum are shown in Figure 12.1. Various noise sources are shown, including atmospheric-noise (dashed) curves A, B,

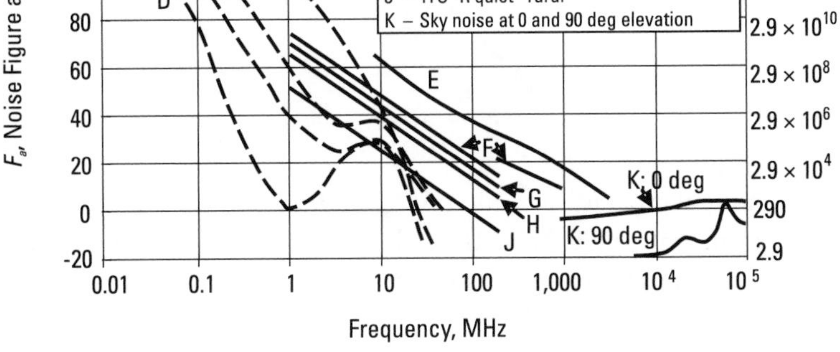

Figure 12.1 Natural and man-made noise.

Table 12.1
Radio Noise Coefficients

Radio Environment	c	d	Notes
Business	44.3	12.3	$200 < f < 900$ MHz
Industrial/business	76.8	27.7	$f < 250$ MHz
Residential	72.5	27.7	$f < 250$ MHz
Rural	67.2	27.7	$f < 250$ MHz
Quiet rural	53.6	28.6	$f < 250$ MHz
Galactic noise	52.0	23.0	—

C, D, and man-made-noise (solid) curves E, F, G, H, J, and K for sky noise at 0 and 90 degrees elevation . The equivalent noise figure is defined as

$$F_a = c - d \log(f) \tag{12.3}$$

where F_a is in decibels above kT_0, and coefficients c and d are from Table 12.1. The table values for industrial, residential, rural, and quiet rural areas are from Rec. ITU-R P.372-8, and the estimates of galactic (cosmic) noise are obtained from CCIR Report 258-4(1982).

Figure 12.1 shows the noise figure on the left ordinate and the noise temperature equivalent on the right ordinate, as related by (12.2). Generally, and as shown earlier by Figure 4.6, noise deceases with increasing frequency, at least up to a few gigahertz, then atmospheric effects due to water, oxygen (curves K of Figure 12.1), and other gases begin to take effect. Our primary focus is from about 100 MHz to just above 10 GHz, where the man-made-noise levels tend to decrease with increasing frequency, as seen in Figure 12.1.

The general form of the expression for equivalent noise temperature in a cascaded receiver system is [2]

$$T_{equiv} = T_1 + \frac{T_2}{G_1} + \frac{T_3}{G_1 G_2} + \frac{T_4}{G_1 G_2 G_3} + \cdots \tag{12.4}$$

where T_n is noise temperature of the nth stage, and G_n is the gain (which can be less than 1) of the nth stage. Since the noise temperature and the noise figure F_e referenced to temperature T_o are related by

$$F_e = \frac{T_e}{T_o} + 1 \tag{12.5}$$

the corresponding overall noise figure is then given by the Friis noise formula:

$$F_{total} = F_1 + \frac{F_2 - 1}{G_1} + \frac{F_3 - 1}{G_1 G_2} + \frac{F_4 - 1}{G_1 G_2 G_3} + \cdots \tag{12.6}$$

where F_n are the noise figures of each stage.

In a cascaded PCD receiver system, the equivalent noise temperature is found using (12.4) to (12.6) in terms of the noise temperature T_{view} in the field of view of the PCD antenna, the overall noise temperature T_{rx} of the receiver, and the antenna losses, A_{loss}. So,

$$T_{equiv} = T_{view} + \frac{T_{rx}}{A_{loss}} \tag{12.7}$$

The antenna loss is here expressed as an efficiency factor less than unity. For example, if antenna losses are 3 dB, $A_{loss} = 0.5$. In a PCD application under Rayleigh faded conditions, the antenna integrates waves weighted by the antenna pattern with a net average directivity of 1. The factor G/T often employed in the design of satellite-based systems is therefore given from (12.4) by $1/T_{equiv}$. In fact, for PCD systems employing Earth-orbiting satellites, (12.7) can be applied with T_{view} just over 290K, as suggested in Figure 12.1. Another useful quantity in PCD applications is the total noise figure,

$$F_{total} = 1 + \frac{T_{view}}{T_o} + \frac{F_{rx} - 1}{A_{loss}} \tag{12.8}$$

written in terms of F_{rx}, the receiver noise figure. Since at the lower frequencies (especially below 130 MHz) T_{view} can be substantially greater than the "room" temperature of 290K, the total system noise figure can easily be dominated by the external noise. At these lower frequencies, small PCD antennas also have substantial losses ($A_{loss} << 1$), so a poor receiver noise figure can be tolerated, especially if it can be traded for another receiver parameter, such as intermodulation immunity. Figure 12.2 shows a typical average gain realized by a PCD antenna at various frequencies, in this case a paging receiver mounted at belt level on people compared with the antenna efficiency. The bump near 60 MHz, as seen earlier in Figure 10.15, is due to the body-enhancement effect for magnetic field antennas. The average enhancement vanishes near 600 MHz and

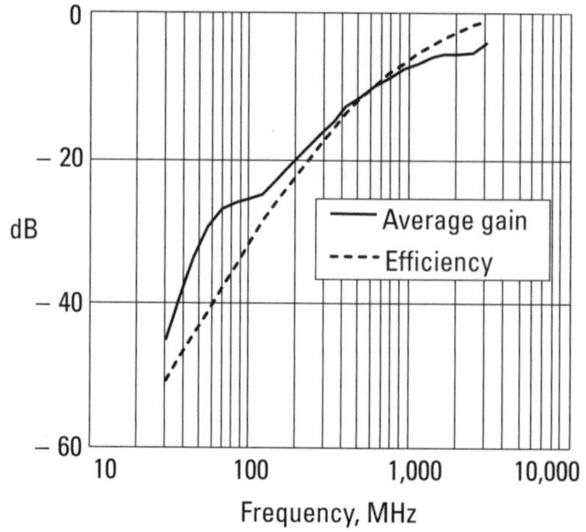

Figure 12.2 Average gain of a body-worn PDC antenna and antenna efficiency.

becomes a slight loss above that frequency. As is typical of small antennas, efficiency decreases rapidly with decreasing frequency.

12.2.3 Noise Asymmetry in Two-Way and Mobile Systems

In two-way applications, it is important to note that the ambient noise level may differ substantially at the PCD location compared with the fixed-site location. Fixed sites often mix receiving antennas with transmitter antennas, so ambient noise at fixed sites can be quite high. Because high-power transmitters may be used, intermodulation products affecting fixed-site receivers can be a serious problem. Consequently, some receiver designs used in fixed-site applications employ passive mixers at the receiver input, which results in receiver noise figures in the range of 7 to 10 dB.

Modern mobile phones are marvels of miniaturization and user features. The digital processing and microprocessor circuitry dominate the designs. That often comes with the cost of self-generated receiver noise that can reduce the effective receiver sensitivity. Recent measurements of mobile phone sensitivities reveal that this desensitization can be as much as 6 to 7 dB for devices in the 1.7–1.9-GHz PCS bands [3].

12.3 Designing a Messaging System Downlink

The designer of a radio system fixed-site transmitter attempts to put enough signal where it is needed with enough isolation from one geographical area to another so that frequencies may be reused on a geographical basis. The tools at the disposal of the system designer are propagation predictions, as well as antenna patterns controlled in azimuth for zone control and in elevation for propagation power law control. For example, a 930-MHz paging and messaging system is designed so that there is a 95% probability that a message will be received on the first floor of a coverage-area building on the contour of coverage. A similar design example could be inferred for a ring alert in a mobile phone system. The coverage contour and the locations of three available sites are shown in Figure 12.3. It is further desired to minimize radiation to the southwest and southeast of Site 0. The available antenna heights are 400m at Site 0, 120m at Site 1, and 100m at Site 2. The three sites will simulcast the transmissions. The methods described in Chapter 9, along with the average gain shown in Figure 12.2, are used to arrive at a design paging field-strength sensitivity of 25 dBμV/m for the particular paging receiver and message. The sensitivity figure is for a call success rate of P_{sens} = 0.99. Suitable sensitivity figures for handheld two-way radios can be found in [4] and were seen in Table 11.4. Coverage into "medium-sized" buildings (according to Table 7.5) is desired so, using Figures 7.18 and 7.19, the loss and standard deviation to consider are 11 dB and 6 dB, respectively. The modified Hata propagation model described in Chapter 7 is to be employed, and the statistical wave description approach [5] will be used to calculate the system performance.

12.3.1 Fixed-Site Antenna Radiation Patterns

The example design uses three sites. Site 0 is a high site with a directional antenna typified by the flat panel patterns shown Figure 2.3. Site 1 is the top of

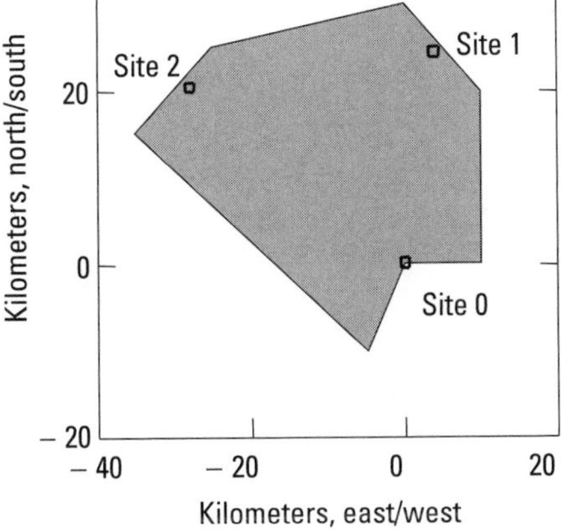

Figure 12.3 Desired area of coverage and site locations.

a building, similar to that shown in Figure 2.14, and Site 2 is a tower-mounted omnidirectional antenna, as depicted in Figure 2.12. The radiation gain patterns from the three sites are computed using the methods of Chapter 2. Figure 12.4 shows the three site radiation patterns, arranged approximately in the relative geographical positions of the three sites. The pattern from Site 0 is shown with no pattern down-tilt. The influence of antennas proximate to the omnidirectional antenna at Site 1 is readily evident, as is the distortion to the tower-mounted omnidirectional antenna of Site 2.

The directional antenna at Site 0 is initially pointed to ϕ = 135° and is tilted down 8° so as to control excess radiation in the northwest direction. The gain pattern in the geographic area from that antenna orientation is shown in Figure 12.5. The peak gain of 16 dBi reaches the ground at a distance of about 400/tan(8°) = 2,900m. The down-tilt in this example is so severe that the antenna side-lobe gain of 5 dBi can be seen in the extreme northwest corner of the coverage zone.

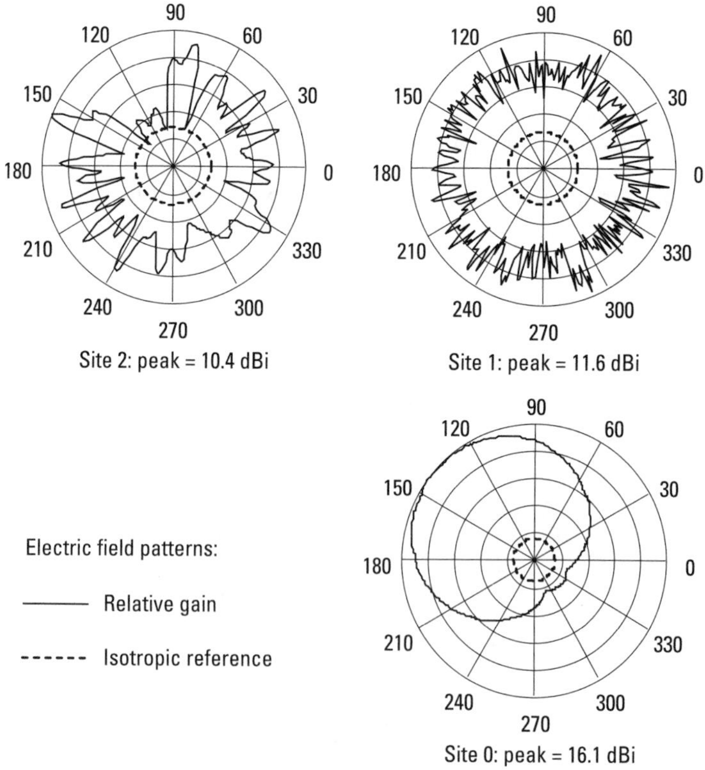

Figure 12.4 Gain amplitude patterns at the three sites.

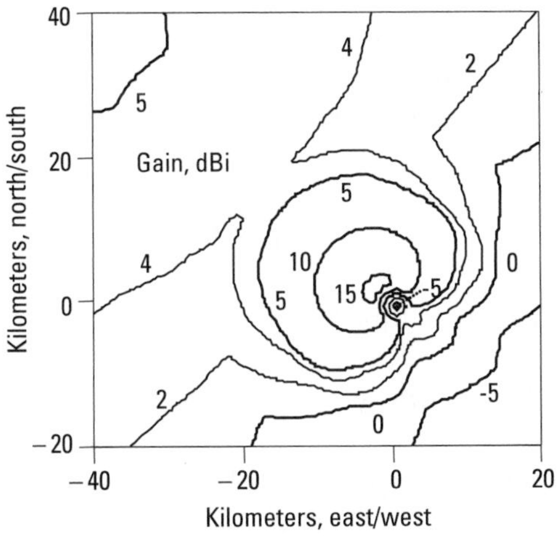

Figure 12.5 Gain footprint (dBi) of an antenna at Site 0 tilted down 8°.

12.3.2 Applying the Statistical Description of Waves

The statistical design method of G. Hagn [5], using the statistical description of waves from Chapter 8, is now applied to calculate the coverage probability in the coverage area. By design, x = 95% coverage is desired for medium-sized buildings at the edge of the coverage area. As will be shown later, this will result in a coverage probability averaged over the area that is higher than 95%. Using the Gaussian approximation (8.18) for the signal statistics, the calling probability P_s is

$$P_s = \frac{1}{2}\left[\operatorname{erf}\left[\frac{z}{\sqrt{2}}\right] + 1\right] P_{sens} \tag{12.9}$$

where the probability of paging sensitivity, or the calling success rate (CSR), is P_{sens}. The factor z is found from (8.18) in terms of the required signal margin M dB of the median signal level compared with the signal required for a calling probability of P_{sens}:

$$z = \frac{M - \sum L_i}{\sigma_{total}} \tag{12.10}$$

where terms in the summation include all losses (11 dB of building losses in the example), and the denominator is the total standard deviation, in decibels, given by (8.19). In the design example, the standard deviation includes a 7.5-dB factor for Rayleigh fading, 4.1 dB for lognormal shadowing, and 6 dB for the building variation. The margin of M dB is the difference between field strength S_r calculated using the propagation model and the required field strength S (25 dBμV/m in the example):

$$M = S_r - S \tag{12.11}$$

The field strength S_r, in (dBμV/m), is

$$S_r = (EIRP) + L + [77.216 + 20\log(f)] \tag{12.12}$$

with frequency f in megahertz and where EIRP, a function of elevation and azimuth angles, is the effective radiated power relative to isotropic radiation in decibels relative to 1 mW (dBm), including the antenna pattern. L is the propagation loss according to a given model, for example, by (7.23) with parameters in Table 12.2, but other models, such as those given in [6–10] and those in Chapter 7, can be used. The term in brackets is the conversion to field strength in decibels above 1 μV/m (dBμV/m).

[12-3a.mcd] Site parameters for a 930-MHz system are as follows: site A transmits from a 400m height with 46-dBm EIRP; site B transmits from 100m with 37 dBm. Calculate the field strength at a 1m height using the modified Hata model for distances up to 40 km.

Figure 12.6 shows the median field strength calculated for each of the three sites along the radial representing the nominal peak gain of the antenna. The power levels at the inputs to the antennas are 30 dBm, 37 dBm, and 37 dBm for sites 0, 1, and 2, respectively. The field strength from site 0 fluctuates with distance because the antenna is tilted downward and the gain pattern on the ground, as shown in Figure 12.5, is not uniform.

12.3.3 Link Margins for Specified Performance

The minimum required margin given the standard deviation and the losses in the design example is found by solving (12.9) for z, given $P_s = 0.95$, which yields $z = 1.74$. Next, the minimum M is found from (12.10), given the losses and total standard deviation. For the example, the minimum $M = 29.2$ dB, of which 18.2 dB is required statistically because of the standard deviation. The minimum desired design field strength for coverage from a single transmitter is therefore $S + 29.2 = 54.2$ dBμV/m. If the three transmitter sites are considered individually, the probability of coverage is then shown in Figure 12.7(a). Coverage was calculated at 0.5-km intervals, or in 0.5-by-0.5-km square grids. These grids are later used to compute the average probability of coverage and in

Table 12.2
Parameters for the Modified Hata Model

Parameter	Definition	Value: (site 0, site 1, site 2)
L_{mh}	Modified Hata propagation, median (dB)	—
H_b	Base antenna height (m)	(400, 120, 100)
H_m	Mobile antenna height (m)	1
U	0 = small/medium, 1 = large city	0.24
U_r	0 = open area, 0.5 = suburban, 1 = urban area	0
B_l	Percentage of buildings on the land (B_l = 15.849 nominally)	15.85
d	Range, not beyond radio horizon (km)	0.5–40
f	Frequency (MHz)	930

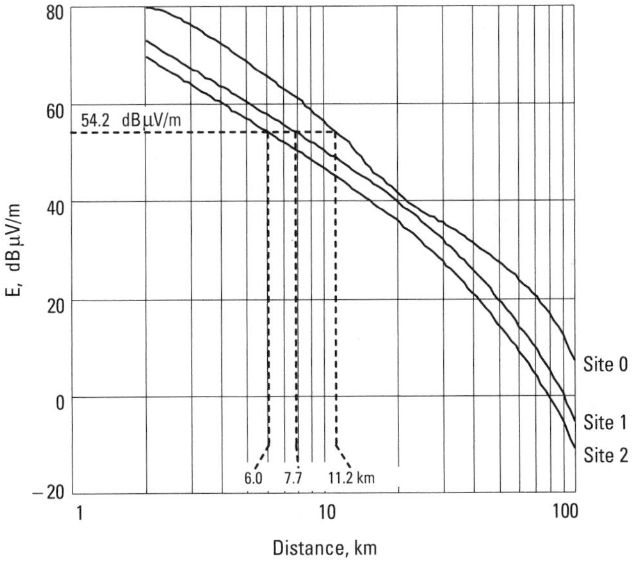

Figure 12.6 Median field strength from sites 0, 1, and 2.

Figure 12.7 to generate contours. The target area is obviously not covered with the required call success probability. The transmitters at the three sites, however, are in simulcast operation, so the actual coverage probability at each geographic point is given by (8.22) since signals from the three sites are uncorrelated as long as simulcast differential-delay interference can be ignored. Figure 12.7(b) shows

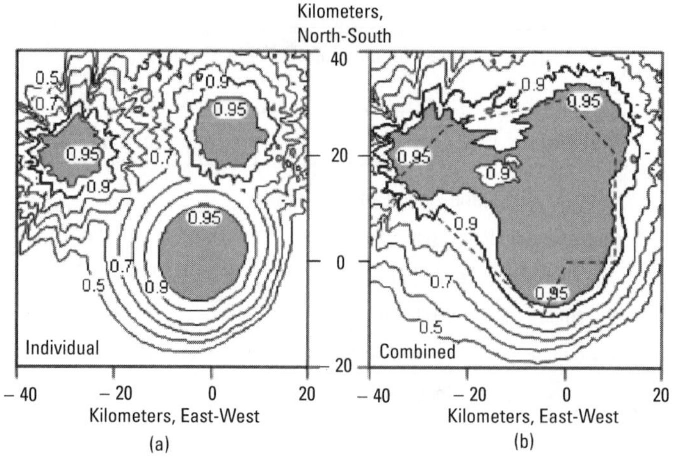

Figure 12.7 (a) Individual and (b) combined coverage probability with 8° down-tilt from site 0.

the combined coverage probability when simulcast differential delay can be ignored. The consequence of the severe down-tilt of the antenna in Site 0 results in inadequate gain, so there is a slight coverage hole nearly centered among the sites.

[12-3b.mcd] Simulcasting sites A and B are 34.7 km apart. Site A transmits from a 400m height with 46-dBm EIRP; site 2 transmits from 100m with 37 dBm. Using the modified Hata model, for a personal communications device with field-strength sensitivity of 25 dBμV/m within buildings having a median 10-dB loss, find (1) the probability of coverage from each site, and (2) the combined probability. The total standard deviation is 10 dB.

The coverage problem noted in Figure 12.7(b) can be corrected by repositioning the Site 0 antenna so that the tilt downward is only 2°. This, however, raises the possibility that the close-in coverage, under the transmitting antenna, will have gaps because of antenna pattern nulls. Figure 12.8 shows the close-in field-strength coverage, and it is seen to exceed the minimum desired field strength of 54.2 dBμV/m. The main antenna lobe points to the ground at about a 3-km distance, and beyond that, the coverage diminishes gradually. Figure 12.9(a) shows individual site coverage, and Figure 12.9(b) shows the simulcast combined coverage for this case where the Site 0 antenna was repositioned. The desired probability of coverage now exceeds the contour of the coverage zone as measured by signal-strength requirements.

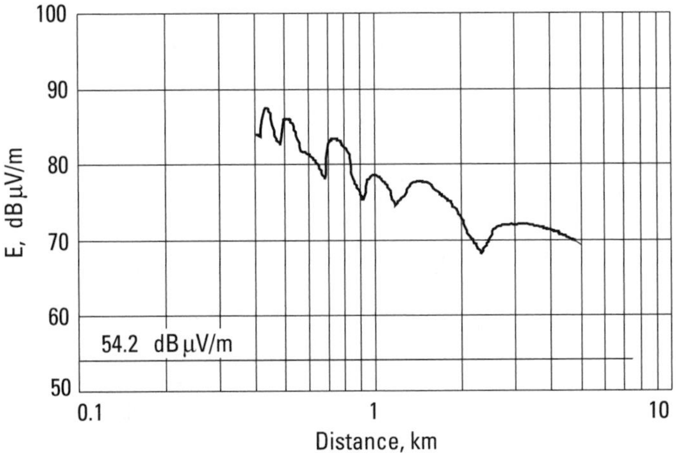

Figure 12.8 Close-in coverage from Site 0 with 2° down-tilt.

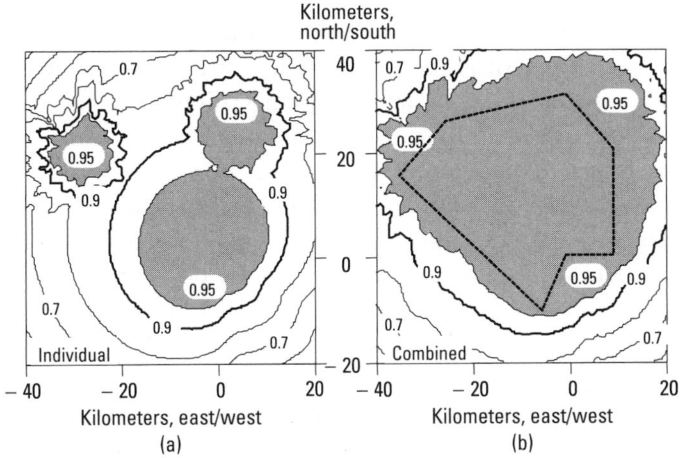

Figure 12.9 (a) Individual and (b) combined coverage probability with 2° down-tilt from site 0.

12.3.4 Simulcast Differential Delay

In simulcast systems, there is a possibility of intersymbol interference (ISI) in digital transmissions due to propagation differential delays from the different sites. Typically, destructive interference occurs when the differential delay is more than 0.15 to about 0.25 of a symbol time and the two signals are within about 2 dB of each other. Figure 12.10 shows the maximums of the three possible differential delays in microseconds among the three sites with no included

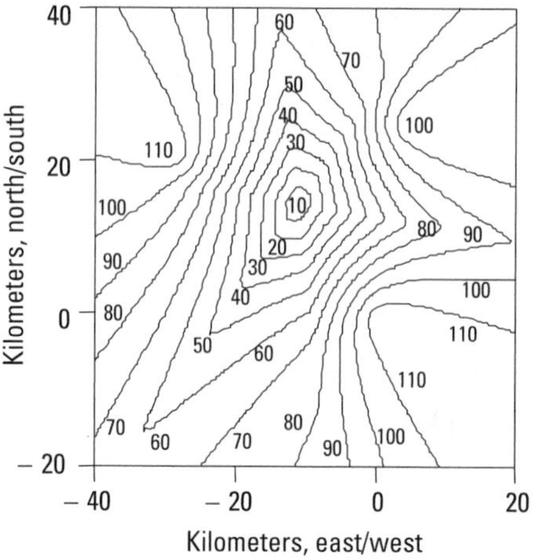

Figure 12.10 Maximum of differential delay (μsec) among the three sites.

transmitter delays. The classical parabola shapes of the equal delay contours are clearly evident. The three sites are at the foci of the three sets of parabolas. Since the median signal strength and propagation delay are known over the geographical area from each site and the total standard deviation of the signal is known, it is possible to compute the probability that two signals are within X dB of each other and at least to identify potential problem areas. The actual probability of missing calls due to simulcast differential delay requires the complete modeling of the modulation and demodulation process, employing knowledge of the signal differential delay, capture, and signal-to-noise performance relative to the desired word error rate, as described by Figure 8.10. The specific detail is beyond the scope of this book.

Referring to Figure 8.6, the Gaussian approximation to the combined Rayleigh and lognormal signal distributions is approximately correct above and below the median as long as σ_{total} is large. The Gaussian approximation to the signal statistics, therefore, will be used to calculate the probability that two Gaussian distributions, in a random draw from with median levels separated by m, are within X dB:

$$P_{overlap} = \frac{1}{2\pi\sigma^2} \int_{-\infty}^{\infty} \exp\left[-\frac{v^2}{2\sigma^2}\right] \int_{v-X}^{v+X} \exp\left[-\frac{(w-m)^2}{2\sigma^2}\right] dw dv$$

The above expression reduces to

$$P_{overlap} = \frac{1}{2}\left[\operatorname{erf}\left[\frac{X+m}{2\sigma}\right] - \operatorname{erf}\left[\frac{X+m}{2\sigma}\right]\right] \qquad (12.13)$$

and, for ($X << \sigma$), can be approximated by

$$P_{overlap} = \exp\left[-\frac{m^2}{4\sigma^2}\right]\frac{X}{\sigma\sqrt{\pi}} \tag{12.14}$$

The probability, based on power levels alone, of having a simulcast problem from two sites, then, depends on the difference in the median levels m in ratio to the standard deviation at a location, as well as on the ratio of X to the standard deviation. The differential delay at that location must also be larger than 0.15 to 0.25 of a symbol time.

Figure 12.11(a) shows a mapping of the combined probability that any two signals will be within 2 dB of each other over the geographical area. The calculation suggests that the propagation delays would be best equalized in the region of highest probabilities: about midway between sites 1 and 2. Accordingly, the transmissions from sites 1 and 2 are delayed 30 and 60 μsec, respectively.

The most suitable method of finding the probability of coverage in a simulcasting system is by a Monte Carlo simulation. Monte Carlo simulation refers to probabilistically based modeling, which relies on randomly selecting system parameters from a presumed distribution and counting the rate of desired outcomes. Such simulations can include a realistic model of the demodulation in the receiver and can include the digital decoding with forward-error correction (FEC). In the absence of a detailed model, it is possible to get an

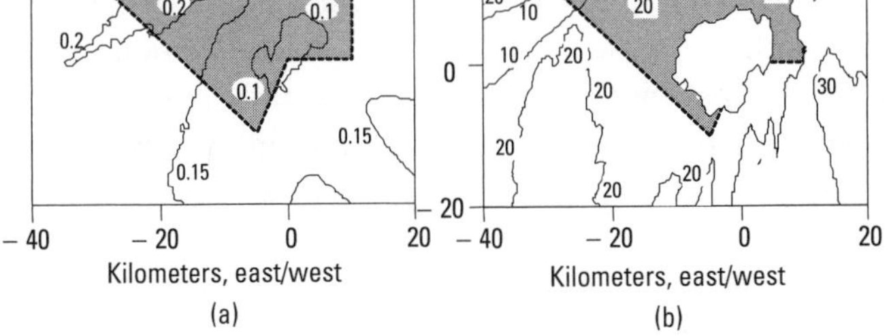

Figure 12.11 (a) Combined probability that signals from any two sites are within 2 dB of each other and (b) weighted differential delay (ms).

indication of potential trouble areas, as shown in Figure 12.10. Another indicator of potential trouble is a calculation based on a power-weighted differential-delay formula, such as (8.10). It is noted that for two equal-power signals, twice the delay spread equals the differential delay. Figure 12.11(b) presents twice the delay spread calculated for the example where Site 1 is delayed 30 μsec and Site 2 is delayed 60 μsec relative to Site 0. A delay-spread formulation, although not the physically correct model, is, in the absence of a full simulation, an indicator of potential trouble spots for simulcast differential-delay intersymbol distortion. The signal-weighted differential delays (with a maximum of 20 μsec) in the coverage area shown shaded in Figure 12.11(b) appear suitable for digital transmissions at symbol rates as high as between 0.15/(20 μsec) = 7.5 kilosymbols/s and 0.25/(20 μsec) = 12.5 kilosymbols/s.

12.4 Designing Two-Way Systems

Two-way personal communications systems can be grouped into two broad categories. The first includes "notification," or paging services, and the second is characterized by cellular telephony operation. Paging systems operate in the range of 10^{-5} Erlangs per user for short messages, so the system economics justify fixed sites that are relatively high and widely spaced. Cellular telephony generally targets 0.026 Erlangs per user, so even with the large cellular frequency bands, economics necessitate relatively small cell areas to meet subscriber capacity demands. Both types of systems must contend with frequency-reuse interference.

12.4.1 Two-Way Paging System

Two-way messaging systems based on paging technology have asymmetric transmitter power capabilities. That is, the fixed-site power level can be in the tens to hundreds of watts, while the miniature PCD is limited to perhaps a watt. In multisite systems the difference can be made up by designing a fixed-site receiver network that is denser than the transmitter network or by using a combination of high-sensitivity fixed-site receivers and fixed-site diversity techniques. Table 12.3 presents the up- and downlink budgets for a balanced two-way paging system having an equal number of fixed transmitter and receiver sites.

The PCD receiver is conventional and, operating here at 930 MHz, exhibits an average field-strength sensitivity [see (9.14)] of 28 dBμV/m. The fixed-site receiver, on the other hand, benefits from digital signal processing techniques to arrive at the improved sensitivity in the table. Furthermore, two horizontally separated fixed-site receiver-antenna combinations are used to exploit space diversity for, on average, an additional gain of 5 dB. The downlink system gain of nearly 163 dB would result in a free space range of about 3.6×10^6 m. Actual terrestrial system range is, however, much smaller. As detailed in the previous section, we require a 29.2-dB margin for desired link success probability. Referring to the universal propagation chart of Figure 7.9, our 54-dBm EIRP to 28 + 29.2 dBμV/m equates with a 62.15-dBm EIRP to 65.35 dBμV/m. The range is 4.3 km for a 66m-high fixed site and 5.4 km for a 100m-high fixed site. Since the down- and uplinks are approximately balanced, this is also the range for the reply link from the PCD. Multiple site reception and simulcasting transmissions will increase this basic range for the same calling probability, as discussed in Section 8.5.

12.4.2 Mobile/Cellular System

Cellular telephone systems operate from fixed sites that are between about 10m and 50m high and employ fixed-site receiver diversity to improve the uplink by up to about 5 dB. The up- and downlink system gains are relatively easily balanced because the up- and downlink transmitter power levels are not greatly different after accounting for the fixed-site diversity. An illustrative system link budget for a cellular mobile telephone system is shown in Table 12.4. The system parameters are not greatly different from GSM or IS-95 parameters.

The mobile receiver in this example includes margin for self-generated interference [3]. The system in this case is uplink limited since the uplink system gain is smaller than the downlink gain. We desire coverage with 90% CSR and need 10 dB for in-building coverage, so the required additional margin is approximately 22 dB. With the help of the universal propagation chart of Figure 7.9, range is about 1.7 km using a 33m-high fixed site. The downlink has a 4-dB advantage in this system, which results in a range of about 2.4 km. For reliable mobile phone system cell coverage, the system operator must design for

Table 12.3

A Two-Way Messaging System Link Budget

	Downlink	**Uplink**
Transmit power at antenna	43 dBm	30 dBm
Fixed-site antenna gain	11 dBi	11 dBi
PCD receiver sensitivity	–120.6 dBm	–130 dBm
PCD antenna gain	–12 dBi	–12 dBi
Diversity gain	0 dB	5 dB
System gain	162.6 dB	164 dB

Table 12.4
Cellular Telephone System Link Budget

	Downlink	**Uplink**
Transmitter power	36 dBm	20 dBm
Fixed-site antenna gain	16 dBi	16 dBi
Receiver sensitivity	–103 dBm	–110 dBm
PDC antenna gain (average)	–5 dBi	–5 dBi
Diversity gain	0 dB	5 dB
90% coverage	12 dB	12 dB
Building penetration	10 dB	10 dB
System gain	128 dB	124 dB

a signal (measured by a 0-dBi antenna) of –103 + 5 + 22 = –84 dBm at the street level.

12.5 Indoor Systems

Personal communications are becoming increasingly pervasive indoors as reflected in Figure 1.1. Wireless local-area networks (WLANs) comprise an access point and portable terminal, usually indoors. Here, we investigate a link budget for a U-NII band 802.11a system and for a personal-area network (PAN) system based on UWB technology.

12.5.1 Wireless Local-Area Networks

WLAN systems operate from a fixed access point to provide coverage for distances up to 100m for various portable and fixed devices. This range diminishes based on the presence and layout of obstacles in an indoor environment. Normally, omnidirectional antennas are used with indoor access points. The link budget in Table 12.5 outlines an example based on the various operating modes of an IEEE 802.11a system [11]. The signal format is orthogonal frequency division multiplexing (OFDM) constructed from 64 subcarriers in a 20-MHz bandwidth. The OFDM symbols are coded and QAM modulated. The average transmit power is about 24 dBm. The individual phenomena that contribute to loss of link margin include multipath losses, fading margin, power amplifier (PA) back off, phase noise, and implementation losses.

In most indoor systems, deep fading affects narrowband signals. Any QAM signal exhibits a nonzero peak-to-average power ratio. Hence, due to the nonlinearities inherent in power amplifiers, there needs to be a power back off to avoid PA saturation. In this example, we assume antennas with 0-dBi gain and an inverse square law propagation of up to 1m; the propagation-loss index increases to 3 using an indoor propagation model like the SBY model of (7.27).

The 802.11a system range increases from about 13m at a 54-Mbps data rate to 45.6m at a 12-Mbps data rate. The range calculation here does not account for system interference due to other 802.11a systems on the same and adjacent channels. Measurements [12] and analysis [13] indicate that, in an 802.11a system, the ergodic capacity hovers at about 25 Mbps per channel for the shortest distances and decreases steadily to about 6 Mbps at maximum range, roughly half of what might be expected from the simple link margin of Table 12.5.

12.5.2 Wireless Personal-Area Networks

Newer technologies are emerging to address the wireless PAN space coverage of up to 10m. We provide an example for a UWB pulsed system operating in the upper part of the unlicensed spectrum band allocated to UWB by U.S. regulations [14]. This example (shown in Table 12.6) presents a spread spectrum (SS) UWB pulsed system with a pulse (or chip) rate of 1,200 MHz. More detail about pulsed UWB signaling and its performance is available in [13]. The maximum allowed emitted power spectral density is –41.3 dBm/MHz under FCC rules, so the average transmit power of a 1,200-MHz-wide signal is approximately –11 dBm. UWB systems tend to be link-margin starved due to low emission limits; hence, they benefit from gain antennas at the receiver. Unlike narrowband systems, a wideband signal does not exhibit deep fading in a multipath environment, as was shown in Section 8.3.6.

Phenomena that contribute to loss of link margin include multipath losses due to ISI and implementation losses. FEC is employed to mitigate losses and improve the energy efficiency of the modulation. In this example, we assumed that the path loss follows an inverse square law that transitions to a propagation index of 2.8 at the break-point distance of 1m, as provided in the SBY propagation model using (7.27).

A UWB system centered at about 8.4 GHz can provide 1.2 Gbps to distances beyond 1m and up to 150 Mbps at distances of 4.3m at transmitter emission levels commensurate with unintentional emission limits when operating under the FCC unlicensed transmitter rules [14].

12.6 System Coverage

There is a number of ways to claim calling success probability in a radio system. One way is to quote the CSR on the outer edge of the coverage area; another is

Table 12.5
Link Budget Example for a WLAN System in the U-NII Band

Parameters	Mode 1	Mode 2	Mode 3	Mode 4	Mode 5
Transmitter					
Carrier spacing (kHz)	312.5	312.5	312.5	312.5	312.5
FFT size	64	64	64	64	64
Channel bandwidth (MHz)	20	20	20	20	20
Number of carriers	52	52	52	52	52
Effective bandwidth (MHz)	16.25	16.25	16.25	16.25	16.25
Coding rate	3/4	2/3	3/4	1/2	1/2
Effective bit rate (Mbps)	54	48	36	24	12
Mean frequency (GHz)	5.15	5.15	5.15	5.15	5.15
Signal bandwidth (MHz)	16.25	16.25	16.25	16.25	16.25
Average power (dBm)	24	24	24	24	24
Receiver					
Noise figure (dB)	7	7	7	7	7
Required E_s/N_0 (dB)	20.6	19.7	15.0	14.4	8.7
Multipath losses (dB)	5	5	5	5	5
Required Power Amplifier back off (dB)	8	7	6	4.5	3.7
Phase noise loss factor (dB)	1.5	1.5	0.8	0.5	0.25
Implementation losses (dB)	2	2	2	2	2
Fading margin (dB)	2	2	2	2	2
Receive power calculations					
Received average power (dBm)	–56.3	–58.2	–63.9	–66.0	–72.5
Average noise power (dBm)	–94.9	–94.9	–94.9	–94.9	–94.9
Maximum path attenuation (dB)	80.3	82.3	87.9	90.0	96.5
Range calculation					
Range (m)	13.1	15.2	23.5	27.6	45.6

to quote the average CSR within the coverage area. For fixed-power terrestrial systems, significantly more signal is available in the vicinity of the transmitters than at the edges of the coverage area, so the CSR averaged over the total area can be significantly higher than the CSR at maximum range. For systems using transmitter power control, the target CSR is maintained within certain limits by reducing transmitter power when possible; thus, the CSR is nearly equal over the coverage area. Similarly, satellite-based personal communications systems may be designed to deliver either a nearly uniform field strength over the coverage footprint, or they may operate with some form of closed-loop power control to deliver a target CSR in the footprint.

12.6.1 Coverage Probability over an Area

The CSR desired in a coverage area, given by (12.9), is a design value at the coverage boundary for a fixed-power system. In the design example, as Figure 12.9 shows, the CSR at the boundary of coverage exceeds the desired value of 0.95. In fact, the average CSR for those points within the coverage boundary is 0.991. The average CSR within the 0.95 contour is calculated by averaging P_s from (12.9) and that average CSR, $P_{s,avg} = 0.984$. W. C. Jakes [15] provides an expression for calculating $P_{s,avg}$ in terms of the lognormal standard deviation σ and the inverse power law n of propagation for a circular coverage area. The average area CSR in terms of the boundary CSR (upon making typographical corrections to [15]) is

$$P_{s,avg} = P_s + \frac{1}{2}\exp\left[\frac{2ab+1}{b^2}\right]\left[1-\operatorname{erf}\left[\frac{ab+1}{b}\right]\right] \tag{12.15}$$

where the constant b is

$$b = \frac{10n\log(e)}{\sigma\sqrt{2}} \tag{12.16}$$

and a is found using numerical methods [16, 17] from

$$P_s = \frac{1}{2}(1+\operatorname{erf}(a)) \tag{12.17}$$

A derivation of an expression equivalent to (12.15) can be found in G. Hess [18]. The average area coverage probability is significantly better than the coverage on the boundary, as was seen above. Expression (12.15) applies to the combination of building and lognormal shadowing standard deviations, so, in the example, $\sigma = 7.27$ with $n = 3$. The average CSR using (12.15) is then $P_{s,avg} = 0.984$, exactly as found by averaging the CSR within the 0.95 coverage boundary of Figure 12.9.

Table 12.6
Link Budget Example for a UWB PAN System in the Upper UWB Band

Parameters	**Mode 1**	**Mode 2**	**Mode 3**	**Unit**
Transmitter				
UWB pulse rate	1,200	1,200	1,200	MHz
FEC rate	1	1/2	1/2	
Spread factor	1	2	4	
Effective bit rate	1,200	300	150	Mbps
Geometric mean frequency	8.38	8.38	8.38	GHz
Bandwidth (3 dB)	1.2	1.2	1.2	GHz
EIRP spectral density limit	–41.3	–41.3	–41.3	dBm/MHz
Tx average power	–10.5	–10.5	–10.5	dBm
Tx antenna gain	0	0	0	dBi
Receiver				
Rx antenna gain	3	3	3	dBi
Noise figure	7	7	7	dB
Minimum required E_b/N_0	12	6	6	dB
Multipath losses	2	2	2	dB
Implementation losses	1	1	1	dB
Receiver power calculations	0	0	0	
Received average power	–61.0	–73.0	–76.0	dBm
Average noise power per bit	–76.0	–82.0	–85.0	dBm
Maximum path attenuation	53.5	65.5	68.5	dB
Range calculation				
Range	1.3	3.3	4.3	m

12.6.2 Proving Measurements

An installed system coverage may be proven by measurements and statistical sampling [19]. It is usually not feasible to measure signal strengths in each predefined grid within the coverage area, so the area is sampled sparsely, but randomly and uniformly. The coverage is then quoted with a confidence factor $P(|p - P_m| < \varepsilon)$, the probability that the difference between the predicted value p and measure value P_m does not exceed ε. Assuming that the sample values are normally distributed, the number N_g of regions that must be sampled is found from [20]

$$P\left(\left|p - P_m\right| \leq \varepsilon\right) = \operatorname{erf}\left[\frac{\varepsilon\sqrt{N_g}}{\sqrt{2\left(R - R^2\right)}}\right] \tag{12.18}$$

Hence,

$$N_g \geq \frac{z^2\left(p - p^2\right)}{\varepsilon^2} \tag{12.19}$$

where z is from

$$P(z) = \operatorname{erf}\left[\frac{z}{\sqrt{2}}\right] \tag{12.20}$$

Other estimates for N_g may be used, such as one based on the Tchebycheff inequality [21], but they inevitably result in a looser bound, therefore a larger required sample size. In the design example, $p = 0.984$, and the desired confidence is $P = 0.99$, giving $z = 2.576$, with $\varepsilon = \pm 0.03$. So, from (12.18), $N_g = 95$. Sampling 95 regions will result in 99% confidence that the measured CSR is within 3% of 0.984.

The coverage area of Figure 12.3 is bounded by (x, y) coordinate pairs: (0, 0), (10, 0), (10, 20), (0, 30), (–25, 25), (–35, 15), (–5, –10), and (0, 0) km. The coverage area is found from the "digital planimeter" formula,

$$Area = \sum_{j=0}^{N} \frac{\left[x_j - x_{j+1}\right]\left[y_{j+1} + y_j\right]}{2} \tag{12.21}$$

where the $j = (N + 1)$ coordinate pair equals the $j = 0$ pair. $Area = 1{,}087.5$ km^2, and if subdivided into regions 0.5 km on an edge, it would comprise 4,350 grid regions. Only 95 randomly and uniformly selected regions need to be sampled to give 99% confidence that the measured CSR is within 3% of 0.984.

The method also applies to bit-error-rate measurements in a stochastic process. For example, in a measurement of BER in AWGN at the 10^{-6} level to ±1% with a confidence of 0.98, we have $p = 10^{-6}$, $\varepsilon = 0.01p$, and $z = 2.326$. Therefore, $N_g = 5.4 \times 10^{10}$ bits must be sampled. Thus, the errors generated by a system operating at 1 Gbps should be averaged over at least $N_g/10^9 = 55$ sec.

The confidence P that a measurement of a Monte Carlo simulation of a stochastic process which generates the error rate p is within $\pm\varepsilon$ after N_g trials can be found from (12.18). For example, in a measurement (or simulation) of error rate p in the 0.01 level, the confidence that the measurement is within $\varepsilon = \pm 0.0001$ with $N_g = 5 \times 10^6$ trials is $P = 0.975$.

12.7 Summary

Personal communications system design was seen as the problem of specifying a desired calling success rate, hence, the manipulation of the available parameters to ensure that the calling rate is satisfied within the boundaries of a specified geographic region. Noise, including thermal noise and interference, was discussed as one limit of PCD receiver performance; simulcast differential delay was discussed as another. A paging system design was carrier through to illustrate the design parameters available at a fixed site. The propagation model illustrated that antenna pattern control can be exploited to alter coverage in a desired geographical area. The statistical design method was used to arrive at a required performance in the coverage area. Simulcast differential-delay characteristics were examined, and design parameters, namely transmitter delays, were altered to improve performance. The link characteristics of two-way paging and cellular telephone systems were explored, as were two indoor systems. One indoor system was based on the IEEE 802.11a standard, and the second system relied on UWB technology to surpass a gigabit-per-second data rate at PAN ranges. Finally, a method of proving designs was presented that involved statistical samplings of the coverage zone, and the confidence of measurements was described.

Problems

12.1 The room temperature sensitivity of a communications receiver is –120 dBm. This receiver is connected to a 6-dBi antenna system located in an RF noisy environment, where the noise temperature is 2,900K. Find the minimum useful signal input to the receiver.

Ans:

The noise floor is increased by 10log(2,900/290) = 10 dB; hence, the required signal must increase by 10 dB to –110 dBm.

12.2 A LEO satellite system services subscribers in suburban residential buildings having an 8-dB average penetration loss. If the sky noise temperature is 60K, what is the noise temperature in the field of view of the PCD?

12.3 A 900-MHz fixed-site transmitter delivers a measured 48 dBμV/m to a street-level point at the edge of an urban coverage boundary characterized by a 6-dB lognormal standard deviation. Find the average CSR in a residential building (loss = 7.6 dB with 4.5-dB standard deviation) on the coverage boundary if the average receiver sensitivity is 29 dBμV/m.

Ans:

The lognormal term is irrelevant; only the Rayleigh and building terms apply.

$M = 48 - 29 = 19$ dB $= \sigma z + 7.6$; $\sigma^2 = 7.5^2 + 4.5^2$, so $z = (19 - 7.6)/8.746 = 1.303$, so the CSR = 0.904.

12.4 A 900-MHz fixed-site transmitter delivers an average 50 dBμV/m to a street-level point at the edge of an urban coverage boundary characterized by a 6-dB lognormal standard deviation. Find the average CSR in residential buildings (loss = 7.6 dB with 4.5-dB standard deviation) near the coverage boundary if the average receiver sensitivity is 29 dBμV/m.

Ans:

The lognormal, Rayleigh, and building terms apply.

$M = 50 - 29 = 21$ dB $= \sigma z + 7.6$; $\sigma^2 = 7.5^2 + 6^2 + 4.5^2$, so $z = (21 - 7.6)/8.746 = 1.263$, so the CSR = 0.897.

12.5 Find the path attenuation to a point where a 50-kW EIRP, 100.1-MHz transmitter delivers a median field strength of 45 dBμV/m.

$L = E - \text{EIRP} - 77.22 - 20\log(f) = 45 - 67 - 77.22 - 40 = -149.2$ dB.

12.6 An FM broadcast station produces a 16.4-kW EIRP at 99.9 MHz from a 400m high antenna. Using Figure 7.9, find the field strength at a 40-km distance.

Ans:

1.64 kW produces 40 dBμV/m at 40 km, so 16.4 kW produces 50 dBμV/m.

12.7 What options are available to improve the up- and downlink system gains in the system of Table 12.3, and what would be the system design implications?

12.8 In a multisite simulcasting paging system, the service area is flooded with signals from multiple transmitters. What system design parameters are available to mitigate intersymbol interference due to differential delays among signals?

12.9 A receiver is in an urban area served by two simulcasting transmitters. The area is generally characterized by a lognormal shadowing standard deviation of 8 dB, and the mean signal strengths from the two transmitters at that location are equal. What is the average magnitude difference in the signal strengths from the two transmitters?

Ans:

In uncorrelated Rayleigh scattered fields, (8.39) applies; hence, the answer is 6 dB.

12.10 A message from a PCD is received at two fixed sites in an urban area generally characterized by a lognormal shadowing standard deviation of 8 dB. Ignoring multipath fading, what is the expected magnitude difference in decibels between the received signals?

Ans:

Expression (8.33) applies.

12.11 A PCD is located 10 km east of a simulcasting transmitter. A second transmitter is 30 km west of the first. Calculate the differential delay, and assuming equal signal strengths from each transmitter, find the maximum digital data transmission rate for the system.

12.12 A GEO direct broadcast satellite system appearing 30° above the horizon is to function with a cluster of terrestrial "fill-in" simulcasting transmitters that each may have a range of 10 km. Find the minimum symbol rate based on simulcast interference criteria if the terrestrial transmitters are more than 20 km apart.

12.13 A system coverage area of 15 km^2 is divided into grid squares 0.5 km on an edge. The system designer is obliged to prove coverage based on a statistical sampling of the grid squares. The design CSR at the limits of the coverage boundary is 0.984. Find the minimum number of grid squares that must be sampled to prove 95% CSR coverage with 99% confidence.

Ans:

$N = 92$, but $15/0.5^2 = 60$. All 60 grid squares must be sampled.

12.14 A system designer randomly samples 80 uniformly distributed squares in a 1,000 km^2 coverage area that was designed to provide 0.90 CSR at the coverage boundary with predicted 0.95 average CSR. The designer reports an average measured CSR of 0.93. What is the confidence level of the measurement?

Ans:

$P = \text{erf}\left(\sqrt{N\varepsilon^2 / p(1-p)}\right)$, $N = 80$, $\varepsilon = 0.05$, $p = 0.95$; so, $P = 0.98$ that the measurement is within 0.05 of 0.95.

References

[1] "Project Cyclops," *NASA Report CR 114445*, NASA/Ames Research Center, Moffett Field, CA, July 1973.

[2] Couch, L., *Digital and Analog Communication Systems*, 4th ed., New York: Macmillan Publishing Company, 1993.

[3] Siwiak, K., "UWB Emissions and EMC," *EMC2006: 18th International Wroclaw Symposium and Exhibition on Electromagnetic Compatibility*, Warsaw, Poland, June 28–30, 2006.

[4] Hill, C., and T. Kneisel, "Portable Radio Antenna Performance in the 150, 450, 800, and 900 MHz Bands "Outside" and In-Vehicle," *IEEE Transactions on Vehicular Technology*, Vol. VT-40, No. 4, November 1991, pp. 750–756.

[5] Hagn, G., "VHF Radio System Performance Model for Predicting Communications Operational Ranges in Irregular Terrain," *IEEE Transactions on Communications*, Vol. COM-28, No. 9, September 1980, pp. 1637–1644.

[6] Maciel, L. R., H. L. Bertoni, and H. H. Xia, "Unified Approach to Prediction of Propagation over buildings for All Ranges of Base Station Antenna height," *IEEE Transactions on Vehicular Technology*, Vol. VT-42, No. 1, February 1993, pp. 41–45.

[7] Furutsu, K., "On the Theory of Radio Propagation over Inhomogenous Earth," *J. Res. NBS*, Vol. 67D, No. 1, 1963, pp. 39–62.

[8] Bullington, K., "Radio Propagation Fundamentals," *Bell System Technical Journal*, Vol. 36, May 1957, pp. 593–626.

[9] Bullington, K., "Radio Propagation for Vehicular Communications," *IEEE Transactions on Vehicular Technology*, Vol. VT-26, No. 4, November 1977, pp. 295–308.

[10] Lee, W. C. Y., *Mobile Communications Engineering*, New York: McGraw-Hill, 1982.

[11] IEEE Std 802.11a-1999 (supplement to IEEE Std 802.11-1999) Part 11: Wireless LAN Medium Access Control (MAC) and Physical Layer (PHY) Specifications: High-Speed Physical Layer in the 5 GHZ Band, ISBN 0-7381-1810-9, New York: Institute of Electrical and Electronics Engineers, September 16, 1999.

[12] Chen, J. C., and Jeffrey M. Gilbert, "Measured Performance of 5-GHz 802.11a Wireless LAN Systems," Atheros Communications, Inc. White Paper, available at www.atheros.com (last accessed December 2001).

[13] Siwiak, K., and D. McKeown, *Ultra-wideband Radio Technology*, London: John Wiley and Sons, 2004.

[14] U.S. 47 C.F.R. Part15, Ultra-Wideband Operations FCC Report and Order, April 22, 2002.

[15] Jakes, W. C., *Microwave Mobile Communications*, American Telephone and Telegraph Co., 1974 [reprinted by IEEE Press, Piscataway, NJ, 1993].

[16] Abramowitz, M., and I. Stegun, (eds.), *Handbook of Mathematical Functions*, New York: Dover Publications, 1972.

[17] *Mathcad User's Guide* (Versions 2.0–7.0 Plus), Cambridge, MA: MathSoft, Inc., 1988–1997.

[18] Hess, G., *Land-Mobile Radio System Engineering*, Norwood, MA: Artech House, 1993.

[19] Hill, C., and B. Olsen, "A Statistical Analysis of Radio System Coverage Acceptance Testing," *IEEE Vehicular Technology Society News*, February 1994, pp. 4–13.

[20] Taub, H., and D. L. Schilling, *Principles of Communication Systems*, 2nd ed., New York: McGraw-Hill, 1986.

[21] Papoulis, A., *Probability, Random Variables, and Stochastic Processes*, New York: McGraw-Hill, 1965.

Appendix A: FORTRAN Programs: The Near Field of Dipoles and Helices

FORTRAN code DIP2.FOR and FIELDS.FOR, complete with all subroutines, are on the CD-ROM included with this book. Additionally included are executable codes DIP2.EXE and FIELDS.EXE, which run under DOS 5.0 or later. In Windows, especially Windows NT or XP and later versions of Windows may give spurious answers. Test your results against the examples below. DIP2 computes the modal current coefficients for a dipole of the specified dimensions. The coefficients are stored in a user-named file. FIELDS then computes the electric and magnetic fields based on currents and charges calculated from the modal coefficients in the user-named file.

Sample (*annotated*) dialog and output from program DIP2, which computes dipole or helical dipole modal excitations and dipole feed-point impedance, are shown below.

```
>DIP2
 INPUT WAVELENGTH (negative value terminates program)...
1                                    (enter the wavelength, m)
    INPUT RADIUS, GAP HALF-HEIGHT, DIPOLE HALF LENGTH...
0.001586,0.00015,0.23629
    INPUT ZERO OR A/P FOR HELIX...(A/P = helix radius to pitch ratio)
0                                   ('0' = for dipole, 'A/P' for helix)
    INPUT NUMBER OF MODES, EVEN AND ODD...
    (-N FOR SELECTED MODES)
1                                   (select # of modes to consider)
```

```
    XMAX  =      254.2321  DX  =      2.5000000E-02
    INPUT NUMBER OF CURRENT POINTS..
```
(# of points to display antenna current)
```
5
  Z, I-AMP, I-PHASE...
  0.0000000      1.5152370E-02 2.426908
  5.9072498E-02 1.3998965E-02 2.426908
  0.1181450      1.3998965E-02 2.426908
  0.1772175      5.7985615E-03 2.426908
  0.2362900      0.0000000     2.426908
  Z(0)  =   (  65.93708,   -2.794604)
```
(feed-point impedance of dipole)
```
  MODAL AMPLITUDES...
     1  AN =   (  7.3589101E-03,   3.1189193E-04)
  ...TO SAVE MODAL AMPLITUDES TYPE YES ...
YES
  ...FILE NAME (4 ALPHA-NUM)...
D100
  ...FILE D100.DAT SAVED
INPUT WAVELENGTH (negative value terminates program)...
1
```
(start another calculation ...)
```
INPUT RADIUS, GAP HALF-HEIGHT, DIPOLE HALF LENGTH...
0.001586,0.00015,0.23629
    INPUT ZERO OR A/P FOR HELIX...
0
```
(dipole selected)
```
    INPUT NUMBER OF MODES, EVEN AND ODD...
    (-N FOR SELECTED MODES)
```
('-' allows selection of modes)
```
-3
    BINARY MODE ENABLE STRING (6(5I1, 1X))...
101
```
(the modes selected are first and third)
```
    XMAX  =      258.4642  DX  =      2.5000000E-02
    INPUT NUMBER OF CURRENT POINTS...
5
  Z, I-AMP, I-PHASE...
  0.0000000      1.3959610E-02 2.329976
  5.9072498E-02 1.3140320E-02 2.329976
  0.1181450      1.0521567E-02 2.329976
  0.1772175      5.9557385E-03 2.329976
  0.2362900      0.0000000     2.329976
  Z(0)  =   (  71.57599,   -2.912296)
  MODAL AMPLITUDES...
     1  AN =   (  7.3589101E-03,   3.1189193E-04)
     2  AN =   (  0.0000000     ,   0.0000000     )
     3  AN =   (  2.2548001E-04,  -2.2127353E-04)
  ...TO SAVE MODAL AMPLITUDES TYPE YES ...
YES
  ...FILE NAME (4 ALPHA-NUM)...
ABCD
  ...FILE ABCD.DAT SAVED
INPUT WAVELENGTH (negative value terminates program)...
-
```
(... terminate program)

Sample (*annotated*) dialog and output from program FIELDS, which computes electric and magnetic fields near the dipole, is presented below.

```
>FIELDS
D001                          (retrieve modal coefficients from D001.DAT)
 WAVELENGTH     =          1.000000
 DIPOLE RADIUS  =      1.586000E-03
 HALF-LENGTH    =      2.362900E-01
 HALF-GAP       =      1.500000E-04
 MODE             1  =        (7.358792E-03,3.132760E-04)
INPUT  RHO (negative value terminates program)...
0.0                                       (r coordinate of field points)
 INPUT  Z...
-.2                                       (z coordinates of field points)
 Z  =    -2.000000E-01
        416 INTEGRATION POINTS, DX  =     9.900000E-03
            (-3.089490,-6.328062)
(4.404556E-01,3.439701E-01)
        (-4.517477E-01,14.092000)
(-4.360922E-02,2.776107E-01)
     (1.884132E-02,-5.838754E-04)
(-4.362090E-03,5.725843E-06)
 E-Z    =          6.544214     -113.878000
                                          (E_z – magnitude and phase)
 E-RHO  =         14.378150       91.974350(E_ρ )
 H-THE  =      1.449077E-02      -2.286581(H_θ )
  INPUT  RHO...
-1                                        (terminate program)
```

The analytical basis for the numerical implementation of this dipole analysis is in [1], and the experimental results are in [2]. The extension of the analysis to helical dipoles, along with experimental results, is in [3].

References

[1] Balzano, Q., O. Garay, and K. Siwiak. "The Near Field of Dipole Antennas, Part I: Theory," *IEEE Transactions on Vehicular Technology*, Vol. VT-30, No. 4, November 1981, pp. 161–174.

[2] Balzano, Q., O. Garay, and K. Siwiak. "The Near Field of Dipole Antennas, Part II: Experimental Results," *IEEE Transactions on Vehicular Technology*, Vol. VT-30, No. 4, November 1981, pp. 175–181.

[3] Balzano, Q., O. Garay, and K. Siwiak. "The Near Field of Omnidirectional Helices," *IEEE Transactions on Vehicular Technology*, Vol. VT-31, No. 4, November 1982, pp. 173–185.

Appendix B: FORTRAN Code: The Near Field of Loops

FORTRAN source codes LOOPK2.FOR and LFLD.FOR, complete with all subroutines, are on the CD-ROM included with this book. Additionally included are executable codes LOOPK2.EXE and LFLDS.EXE, which run under DOS 5.0 or later. In Windows, especially Windows NT or XP and later versions of Windows may produce spurious results. Test your answers against the examples below. LOOPK2 computes the current densities and charge distributions around a loop of the specified dimensions. The data is stored in a user-named file as a function of angle ϕ. LFLD then computes the electric and magnetic fields based on current densities and charges from the user-named file. Since the assumed loop gap voltage is 1V, the loop impedance can be found from the inverse of the first current density pair (real and imaginary) divided by the loop wire circumference, $2\pi a$.

Sample (*annotated*) dialog and output from program LOOPK2 is shown below.

```
>LOOPK2
FMHZ....(negative value terminates program)
300.
A,B (meters)....
0.001, 0.1
   MAS...(> = 6)...                        (enter # of circumferential modes)
16
 ... TO SAVE MODES TYPE   YES   ...
YES
 ...FILE NAME (4 ALPHA-NUM)...             (currents and charges stored)
```

```
L001
Stop - Program terminated.
```

Sample (*annotated*) dialog and output from program LFLD, which computes fields, is presented below.

```
>RUN LFLD                                         (start program LFLD)
FILE NAME FOR MODES...
L001
   OMEGA  =        12.886000     NN  =          0
      A   =     1.000700E-03
      B   =     1.000700E-01
      K   =         6.283200
 Field points: RHO,PHI,Z ...   (input ρ, φ (deg) and z of field point)
ENTER RHO  (negative value terminates program)
0.2
ENTER PHI(deg)
0
ENTER Z
0
(Eρ, Eφ, Eρ, Hρ, Hφ, Hz magnitude and phase-deg:)
        Er          Ep          Ez          Hr          Hp
Hz
amplitude: 2.835E-08 5.113E-01 1.194E-10 8.892E-14
3.287E-14 1.297E-03
    phase: 9.582E+01-1.669E+02
4.234E-01-1.235E+02-9.330E+01-1.229E+02
ENTER RHO  (negative value terminates program)
-1                                             (terminate program)
The user data file has the following structure:
.12886E+02   .10007E-03   .10007E-01   .62832E+01
 181
   .47106E-03  -.15206E+02  -.10692E-04   .73554E+00
   .00000E+00   .00000E+00   .00000E+00   .00000E+00
   .47106E-03  -.15213E+02  -.10692E-04   .73528E+00
   .59102E+02  -.12848E-04   .76934E+00  -.20782E-06
   .47105E-03  -.15221E+02  -.10692E-04   .73551E+00
   .40491E+02  -.25693E-04  -.19455E+01  -.41557E-06
```

... and so on, where the first line contains Storer's parameter $\Omega = 2\ln(2\pi b/a)$, loop wire radius a, loop radius b, and wave number k. The second line indicates that 181 pairs of lines are to follow. The first line in the pair of lines is Re{J_ϕ}, Im{J_ϕ} for the first mode, and Re{J_ϕ}, Im{J_ϕ} for the amplitude of the cosinusoidal current density variation around the wire. The second line in the pair contains the first and second modes of the charge density. The loop feed-point impedance in this example is

$$Z = 1/(0.00047106 - j15.206)(2\pi\ 0.0001) = 0.0032 + j104.67$$

The analytical basis for the numerical implementation of this loop analysis can be found in [1, 2], along with experimental measurements of the close near fields around both thin and thick wire gap-fed loops.

References

[1] Balzano, Q., and K. Siwiak. "Radiation of Annular Antennas," *Correlations* (Motorola Engineering Bulletin, Schaumburg, IL), Vol. 6, No. 2, Winter 1987.

[2] Balzano, Q., and K. Siwiak. "The Near Field of Annular Antennas," *IEEE Transactions on Vehicular Technology*, Vol. VT-36, No. 4, November 1987, pp. 173–183.

Appendix C: Digital Communications Codes and Character Sets

Morse Code

Morse code is listed here because a form of this code was used as the first digital communications code. The timing is such that the dot ("dit") is equal to one unit in time followed by a one unit space. The dash ("dah") is three time units followed by a space unit. Spaces between characters are an additional two units; spaces between words are an additional four units. Thus, the standard-length word "PARIS" (including the trailing space) is 50 time units long, the same length as a string of 25 dits.

Table C.1
International Morse Code (ITU-R Recommendation M.1677)

A	•–	N	–•	1	•– – – –
B	–•••	O	– – –	2	••– – –
C	–•–•	P	•– –•	3	•••– –
D	–••	Q	– –•–	4	••••–
E	•	R	•–•	5	•••••
F	••–•	S	•••	6	–••••
G	– –•	T	–	7	– –•••
H	••••	U	••–	8	– – –••
I	••	V	•••–	9	– – – –•
J	•– – –	W	•– –	0	– – – – –
K	–•–	X	–••–		
L	•–••	Y	–•– –	@	•– –•–•
M	– –	Z	– –••		

Period [.]	•–•–•–	Wait sign (AS)	•–•••
Comma [,]	– – ••– –	Double dash (break)	–•••–
Question mark [?]	••– –••	Error sign	••••••••
Quotation mark [" "]	•–••–•	Fraction bar (/)	–••–
Colon [:]	– – –•••	End of message (AR)	•–•–•
Parenthesis [(]	–•– –•–	End of transmission (SK)	•••–•–
Parenthesis [)]	–•– –• –	International distress (SOS)	•••– – –•••

On May 24, 2004, the 160th anniversary of the first telegraphic transmission, the ITU added the "@" ("commercial at" or "commat") to the Morse character set to facilitate sending electronic mail addresses by Morse code. Notably, this is the first official addition to the Morse set of characters since World War I.

The Morse code exhibits characteristics that today are associated with Huffman codes. It is a variable length code with the shortest-length characters generally appearing the most often in normal language text messages.

Digital Paging Codes

Paging signaling uses a 4-bit code for numeric data. Some standard numeric codes are listed in Table C.2.

Table C.2
Paging 4-Bit "Numeric-Only" Character Set

4-Bit Combination: Bit Number: 4 3 2 1	**Displayed Character**	
0 0 0 0	0	
0 0 0 1	1	
0 0 1 0	2	
0 0 1 1	3	
0 1 0 0	4	
0 1 0 1	5	
0 1 1 0	6	
0 1 1 1	7	
1 0 0 0	8	
1 0 0 1	9	
1 0 1 0	Spare	A
1 0 1 1	U	B
1 1 0 0	Space	Space
1 1 0 1	-	C
1 1 1 0	]	D
1 1 1 1	[	E
	(1)	*(2)*

The last six entries in the table can have alternative renditions of this code: the column labelled *(1)* is used with POCSAG and FLEX signaling, and *(2)* is an alternate FLEX rendition.

The 7-bit character set shown in Table C.3 is based on the American National Standard Code for Information Interchange (ASCII) and ISO 646-1983E. It has been adopted for character-based paging in the United States and elsewhere using, for example, the FLEX signaling format. Some of the characters, such as "FS," "GS," and "RS," may be reserved for special control functions in paging signaling. Others from the character group #$@[\]^_’{|}~ and "DEL" are available for special purposes and for mapping other symbols from other alphabets as needed.

Table C.3
The ISO 646-1983E 77-Bit Paging Character Set.

Bit Position				**7**	0	0	0	0	1	1	1	1
				6	0	0	1	1	0	0	1	1
4	**3**	**2**	**1**	**5**	0	1	0	1	0	1	0	1
0	0	0	0		NUL	DLE	SP	0	@	P	\	p
0	0	0	1		SOH	DC1	!	1	A	Q	a	q
0	0	1	0		STX	DC2	"	2	B	R	b	r
0	0	1	1		ETX	DC3	#	3	C	S	c	s
0	1	0	0		EOT	DC4	$	4	D	T	d	t
0	1	0	1		ENQ	NAK	%	5	E	U	e	u
0	1	1	0		ACK	SYN	&	6	F	V	f	v
0	1	1	1		BEL	ETB	'	7	G	W	g	w
1	0	0	0		BS	CAN	(	8	H	X	h	x
1	0	0	1		HT	EM	)	9	I	Y	i	y
1	0	1	0		LF	SUB	*	:	J	Z	j	z
1	0	1	1		VT	ESC	+	;	K	[	k	{
1	1	0	0		FF	FS	,	<	L	\	l	:
1	1	0	1		CR	GS	-	=	M	]	m	}
1	1	1	0		SO	RS	.	>	N		n	~
1	1	1	1		SI	US	/	?	O	_	o	DEL

ACK	Acknowledge	ENQ	Enquire	RS	Record separator
BEL	Bell or alarm	EOT	End of transmission	SI	Shift in
BS	Backspace	ESC	Escape	SO	Shift out
CAN	Cancel	ETB	End of transmission block	SOH	Start of heading
CR	Carriage return	ETX	End of text	SP	Space
DC1	Device control 1	FF	Form feed	STX	Start of text
DC2	Device control 2	FS	File separator	SUB	Substitute
DC3	Device control 3	GS	Group separator	SYN	Synchronous idle
DC4	Device control 4	HT	Horizontal tab	US	Unit separator
DEL	Delete	LF	Line feed	VT	Vertical tab
DLE	Data link escape	NAK	Negative acknowledgment		
EM	End of medium	NUL	Null, or all zeros		

The FLEX specification defines many additional character sets, including ones for the Japanese, Chinese, Russian, and other languages. The code of Table C.3, when used in wire-line communications, can include an eighth parity bit, which may be odd or even parity, depending on the particular selection made.

Appendix D: HF Propagation Models

The CD-ROM included with this text contains software programs: VOACAP (as well as ICEPAC and REC533) from NTIA, and HamCAP from DXatlas. These are used for HF propagation examples in Chapter 4. They require a computer with at least

600 MHz Pentium II CPU;
MS-Windows 95/98/NT/2000/XP;
64 MB RAM;
16-bit or 32-bit color video card.

The signal-to-noise ratio in decibels per 1 Hz bandwidth is calculated in VOCAP and includes man-made noise terms from Rec. ITU-R P.372-8 for various radio noise environments, as seen in Chapter 12. HamCAP calls out the residential man-made noise level given by

$$N_a = 67.2 - 27.7\log(f_{MHz}) - 204 \text{ dBW/Hz} \qquad \text{(D-1)}$$

which at 3 MHz equals –144.7 dBW/Hz. The propagation prediction is for a circuit reliability of 50%. That is, the prediction is exceeded for half the time and is below the predicted value for half the time in any given month. For reference, a single-sideband voice circuit requires a 49-dB SNR/Hz for 90% intelligibility, and a Morse telegraphy circuit needs an SNR of 33 dB/Hz. The HF broadcast service, on the other hand, requires a 73-dB SNR/Hz to assure a usable circuit 90% of the time. The programs are used here to illustrate examples of HF propagation phenomena. When fully understood and carefully used, these programs can provide valuable predictions of HF propagation.

VOACAP, ICEPAC, REC533

This HF planning software was developed by an agency of the U.S. government:

U.S. Department of Commerce
National Telecommunications and Information Administration
Institute for Telecommunication Sciences (NTIA/ITS)
325 Broadway
Boulder, Colorado 80305

NTIA/ITS has no objection to the use of this software for any purpose since it is not subject to copyright protection in the United States. No warranty, expressed or implied, is made by NTIA/ITS or the U.S. government as to the accuracy, suitability, or functioning of the program and related material; nor shall the fact of distribution constitute any endorsement by the U.S. government. ICEPAC, VOACAP, and REC533 can be obtained for free from the Internet site http://elbert.its.bldrdoc.gov/pc_hf/hfwin32.html.

Later versions of the program may be found at www.greg-hand.com/hf.html.

The file included on the CD-ROM, itshfbc_060418.exe (used with permission), is identified by the year, month, and day it was created. Place this file in a temporary directory and install it by executing it. VOACAP and the other executable programs may be run from the executable files in the installation folder, or they may be invoked by the user-friendly interface program HamCAP. The examples in Chapter 4 rely on the HamCAP interface program.

HamCAP

HamCAP was written by Alex Shovkoplyas for use in the Radio Amateur Service and is available free of charge from www.dxatlas.com/hamcap.

Version 1.4 (used with permission) is included on the CD-ROM. HamCAP is a compact freeware interface program to VOACAP, incorporating both graphical point-to-point and area coverage predictions. The program is self-contained and needs, at most, the smoothed International Sunspot Number from the NGDC at (www.ngdc.noaa.gov/stp/SOLAR/ftpsunspotnumber.html) site. It is installed by running the setup.exe executable program.

The HF propagation examples in Chapter 4 are based on the contents of the file HamCap.ini included on the CD-ROM. After HamCAP is installed, the predefined HamCap.ini file included on the CD-ROM can be written over the installed HamCap.ini to set the initial conditions used in the examples. All parameters that are saved in the .ini file may be controlled from the program. Please note that VOACAP must be installed before HamCAP can be used. HamCAP capabilities and pointers to more detailed explanations are available at the dxatlas.com Web site.

Figure D.1 shows the opening screen in HamCAP, which uses VOACAP as the calculation engine. May 2006 is shown selected, and VR2, the designation for Hong Kong, has been entered in the "DX Call" box. Figure D.2 shows the 50% probability propagation prediction on the "Map" screen, that is, the probability that, on any given day in that prediction month, the predicted SNR will be achieved. A cursor has been placed in the South Pacific region. The bottom of the screen shows its geographical location and the computed SNR at that location. The SNR is in decibels per hertz. A great circular path from Florida, northwest past Alaska, to Hong Kong is also visible on the map.

Finally, the "Chart" tab in Figure D.3 shows a calculation of the 50% probability of the maximum usable frequency (MUF) for the great circular path between Florida and Hong Kong for a 24-hour period on any given day in May 2006. A white vertical bar at 22:00 UTC corresponds to the time selected on the SNR prediction map.

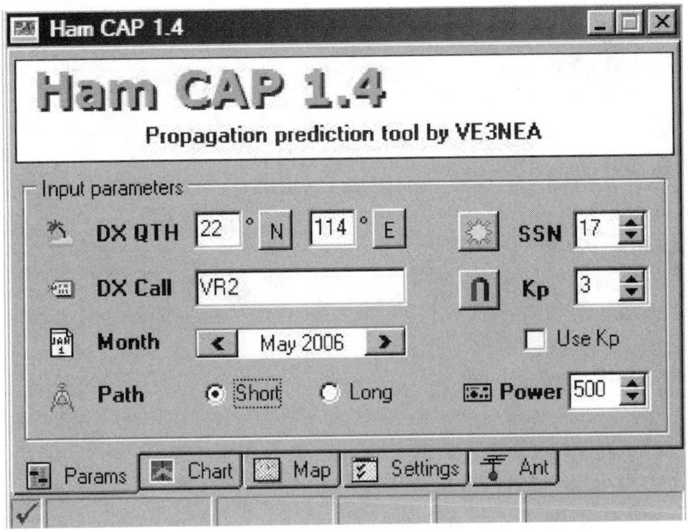

Figure D.1 Opening screen in HamCAP.

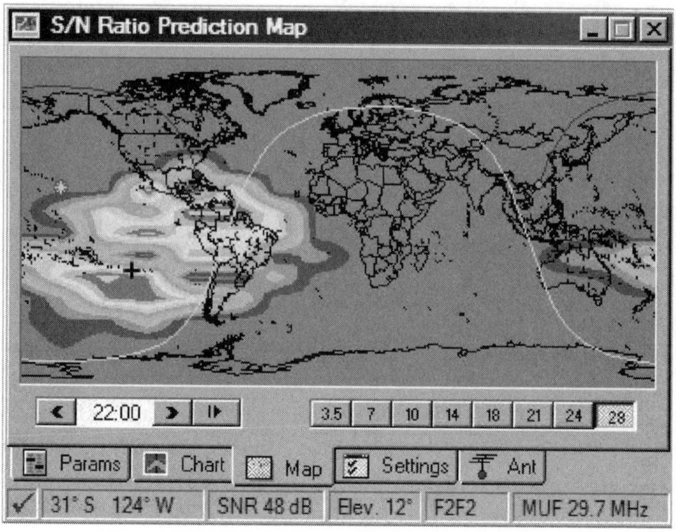

Figure D.2 The "Map" tab screen in HamCAP.

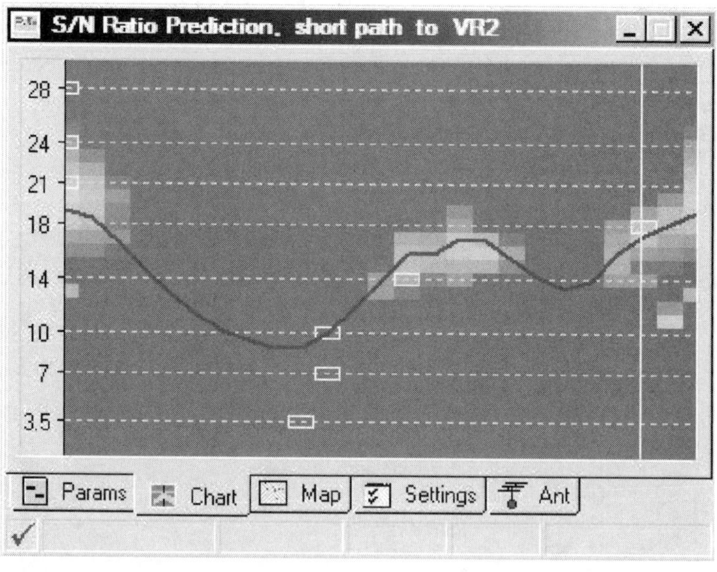

Figure D.3 The "Chart" tab screen in HamCAP.

About the Authors

Kazimierz "Kai" Siwiak is founder of TimeDerivative, Inc., a wireless consultancy helping business and technical clients with UWB, MIMO techniques, antennas, propagation, and EMC. While working at Motorola, he was named a Dan Noble Fellow and received the Silver Quill Award. He holds more than 30 U.S. patents and has published extensively. A senior member of the IEEE, he served on ETSI and IEEE 802 committees tasked with developing radio telecommunications standards. Kai received his BS.E.E and M.S.E.E from the Polytechnic University of Brooklyn and his Ph.D. from Florida Atlantic University.

Yasaman Bahreini is a consultant focused on technology development and business executive leadership in the communications industry. Specializing in physical layer (PHY) architectures, she has served on various industry standards committees developing specifications for 3G and wireless local- and personal-area systems. Her most recent contributions have been on UWB activities within various standards groups. She was a U.S. delegate to the ITU-R on UWB technology. Yassi is a senior member of the IEEE and holds a B.S. and M.S. in electrical engineering from George Washington University.

List of Symbols

NOTE: E (electric) and *H* (magnetic) fields in this text are generally root mean square (RMS) fields, not peak values, as is common in many electromagnetics texts. This matters only when the fields quantities are related to power or power density. This is the usual method of specification in the telecommunications industry.

In this text, quantities such as c and μ_0 are usually stated to their full known precision, not to imply that results calculated using that precision results in any sort of enhanced accuracy, but because the full precision is (a) correct and (b) often very useful in debugging numerical calculations carried out using numerical methods on modern computers. In some applications, such as the calculation of satellite orbits, the full precision is necessary.

Not every symbol used in the text is listed here. Some symbols are reused in the text, and their meaning is clearly obvious form their context. Vector quantities, such as **E** and **H**, are in bold, and listed at the end. The physical constants are from the 2002 CODATA recommended values; they are generally recognized worldwide for use in all fields of science and technology. The values became available in December 2003 and replaced the 1998 CODATA set. They are based on all of the data available through December 31, 2002. A detailed description of the data and the analysis that led to these values is published in the *Reviews of Modern Physics* 77 (2005), pp. 1–107. The constants are also available at http://physics.nist.gov/cuu/Constants.

Symbol	Value and Units
A	surface wave field factor
A_{area}	physical aperture area of an antenna

A_g	$\alpha_g \times 20\log(e) = \alpha_g \times 8.6859$ dB/m, attenuation
A_e	effective antenna aperture (m^2)
AF	antenna factor, ratio of field strength E to voltage across the receiver input terminals V_r (dB)
AR	axial ratio of polarization; see Section 1.4
B	magnetic flux density (tesla)
B_0	$\approx 5 \times 10^{-5}$ tesla, Earth's magnetic field
BW	bandwidth
c	$(\mu_0 \varepsilon_0)^{-1/2} = 299{,}792{,}458$ m/s, speed of light in vacuum (exact)
C	capacitance (F)
$D(\theta,\phi)$	antenna directivity, directive gain, expressed as a power ratio, or in decibels relative to isotropic radiation
D	directivity in the direction of maximum radiation density
D_d	ambipolar diffusion coefficient (m^3/s)
D_ε	$\sigma/(\omega\varepsilon_0\varepsilon_r) = \varepsilon_i/\varepsilon_r$ dissipation factor for dielectrics
e	2.718281828459045 ...
e_e	$1.60217653 \times 10^{-19}$ C, electron charge
E	electric field strength (V/m)
f	frequency (Hz, unless otherwise specified)
F	free space loss (dB), increasingly positive with more loss
$F_{\text{“}n\text{”}}$	noise figure of stage “n”
G	conductance (S/m)
$G^{\pm}$	diffraction coefficient; see Chapter 5
h	$6.6260693 \times 10^{-34}$ J sec, Planck constant
$h_{\text{“}u\text{”}}$	polarization amplitude for “u”-oriented polarization
H	magnetic field strength (A/m)
I	current (A)
J_s	linear surface current (A/m)
k	$2\pi/\lambda = \omega/c$ rad/m, wave number
k_b	$1.3806505 \times 10^{-23}$ J/K, Boltzmann constant
K	atmospheric refraction factor, commonly equal to 1.333
K	degrees kelvin, or kelvins
$L_{\text{“}x\text{”}}$	path attenuation, model “x” dB, increasingly negative with more loss
m_e	$9.1093826 \times 10^{-31}$ kg, electron rest mass

M	equivalent "magnetic current" (V/m^2)
n_e	linear electron density, electrons per meter
N	atmospheric refraction unit; see Chapter 4
$N(z)$	free electron density profile along path z per cubic meter
N_e	electron density per cubic meter
N_{max}	electron density per cubic meter associated with the critical frequency
P_r	power density (W/m^2)
$P_{\text{“event”}}$	probability of "event"
PF	$\sin[\tan^{-1}(D_\varepsilon)]$ dielectric power factor
Q	$2\pi \times$ ratio of energy stored to energy dissipated in one cycle
R_e	6,378.145 km, mean equatorial Earth radius
r_e	$2.817940325 \times 10^{-15}$ m, classical electron radius
s	9,192,631,770 periods between the two hyperfine levels of the ground state of the cesium-133 atom = 1 sec
SAR	specific absorption rate (W/kg)
T	orbital period (sec, unless otherwise specified)
T_b	background sky temperature, 2.725K
T_0	290K, standard room temperature
T_s	86,164.0905 sec, Earth's sidereal day
V_r	voltage across input terminals of a receiver (V)
Z	impedance (Ω)
α_g	nepers per meter, attenuation
δ_s	$1/\alpha_g$ [$\approx(2/\omega\mu_0\sigma)^{1/2}$, good conductors], skin depth in meters in lossy medium
ε	$\varepsilon_0\varepsilon_d$ complex permittivity (F/m)
ε_d	$\varepsilon_r(1 - jD_\varepsilon) = \varepsilon_r - j\varepsilon_I$ complex relative dielectric constant
ε_0	$1/\mu_0 c^2$ (exact) = $8.854187817 \ldots \times 10^{-12}$ F/m
ε_r	real part of relative dielectric constant
ε_i	$\varepsilon_r D_\varepsilon$ imaginary part of relative dielectric constant
Γ	reflection coefficient
Γ_v	reflection coefficient, vertical polarization
Γ_h	reflection coefficient, horizontal polarization
η	efficiency

η_0	$c\mu_0$ (exact) = 376.730313 ... ≈ $120\pi\,\Omega$, intrinsic free space impedance
λ	wave length (m)
μ	$\mu_0\mu_d$ complex permeability (H/m)
μ_d	complex relative permeability
μ_0	$4\pi \times 10^{-7}$ H/m, free space permeability (exact)
μ_r	real part of relative permeability
$\mu_\oplus$	398,600.441 km^3/s^2, geocentric gravitational parameter for Earth
π	3.141592653589793 ...
ρ	charge density (C/m^2)
σ	conductivity (S/m)
$\sigma_{``x"}$	standard deviation associated with "x"
σ_{Stefan}	$2\pi^5 k_b^4 c^{-2} h^{-3}/15 = 5.670400 \times 10^{-8}$ $W/m^2/K^4$, Stefan-Boltzmann constant
τ	time constant (sec, unless specified otherwise)
τ_{pol}	polarization tilt, magnitude of angle between θ direction and major axis of polarization ellipse
ω	$2\pi f$ rad/s, radian frequency

Vector quantities

Symbol	Value and Units
A	vector potential (electric sources) (Wb/m)
B	magnetic flux density vector (tesla)
D	electric displacement field vector (C/m^2)
E	electric field vector (V/m)
F	vector potential (magnetic sources) (C/m)
H	magnetic field vector (A/m)
$\mathbf{h}_a$	complex polarization amplitude vector
J	current density vector (A/m^2)
M	equivalent vector "magnetic current" (V/m^2)
x	unit vector parallel to the x-axis
y	unit vector parallel to the y-axis
z	unit vector parallel to the z-axis
r	unit vector parallel to the radial direction

R	vector from origin to field point (m)
S	complex Poynting vector (W/m^2)
θ	unit vector in the θ direction
ϕ	unit vector in the ϕ direction

Index

Recent Titles in the Artech House Mobile Communications Series

John Walker, Series Editor

3G CDMA2000 Wireless System Engineering, Samuel C. Yang

3G Multimedia Network Services, Accounting, and User Profiles, Freddy Ghys, Marcel Mampaey, Michel Smouts, and Arto Vaaraniemi

802.11 WLANs and IP Networking: Security, QoS, and Mobility, Anand R. Prasad, Neeli R. Prasad

Advances in 3G Enhanced Technologies for Wireless Communications, Jiangzhou Wang and Tung-Sang Ng, editors

Advances in Mobile Information Systems, John Walker, editor

Advances in Mobile Radio Access Networks, Y. Jay Guo

Applied Satellite Navigation Using GPS, GALILEO, and Augmentation Systems, Ramjee Prasad and Marina Ruggieri

CDMA for Wireless Personal Communications, Ramjee Prasad

CDMA Mobile Radio Design, John B. Groe and Lawrence E. Larson

CDMA RF System Engineering, Samuel C. Yang

CDMA Systems Capacity Engineering, Kiseon Kim and Insoo Koo

CDMA Systems Engineering Handbook, Jhong S. Lee and Leonard E. Miller

Cell Planning for Wireless Communications, Manuel F. Cátedra and Jesús Pérez-Arriaga

Cellular Communications: Worldwide Market Development, Garry A. Garrard

Cellular Mobile Systems Engineering, Saleh Faruque

The Complete Wireless Communications Professional: A Guide for Engineers and Managers, William Webb

EDGE for Mobile Internet, Emmanuel Seurre, Patrick Savelli, and Pierre-Jean Pietri

Emerging Public Safety Wireless Communication Systems, Robert I. Desourdis, Jr., et al.

The Future of Wireless Communications, William Webb

Geographic Information Systems Demystified, Stephen R. Galati

GPRS for Mobile Internet, Emmanuel Seurre, Patrick Savelli, and Pierre-Jean Pietri

GPRS: Gateway to Third Generation Mobile Networks, Gunnar Heine and Holger Sagkob

GSM and Personal Communications Handbook, Siegmund M. Redl, Matthias K. Weber, and Malcolm W. Oliphant

GSM Networks: Protocols, Terminology, and Implementation, Gunnar Heine

GSM System Engineering, Asha Mehrotra

Handbook of Land-Mobile Radio System Coverage, Garry C. Hess

Handbook of Mobile Radio Networks, Sami Tabbane

High-Speed Wireless ATM and LANs, Benny Bing

Interference Analysis and Reduction for Wireless Systems, Peter Stavroulakis

Introduction to 3G Mobile Communications, Second Edition, Juha Korhonen

Introduction to Communication Systems Simulation, Maurice Schiff

Introduction to Digital Professional Mobile Radio, Hans-Peter A. Ketterling

Introduction to GPS: The Global Positioning System, Ahmed El-Rabbany

An Introduction to GSM, Siegmund M. Redl, Matthias K. Weber, and Malcolm W. Oliphant

Introduction to Mobile Communications Engineering, José M. Hernando and F. Pérez-Fontán

Introduction to Radio Propagation for Fixed and Mobile Communications, John Doble

Introduction to Wireless Local Loop, Second Edition: Broadband and Narrowband Systems, William Webb

IS-136 TDMA Technology, Economics, and Services, Lawrence Harte, Adrian Smith, and Charles A. Jacobs

Location Management and Routing in Mobile Wireless Networks, Amitava Mukherjee, Somprakash Bandyopadhyay, and Debashis Saha

Mobile Data Communications Systems, Peter Wong and David Britland

Mobile IP Technology for M-Business, Mark Norris

Mobile Satellite Communications, Shingo Ohmori, Hiromitsu Wakana, and Seiichiro Kawase

Mobile Telecommunications Standards: GSM, UMTS, TETRA, and ERMES, Rudi Bekkers

Mobile Telecommunications: Standards, Regulation, and Applications, Rudi Bekkers and Jan Smits

Multiantenna Digital Radio Transmission, Massimiliano "Max" Martone

Multiantenna Wireless Communicatio ns Systems, Sergio Barbarossa

Multipath Phenomena in Cellular Networks, Nathan Blaunstein and Jørgen Bach Andersen

Multiuser Detection in CDMA Mobile Terminals, Piero Castoldi

OFDMA for Broadband Wireless Access, Sławomir Pietrzyk

Personal Wireless Communication with DECT and PWT, John Phillips and Gerard Mac Namee

Practical Wireless Data Modem Design, Jonathon Y. C. Cheah

Prime Codes with Applications to CDMA Optical and Wireless Networks, Guu-Chang Yang and Wing C. Kwong

QoS in Integrated 3G Networks, Robert Lloyd-Evans

Radio Engineering for Wireless Communication and Sensor Applications, Antti V. Räisänen and Arto Lehto

Radio Propagation in Cellular Networks, Nathan Blaunstein

Radio Resource Management for Wireless Networks, Jens Zander and Seong-Lyun Kim

Radiowave Propagation and Antennas for Personal Communications, Third Edition, Kazimierz Siwiak and Yasaman Bahreini

RDS: The Radio Data System, Dietmar Kopitz and Bev Marks

Resource Allocation in Hierarchical Cellular Systems, Lauro Ortigoza-Guerrero and A. Hamid Aghvami

RF and Baseband Techniques for Software-Defined Radio Peter B. Kenington

RF and Microwave Circuit Design for Wireless Communications, Lawrence E. Larson, editor

Sample Rate Conversion in Software Configurable Radios, Tim Hentschel

Signal Processing Applications in CDMA Communications, Hui Liu

Smart Antenna Engineering, Ahmed El Zooghby

Software Defined Radio for 3G, Paul Burns

Spread Spectrum CDMA Systems for Wireless Communications, Savo G. Glisic and Branka Vucetic

Third Generation Wireless Systems, Volume 1: Post-Shannon Signal Architectures, George M. Calhoun

Traffic Analysis and Design of Wireless IP Networks, Toni Janevski

Transmission Systems Design Handbook for Wireless Networks, Harvey Lehpamer

UMTS and Mobile Computing, Alexander Joseph Huber and Josef Franz Huber

Understanding Cellular Radio, William Webb

Understanding Digital PCS: The TDMA Standard, Cameron Kelly Coursey

Understanding GPS: Principles and Applications, Second Edtion, Elliott D. Kaplan and Christopher J. Hegarty, editors

Understanding WAP: Wireless Applications, Devices, and Services, Marcel van der Heijden and Marcus Taylor, editors

Universal Wireless Personal Communications, Ramjee Prasad

WCDMA: Towards IP Mobility and Mobile Internet, Tero Ojanperä and Ramjee Prasad, editors

Wireless Communications in Developing Countries: Cellular and Satellite Systems, Rachael E. Schwartz

Wireless Communications Evolution to 3G and Beyond, Saad Z. Asif

Wireless Intelligent Networking, Gerry Christensen, Paul G. Florack, and Robert Duncan

Wireless LAN Standards and Applications, Asunción Santamaría and Francisco J. López-Hernández, editors

Wireless Technician's Handbook, Second Edition, Andrew Miceli

For further information on these and other Artech House titles, including previously considered out-of-print books now available through our In-Print-Forever® (IPF®) program, contact:

Artech House
685 Canton Street
Norwood, MA 02062
Phone: 781-769-9750
Fax: 781-769-6334
e-mail: artech@artechhouse.com

Artech House
46 Gillingham Street
London SW1V 1AH UK
Phone: +44 (0)20 7596-8750
Fax: +44 (0)20 7630-0166
e-mail: artech-uk@artechhouse.com

Find us on the World Wide Web at: www.artechhouse.com